International Yearbook of Soil Law and Policy

The *International Yearbook of Soil Law and Policy* is a book series that discusses the central questions of law and policy with regard to the protection and sustainable management of soil and land. The Yearbook series analyzes developments in international law and new approaches at the regional level as well as in a wide range of national jurisdictions. In addition, it addresses cross-disciplinary issues concerning the protection and sustainable management of soil, including tenure rights, compliance, food security, human rights, poverty eradication and migration. Each volume contains articles and studies based on specific overarching topics and combines perspectives from both lawyers and natural scientists to ensure an inter-disciplinary discourse.

The *International Yearbook of Soil Law and Policy* offers a valuable resource for lawyers, legislators, scholars and policymakers dealing with soil and land issues from a regulatory perspective. Further, it provides an essential platform for the discussion of new conceptual approaches at the international, national and regional level.

More information about this series at http://www.springer.com/series/15378

Harald Ginzky • Elizabeth Dooley •
Irene L. Heuser • Emmanuel Kasimbazi •
Robert Kibugi • Till Markus • Tianbao Qin •
Oliver Ruppel
Editors

International Yearbook of Soil Law and Policy 2019

Springer

Editors
Harald Ginzky
German Environment Agency
Dessau, Germany

Irene L. Heuser
Specialist Group on "Soils, Desertification and Sustainable Agriculture"
IUCN World Commission on Environmental Law
Kleinmachnow, Germany

Robert Kibugi
School of Law
University of Nairobi
Nairobi, Kenya

Tianbao Qin
Research Institute of Environmental Law
Wuhan University
Wuhan, China

Elizabeth Dooley
University of Exeter
Exeter, UK

Emmanuel Kasimbazi
School of Law
Makerere University
Kampala, Uganda

Till Markus
Helmholtz-Centre for Environmental Research
UFZ
Leipzig, Germany

Oliver Ruppel
Development and Rule of Law Programme [DROP]
University of Stellenbosch
Stellenbosch, South Africa

ISSN 2520-1271 ISSN 2520-128X (electronic)
International Yearbook of Soil Law and Policy
ISBN 978-3-030-52316-9 ISBN 978-3-030-52317-6 (eBook)
https://doi.org/10.1007/978-3-030-52317-6

This Springer imprint is published by the registered company Springer Nature Switzerland AG.
The registered company address is: Gewerbestrasse 11, 6330 Cham, Switzerland

Foreword

It is with great pleasure that I write this foreword for the volume of the *International Yearbook of Soil Law and Policy* that includes as the theme the presentations held at the workshop "Implementing LDN in Africa: Means, legal instruments and institutional challenges?". This workshop was co-organised in cooperation between the German Environment Agency (UBA), the Konrad-Adenauer-Stiftung—Climate Policy and Energy Security Program for Sub-Saharan Africa, the GIZ and the University of Nairobi and held at Windsor Hotel, Nairobi, in August 2018.

The objective of the workshop was to enable an effective implementation of policies on land degradation and soil protection, in particular in Africa. It was the second conference of its kind, the first one being held in September 2017 in Kampala, Uganda. To this end, the organisers not only managed to assess how an enabling environment for the implementation of sustainable land use practices could be established, with respect to legislation, as well as institutional and organisational necessities, but they also investigated how international cooperation could support the creation of such an enabling environment.

One of the outcomes from the workshop was that African governments must handle the challenge of adapting and coping with the consequences of climate change on soils, biodiversity and ability to sustainably manage our land and natural resources. The law plays an important role in this regard.

Africa requires sufficient technical, technological, financial and human resources to face the soil challenge in a coordinated, synergistic manner involving all stakeholders. The Pan-African Parliament (PAP) could play a key role in this endeavour. As set out in Article 17 of the African Union Constitutive Act, the PAP has "to ensure the full participation of African peoples in the development and economic integration of the continent". The Parliament can thus serve as a platform for the people from all African states to be involved in the discussion and decision-making towards more soil protection, and the harmonisation and coordination of Member States' laws.

Pan-African Parliament, Johannesburg, South Africa

Janet Oongera

Preface

The International Panel on Climate Change (IPCC) has published two reports in the past 2 years that highlight the enormous challenges for land and soil protection and submit recommendation for future actions. The special report of 2018 stated that scientific evidence shows that temperature increase has to be held by 1.5 degrees above pre-industrial levels. Otherwise, humanity will face the risk of disastrous effects caused by tipping points of the climate system. Moreover, the report stressed the need for immediate and far-reaching emission's reduction (80–95% worldwide by 2050). In addition, the IPCC clearly stated that appropriate measures to achieve the so-called negative emissions are inevitably necessary as the 1.5-degree objective could not be met by emission's reductions alone. The IPCC called for sustainable agriculture and forestry as one important means to achieve these "negative emissions".

Another IPCC report, published in August 2019, focused on "climate change and land". Herein the IPCC stresses that food production is a relevant source of CO_2 emissions and that land use, climate change and food security are closely interlinked and interdependent. The IPCC strongly advocates for sustainable land use, for substantive reductions of food waste and for more sustainable consumption patterns, as well as for sustainable agriculture and forestry. Otherwise, the challenge both to ensure food security for all human beings on the earth and to cope with climate change requirements will fail—causing effects one hardly dares to envisage.

What does that mean for soils and for soil governance? What does that mean for the "International Yearbook of Soil Law and Policy"? In essence, the two aforementioned IPCC reports underline the importance of soils as the "melting pot of all ecological functions"—not neglecting a similar importance of the world's oceans. If humanity could find the ways and means to use soils sustainably—throughout the continents and on a permanent scheme—it would contribute significantly to ensuring food security and combatting climate change and preventing major climate-induced migration streams and political conflicts.

"Soil Law and Policy" is—understood broadly—a means to establish sustainable soil management. "Law and Policy" not only needs to encompass legal provisions, statutes and ordinances but also equally needs to address scientific issues, enabling institutional arrangements as well as sufficient financial resources. Moreover, "Law and Policy" has to ensure the involvement of all relevant stakes and stakeholders.

As sustainable soil management is critically important for humanity's needs and survival, the *International Yearbook of Soil Law and Policy* is deemed to play a crucial role by providing a platform for discourse on the best practices, strategies and legal concepts ensuring increased sustainable soil management—for academics, legislators and policymakers.

This fourth volume attempts to cope with these high expectations. The "theme" of the volume is coined to address how an "enabling environment for effective implementation of soil governance in Africa" can be achieved more effectively. This volume consists of about 19 chapters, which can inter alia be considered as an outcome of a two-day workshop held in Nairobi, Kenya, in August 2018. This workshop was successfully co-organised by the German Environment Agency, the Konrad-Adenauer Foundation and the University of Nairobi. Both the workshop and the chapters on this "theme" in Part I are intended to contribute to a more focused debate in Africa on the interplay of soils as a natural resource, food security and climate change.

Part II—"Recent International Developments"—entails a discussion on the future of international soil governance. Bodle et al. provide an overview of the current legal frameworks at international level and present the recommendations to develop this further. Fowler and Hannam's chapter entails meaningful comments on these recommendations. The chapter by Rodriguez elucidates, in particular, the role of the Food and Agriculture Organization (FAO) in the aforementioned context.

Part III—"Regional and National Reports" —provides insights on the soil protection governance in Mexico (Gonzales), California (Desai), Brazil (Silveira et al), Germany (Ginzky) and the European Union (on Common Agriculture Policy—Raffelsiefen).

In Part IV—"Cross-cutting Issues", numerous topics are addressed. First, the perception of soils within the Catholic Church is explained (Voigt). Another chapter by Bartke et al. addresses the challenges of the urban and rural linkages with regard to sustainable soil management and works out the ways and means of how to solve them. Finally, the chapter by Hannam addresses the aspects of a legislative and policy framework to manage soil carbon sequestration.

In conclusion, we as the editors hope that by including a variety of perspectives ranging from international to regional to national level, the diversity of topics such as Christian soil perspectives or the land--urban linkages and the involvement of authors from all regions of this planet, the volume of *International Yearbook of Soil Law and Policy* will contribute in addressing the challenges expressed by the IPCC.

Finally, we would like to express our deepest gratitude to all authors of this volume for their engagement, commitment and contributions, to the members of the advisory board for providing their continued insight during the review process, and last but definitely not the least, to the publishing house SPRINGER, in particular Anke Seyfried, for the ongoing operational support and their technical assistance.

Dessau, Germany — Harald Ginzky
Exeter, UK — Elizabeth Dooley
Kleinmachnow, Germany — Irene L. Heuser
Kampala, Uganda — Emmanuel Kasimbazi
Nairobi, Kenya — Robert Kibugi
Leipzig, Germany — Till Markus
Wuhan, China — Tianbao Qin
Stellenbosch, South Africa — Oliver Ruppel

Contents

Part I
Enabling Environment for an Effective Implementation of Soil Governance in Africa

Challenges to Soil Protection and Sustainable Management in Africa

Francis Marthy Tetteh

1 Introduction

Soil is the main resource base for many people in Africa and the sustainable management of soil resources in Africa is a formidable challenge, crucial for the survival of over one billion people. With an estimated population growth for Sub-Saharan Africa (SSA) from the current 900 million to 1.4 billion in 2030, the region's soils will experience increasing pressure as a natural resource to provide for the needs of its people. Food production in Sub-Saharan Africa today is inadequate to feed the fast growing population. Smallholder farmers in Ghana, as elsewhere in Sub-Saharan Africa, are widely considered to be the largest as well as the most vulnerable component of the rural sector. Smallholder farmers make up approximately 70% of the estimated 5 million farming households. Sixty percent (60%) of all farms in Ghana are less than 1.2 ha in size, 25% are between 1.2 and 2.0 ha and 15% above 2.0 ha.

Only approximately 8% of the African continent is covered by soils that 'are relatively free of natural constraints for agriculture', and much of the presently cultivated land in Africa occurs on areas that are deemed unsuitable while other areas appear to be suitable but are not cultivated (Laker 2015). Throughout the continent, many areas now experience land degradation from inappropriate cultivation practices which are driven by a desire for quick economic return but ignore the capacity of the soil to support them, exacerbated by high rural poverty. About 65% of arable lands, 30% of grazing lands and 20% of forests are already degraded in Africa.[1] The degradation may be from single or multiple factors in combination such

[1]FAO (2015), p. 16.

F. M. Tetteh (✉)
African Soil Partnership, Soil Science Society of Ghana and CSIR-Soil Research Institute, Kumasi, Ghana

H. Ginzky et al. (eds.), *International Yearbook of Soil Law and Policy 2019*, International Yearbook of Soil Law and Policy 2019,
https://doi.org/10.1007/978-3-030-52317-6_1

as soil compaction, soil erosion, soil fertility decline, reduction of biomass, salinity, loss of biodiversity and other physical and chemical alterations as a result of inadequate drainage and misuse of soils, as well as loss of soil biodiversity.

There is therefore a great opportunity for Sub-Saharan Africa to position itself as champion in terms of increasing food production and security, achieving land restoration and increasing agricultural resilience to climate change.

Currently in West Africa vast agricultural lands are also being destroyed through illegal artisanal mining. These processes and activities pose a major threat to environment conservation and sustainable development of the soil resource base, a real challenge to national development goals.

Research indicates that 1% growth in the agricultural sector translates into 2.5% growth in income for the poor people in the region. However such growth is still low and per capita food production has been insufficient and decreasing over time because it is on smallholder agriculture that often provides insufficient incentives to use land resources sustainably. Compounding problems include volatility in input prices, markets and climate, declining farm sizes as a result of rising population pressures, weak extension services, low adoption of improved technologies and limited government investments.

Sub-Saharan Africa uses one-tenth the level of fertilizer as compared to the rest of the world due to the following: (1) Smallholder farmers can neither access nor afford to buy locally appropriate fertilizers. (2) Farmers lack the knowledge of how to efficiently combine mineral fertilizers and organic inputs to maximize the return on their investment. (3) Counter-productive policies have corrupted the fertilizer market and decimated the systems needed to deliver knowledge to poor farmers.

2 What Is Soil Protection and Sustainable Soil Management?

Soil protection means to sustainably secure or restore and improve soil functions and other ecosystem services. Soil protection is achieved by addressing a number of threats to soil functions. The most important threats to soil function globally are soil erosion, loss of soil organic carbon and nutrient imbalance. Human pressures on soil resources are reaching critical limits. If soils are not protected sustainably, there will be further loss of productive soils leading to increased food-price volatility, thus sending millions of people into poverty.

The SSM definition drawn directly from the 2015 World Soil Charter states as follows:

> Soil management is sustainable if the supporting, provisioning, regulating, and cultural services provided by soil are maintained or enhanced without significantly impairing either the soil functions that enable those services or biodiversity.

Sustainable Soil Management has different meanings for different people. For some, it means continuing present soil management practices. For others, the focus is on ecological integrity at the expense of any other concern.[2] Everyone assumes that agriculture must be sustainable. But we differ in the interpretations of conditions and assumptions under which this can be made to occur.[3] A great deal of effort has gone into trying to define sustainability in absolute terms. To some people, sustainability implies persistence and the capacity of something to continue for a long time. With regard to the soil and environment, it involves not damaging or degrading natural resources. To others, it implies resilience and the ability to bounce back after unexpected difficulties. It is also important to clarify what is being sustained, for how long, for which benefit and at whose cost, over what area and measured by what criteria. For people in SSA, one of the root causes of stagnating or declining agricultural productivity is soil degradation which is manifested in the form of soil erosion. Other related problems that reduce agricultural land productivity include soil organic matter loss, nutrient depletion, soil biodiversity loss, etc. Sustainable soil management for every country in SSA therefore will involve rehabilitation of degraded land with focus on four main types of soil change that are considered as major threats, which are soil erosion, soil organic matter depletion, soil nutrient depletion and loss of biodiversity.

In many parts of Europe today, anthropogenic pressures are the main reason for soil degradation. In addition the western part of the European region, unlike other regions of the world, has a history of over 200 years of industrialization, putting additional pressure on the soil particularly through contamination. Soil sealing and land take by infrastructure and housing are the predominant threats to Western European soils while soil salinization and sodification are the predominant threats to Eastern European soils. Common to the entire region is the concern for soil contamination, mainly related to a large number of contaminated sites that are the heritage of a very long history of industrialization. Comparing soil threats in Europe and SSA, the emphasis is different and therefore the interpretation of sustainable soil management in these contrasting regions may be different.

Improving soils has proved to be one of the hardest challenges in agricultural development, with efforts plagued by short-term solutions, conflicting ideologies about entry points and inadequate understanding of market development and of farmer behaviour.

It is also reported that 'Africa's climate has made agricultural improvement difficult'. Most of these soils have limitations for crop production and they require careful management if they are to be used sustainably.[4]

The 2015 Soil Charter identified the ten threats to soil functions as soil erosion, soil organic carbon loss, nutrient imbalance, soil acidification, soil contamination, water logging, soil compaction, soil sealing, salinization and loss of biodiversity.

According to the Global Soil Partnership (GSP) (2016), sustainable soil management is vital to achieving increasing food production and security, land restoration

---

[2]Eswaran (1995).

[3]Francis and Hildebrand (1989).

[4]Jones et al. (2013).

and increasing agricultural resilience to climate change. There are soil-related strategies for increasing food supply while minimizing harmful environmental impacts. Restoring productivity to soils that have previously experienced productivity losses and reducing future productivity losses due to degradation through agricultural intensification, increasing the efficiency with which agricultural inputs such as irrigation, fertilizers and pesticides are used in farming systems have implications for increased food supply.

There are proven soil management practices that can help farmers adapt to adverse effects of increasing weather variability and climate change. These practices also contribute to reducing greenhouse (carbon dioxide, methane and nitrous oxide) gas emissions from agriculture. In SSA farming systems, climate change and variability may affect soil health through reduced and erratic rainfall and more frequent and severer periods of droughts, that reduce the capacity of soils to supply water and nutrients to crops and intense rainfall and storms, that increase the risk of soil erosion and increase soil surface temperatures, leading to increased rates of carbon dioxide loss into the atmosphere.

The African Soil Partnership (AfSP), which caters for Sub-Saharan Africa, identified the key priority areas as:

- Protect and conserve healthy soils and restore degraded soil.
- Promote sustainable management of soil resources at all levels and in all land uses.
- Enhance soil information by using state of the art methods of digital soil mapping and advocate for having effective national soil information systems.
- Encourage investment in the promotion of sustainable soil management to achieve sustainable development.

3 Key Constraints to Solving Africa's Soil Health Problems

Improving the fertility of Africa's soils is one of the most challenging problems in developing agriculture and yet presents one of the most powerful leverage points for stable agricultural growth.

The critical issues are:

3.1 Inappropriate Technologies for Farmer Needs

Soil fertility technologies that emphasized low inputs, but did not take into account the labour involved and income foregone in systems where nitrogen fixing plants or trees were not marketable, were on the research oriented agenda of some donors a couple of years ago. Purely organic approaches to African soil fertility are not sufficient to address the problem of poor farmers. Lack of fertilizers in Africa and the attractiveness of a sustainable paradigm led to significant experimentation at scale with 'low-input' and organic farming methods. An example is the 'alley farming' with leguminous trees, improved fallows with leguminous trees and indigenous legumes, biomass transfer systems, green manure cover crops and animal manure and composting. These technologies are weak entry points to adoption of sustainable soil management technologies, since benefits accrue over long time periods and cannot be easily assessed by farmers, just like education and preventive medicine, have a public good character. Each of these methods has achieved some isolated successes and may be part of a local integrated soil fertility management strategy, but has not been adopted at large scale. The reasons for low adoption were linked to the fact that most of the 'low-input methods are also characterized by 'low out-put' systems due to the low quantity and quality of nutrient provision. Also, these methods are labour and land intensive and require famers to dedicate resources to producing and retaining biomass other than staple and cash crops. This could be addressed by rooting fertilizer use in integrated soil fertility management that is appropriate to local farmer knowledge, practice and agro-ecology. Sustainable soil management practices can be promoted and adopted when the aims are (1) to improve fertilizer efficiency and (2) to increase yields of a marketable crop.

3.2 Smallholder Incentives for Soil Fertility Management

Adoption of fertilizers by small holder farmers is conditioned by risk, market demand for farm products and access to affordable credit. Smallholders live close to the subsistence margin. They need short term as well as long-term benefits for whatever investments they make.

Soil management practices for maintenance of soil quality, structure, nutrients and proper chemistry can be a partial alternative to the use of the mineral fertilizers but alone cannot meet nutrient demands. Fertilizers alone cannot create big yield increases in degraded soils and low fertilizer use efficiency has proved to be too expensive for poor farmers and is environmentally unsustainable as well. There is a consensus in the scientific community that the highest and most sustainable gains in crop productivity per unit nutrient are achieved from mixtures of fertilizer and organic inputs.[5] There is substantial evidence that when organic and mineral sources are used together they are complementary; organic materials increase the agronomic efficiency of fertilizer and fertilizer helps increase the returns on organic material application through positive interactions on soil properties.

[5]Giller et al. (1998).

3.3 *Environmental Degradation*

Many traditional 'slash and burn' agricultures relied on long-term fallows to restore fertility and were carried out in environments where significant tree cover remained intact. As agriculture has become more intensive the fallow periods have shortened and the trees needed to keep soil in place and recycled nutrients from deeper soil profiles have disappeared and thus created a cycle of increasing degradation. Solutions require new ecosystem- and landscape-wide approaches that take into account the entire soil and water landscape and incorporate productive trees as part of the farm.

4 Challenges for Soil Protection and Sustainable Management

Several challenges often lead African countries to resist adopting the concept and practices of soil protection and sustainable soil management.

4.1 *Inadequate Capacity, Knowledge and Experience*

In most SSA countries some of the major challenges in soil protection are inadequate human and institutional capacity, knowledge and experience to plan and implement SSM and optimally manage, mitigate and monitor the productive and degradative status of the soils; especially under intensive cultivation. Research and development in agriculture are inadequate and suffer from lack of trained personnel and facilities (inadequate soil scientists, inadequate laboratory equipment and field testing facilities).

Donor-supported research is common, but this does not help African countries to build satisfactory research traditions and local expertise.

4.2 *Lack of Economic Incentives for Smallholders*

Adoption of sustainable soil management technologies by smallholder farmers is conditioned by risk, market demand for farm produce and access to affordable credit. Smallholders live so close to the subsistence margin, they need short term as well as long term benefits for whatever investments they make. Reduction in soil loss or long-term environmental degradation are not tangible inducements for small farmers to adopt soil protection practices when their immediate concern is simply feeding their families. A farmer's willingness to buy fertilizer is affected by output prices, the yield impact of fertilizer and competing household demands for that money. This decision occurs against the risks and uncertainty of market prices, attainable yields uncertain factors (seed quality, rains, etc.) and other sources of income.

In Sub-Saharan Africa and for that matter Ghana, tenant farmers may not adopt soil protection measures with long gestation period for fear of ejection from the land. The tenant farmers may not even use mineral fertilizer as long as the value of crops harvested will cover their operation costs. They will continue to mine the nutrients in the soil and cause further loss in the fertility of the soils.

4.3 Lack of Education and Awareness

Soil awareness means developing a responsible behaviour towards soils and soil management and protection, based on knowledge and attitude. Knowledge of soil and land resources is the foundation of achieving sustainable soil management. It should be integrated into formal education preferably at all levels of schooling. There is no soil component in the education of most people. It is important that we teach the importance of soil to the society at large; from young children, school teachers, farmers, gardeners and industrialist to planners and politicians. Most soil scientists, however, have very limited interaction with public education activities.

The need to raise awareness and understanding of the soil and soil functions is being recognized increasingly as a critical step in the protection and sustainable use of soils not only at the farm level but also at higher levels in the society. In most African countries, soil protection is not explicitly included in the legislation. Therefore increasing everybody's knowledge about soil through education and research is the first logical step to allow society as a whole to understand and appreciate soil and its importance.

Even if the farmers are willing and able, extension services are poor or non-existent in most Sub-Saharan African countries. Most extension technological packages to farmers lack soil protection and sustainable soil management information.

4.4 Institutionalization of Systems

Sustainable soil resource management cannot be the domain of one ministry or institution. It is important therefore that in formulating an action plan on soil protection and fertility enhancement, all related ministries and institutions dealing directly with the management of the resource base should collaborate and directly contribute to its success. The National Agricultural Research Institutions (NARS), Ministries of Food and Agriculture, Ministries responsible for environment, Universities, Farmers Organization, Banks, Non-Governmental Organizations and traditional leaders should be directly involved in programs for soil resource conservation and management for sustainable agricultural production. In many African countries research and development in agriculture are inadequate and suffer from lack of trained personnel, facilities and motivation. In most African countries, donor-supported research is still the rule. This makes it difficult for a country to build satisfactory research traditions and local expertise; consequently, the benefits are limited to individual projects. The management technologies developed by the projects cease to be adopted at the end of the project.

Numerous stakeholders including the public sector, development agencies, NGOs, research institutions, private companies and farmer organizations play significant, but isolated roles in improving agricultural production. However, poor coordination between all the sectors that control agricultural production inputs, services and infrastructure reduces the potential for the adoption of innovative technologies. There are many initiatives, interventions and investments seeking to improve adoption of sustainable soil management technologies by smallholder farmers at village, district, country and regional levels. Some efficiency can be gained by better coordination between the myriad of players and initiatives by building synergies and reducing unnecessary duplications. In West Africa today, several interventions are being implemented by development agencies and NGOs as well as national research institutions in sustainable soil management. These efforts are very commendable. However, these sustainable soil management technologies are not accessible to the farmer for adoption or technology adoption is constrained by some institutional factors.

Some of the institutional factors affecting use of fertilizer by farmers include property rights and land tenure arrangements, research extension linkage, institutional lending, distribution of inputs and input and output market. Migrant or settler farmers will have to rent land for crop production for one or two cropping seasons or practice share cropping, where the farmer will share the produce from the farm with the land owner at harvest. In these situations, there is no commitment on the part of the farmer to improve the quality of the soil through, for example, integrated soil fertility management, because he is not sure of his continuous stay on the land. He might be ejected from the land by the land owner. In some communities, especially in the northern part of Ghana, women do not have ownership right to land.

4.5 Property Rights and Land Tenure Arrangements

Land is key asset for rural poverty alleviation. Land tenure embodies the rules and procedure governing the rights and responsibilities of individuals and groups in the use and control of land. For all practical purposes, in Ghana, Mali, Zambia, Malawi, Burkina Faso and in large parts of Sierra Leone, Nigeria, among others, land is predominantly regulated by customary rather than statutory laws. Channels of access to land in Ghana include the family, spouses, sharecropping, lease and purchase, or gift transfers. Land is becoming increasingly scarce due to a variety of pressures, including demographic growth.

4.6 Information Base (Data Base)

Implementation of sustainable soil management practices assumes that:

(a) Reasons for non-sustainability are known
(b) There is sufficient information on the resource base to target activities that will foster sustainable soil management
(c) The resource base can be monitored to evaluate sustainability

Few African countries have a systematic, detailed soil resources inventory program. Agronomic research programs, including those by western expatriates have been conducted and are still being conducted on soils about which little is known. In the absence of information on the resource base, it is usually a waste of time and effort to try and institute sustainable soil management.

4.7 Diverse Agro-Ecologies and Crop Systems Demand Local Adaptation

Africa's staple crops are highly diverse (Africa's food base is made up of at least eight staple crops as opposed to just two crops in Asia) and African production environments are highly diverse, with significant difference in local pests, diseases and rainfall patterns and soil properties. Sustainable soil management practices such as erosion control, rotation, intercropping, planting patterns and water harvesting must be adapted to local conditions and incentives. A major problem for effective utilization of fertilizers has been 'blanket' recommendations that fail to take into consideration differences in farmer's resource endowment. Past fertilizer recommendations have been based on single crops such as maize, cotton, etc and did not take into account complex farming systems involving crop rotations, intercropping and conservation agriculture that are characteristic of most smallholder farming systems in Africa.

4.8 Ensuring Inter-Generational Equity

Population pressures on soil resources have reached critical limits to the point that ensuring inter-generation equity is becoming more difficult. Most traditional cultures and systems of family farming have strong cultural norms that ensure tribal land or family farms are passed to the next generation in the same or better condition than when they were inherited.

4.9 Policy Issues

In many countries, policies regulating soil use are lacking.

Where regional and national SSM policies exist, financing is often not a priority and/or implementation can be ineffective due to lack of political will or a lack of implementation capacity (Pretty 1995).

For policy makers, one very significant issue is the disconnection between the increasingly urbanized population and the soil. Most people especially in the urban areas access food everyday but forget that the food comes from the soil. Human societies depend more than ever before on products from the soil and the services it provides for the maintenance of the biosphere. There is therefore the need to make clear these essential connections between human well-being and the soil.

Achieving sustainable soil management and ensuring success with related policies (e.g. food security, conservation of biodiversity, climate change adaptation and mitigation) requires an understanding of the interconnectedness of policies and the consequences of interactions.

4.10 Socio-Cultural Constraints

This is the socio-cultural environment in which soil management technology transfer and adoption function. Sustainable soil management technology transfer needs to operate at the local community level. The packages of soil management technologies have to be delivered to the farmer in his locality easily and these technologies must be visible to the farmer. In Ghana the farmer/extension ratio is about 1500/1. In a situation as in Ghana and other Sub-Saharan African countries, there cannot be effective extension by implementing 'face to face' extension approach. This therefore, means quite a number of farmers are not reached.

Socio-cultural norms will also dictate to a significant extent the manner in which females are considered to be target of extension's educational programme. In some communities in Northern Ghana it is unacceptable to hold meetings with women without the presence of their husbands.

The role of Non-Governmental Organizations (NGOs) in rural development and for that matter technology transfer cannot be overemphasized. The ability of NGOs to mobilize local people would form a crucial factor in sensitization and awareness creation within farming groups and communities. Over 90% of all NGO activities have direct or indirect relation to the soil.[6] In some areas, the NGOs have tended to improve the farming system of the localities, yet in other areas, the NGOs have openly rejected the use of inorganic fertilizers 'en masse'.

Sub-Saharan Africa has the tragic distinction of being the only region in the world where overall food security and livelihoods are deteriorating, rather than improving.

[6]AGRA (2013), p. 198.

Africa is not producing enough food to keep pace with its needs. Cereal yields in Sub-Saharan Africa averaged less than 1.3 metric tonnes per hectare in 2010 as compared to yields in Asia of about 3 metric tons per hectare.[7] Africa's food production lags because its soils are low in nutrients, low in organic matter and have poor water holding capacity. Africa's soils will continue to degrade and its food situation will continue to deteriorate until those conditions are reversed.

5 Conclusion

Several constraints often lead to resist adoption of sustainable soil management practices.

- Sustainable soil management calls for educating farmers, emphasizing the long-term consequences of their traditional methods of agriculture and helping them implement innovative, appropriate soil protection practices with appropriate incentives.
- Without financial and technical assistance, SSM in Africa will be untenable in the immediate future.
- Awareness creation. The ability of NGOs to mobilize local people would form a crucial factor in sensitization and awareness creation within farming groups and communities.
- Institutional and societal barriers constraining application of SSM technologies that reduce land degradation exist. Lasting solutions can be found through social and economic reforms.

Traditional practices (e.g. bush fallow) have not been replaced by new methods of soil management and cropping systems due to lack of knowledge, lack of essential inputs and lack of incentives.

References

AGRA (2013) Africa Agriculture Status report. Focus on staple crops. 198

Eswaran H (1995) Sustainable agriculture in developing countries: challenges and U.S role. Agricult Environ 27:199–204

FAO (2015) Status of the World Soil Resources - Technical Summary

Francis CA, Hildebrand PF (1989) Farming systems research- extension and the concepts of sustainability. FSRE Newsl 3:6–11

Giller KE, Cadisch GEORG, Mugwira LM, Tagwira F (1998) Potential benefits from interactions between mineral and organic nutrient sources. Soil fertility research for maize-based farming systems in Malawi and Zimbabwe. Soil Fertility Network and CIMMYT-Zimbabwe, Harare, 155–158

---

[7] AGRA (2013), p. 198.

Global Soil Partnership (GSP) (2016) Regional implementation plan for the African soil partnership. Summary of activities. FAO. p 16

Jones A, Breuning-Madsen H, Michel B, Dampha A, Deckers J, Dewitte O, Gallali T, Hallett S, Jones R, Kilasara M, Le Roux P, Micheli E, Montanarella L, Spaargaren O, Thombiano L, Van Ranst E, Yemefack M, Zougmoré R (2013) Soil atlas of Africa. Union Européenne, Luxembourg, p 176. (Joint Research Centre Soil Atlas Series). ISBN 978-92-79-26715-4

Laker MC (2015) Understanding sustainable soil management in Africa. Nat Faune 30(1):3–6

Pretty JN (1995) Regenerating agriculture: policies and practice for sustainability and self-reliance. Earthscan Publications

Local Agricultural Intensification Practices and Soil Degradation in Momo Division, North West Cameroon

Tassah Ivo Tawe

1 Introduction

Humans depend on soil and the interactions surrounding soil for existence and well-being. The over 7.1 billion people on planet earth depend on the biological products derived from the soil for food. This is provided by the 2.6 billion farmers whose livelihoods largely depend on generating food from the soil.[1] However, many people believe humanity has been given stewardship in the sense of supreme authority over nature, and it has resulted in environmental degradation and soil deterioration as well as over exploitation of the land and other natural resources with resultant climatic implications. Meanwhile, the land has to continue taking care of the ever increasing demands of man, despite the fact that there is a diminishing ability to continue producing.

The situation in Cameroon shows that agriculture is the main source of income to approximately 70% of the population and contributes about 20% to the GDP.[2] Cameroon is also one of the Sub-Saharan Africa countries with substantial potentials for growth, profitability, and sustainability in the agricultural sector, owing to its diversity in soils and ecosystems that support the growth of diverse crop types.[3] Despite the numerous opportunities in the natural resource base of the country, the sector has been undervalued leading to underperformance and declining productivity. Studies indicate that Cameroon has gradually moved from a country of food abundance to a country experiencing food shortages, forcing her to receive her first

[1]Chasek et al. (2014).

[2]Central Intelligence Agency (2012).

[3]Dewbre and Borot de Battisti (2008).

T. I. Tawe (✉)
Ministry of Scientific Research and Innovation Yaoundé, Department of Economics and Environmental Sciences, Yaoundé, Cameroon

H. Ginzky et al. (eds.), *International Yearbook of Soil Law and Policy 2019*, International Yearbook of Soil Law and Policy 2019,
https://doi.org/10.1007/978-3-030-52317-6_2

donation of 2600 metric tonnes of food from the United Nations and the Food and Agricultural Organization (FAO) in 1991.[4] These revelations contradict sharply with the economic development policy adopted immediately after independence and reunification of the country in 1961. After independence however, the government of Cameroon recognised agriculture as the main engine for economic development. In line with this vision, Cameroon subsidised agricultural development for decades prior to the Structural Adjustment Programmes (SAP) reforms in the 1990s.

The SAP reforms provided by the Bretton Woods institutions, notably the World Bank, the International Monetary Fund (IMF) and other external agencies were aimed at stimulating economic recovery in the newly independent African countries and Cameroon in particular after the economic crunch that hit the world in the 1980s and resulted in a serious balance of payment deficit. The inability to withstand shocks from the economic crunch of the 1980s prompted the Bretton Woods institutions to implement short term austerity measures for economic recovery in these newly independent African countries and Cameroon in particular. In an attempt to service foreign debts and stimulate economic growth in Cameroon, it was noted that the situation only helped worsen things further, as living standards fell drastically with rising poverty levels.[5] This was mainly due to the conditionalities attached for receiving assistance from the IMF and the World Bank, which in themselves had untold consequences for the government of Cameroon and its citizenry.

Some of the consequences of the conditionalities were the massive retrenchment in the public service, huge salaries cuts, liberalization of agriculture and the withdrawal of subsidies from farmers. After SAP Policies eliminated agricultural subsidies, the rural farmers were plunged into the abyss of poverty and untold misery, as a significant proportion of the retrenched workers from the public service had to add to the already huge mass of small scale farmers, who were exploiting the land with little or no assistance from the government using unsustainable local practices like bush burning, the *ankara* system, slash and burn among others, which in themselves are not suitable for the health of the soil. This marked the era of wide spread environmental deterioration and soil degradation especially for the small scale rural farmers who had no other means to improve on farm outputs.[6] The soil which constitutes the natural resource base became vulnerable to the wimps and caprices of unsustainable local agricultural intensification practices, leading to declining outputs, unfavourable climatic cycles and environmental deterioration. According to studies by Aroldo,[7] soil is at the center of the main challenges of the planet, food, fibre and energy production as well as environmental services. He further noted that as soils are responsible for 95% of global food production, its sustainable management could increase food production by as much as 58% and help reducing in climate change.

[4]Fonjong (2004).

[5]Ibid: 4.

[6]Birner et al. (2006).

[7]de Oliveira (2016).

In order to meet up with the rising food demand amidst the rapid demographic increase in the population of Cameroon, the local farming populations had to resort to unsustainable local practices to enhance production. It has been observed in many developing countries that these local practices of land exploitation for food production do not guarantee soil improvement and sustainability. Studies by Aroldo[8] indicate that soils make up one quarter of the planets biodiversity and account for the second largest global carbon stock, after the oceans. He further stressed that approximately one third of this natural resource is degraded worldwide due to erosion, compaction, sealing, acidification, pollution and depletion of nutrients caused by unsustainable management practices and therefore creating a future challenge for generations to come.

2 The Study Site and Methodology

Momo Division is one of the seven divisions that make up the North West Region in the Western Highlands of Cameroon (see Fig. 1). It is located between longitudes $9^0 58''$ and $28^0 02''$ East of the Greenwich meridian and latitudes $5^0 54''$ and $1^0 91''$ North of the equator. Relatively, the division shares its borders with Menchum Division in the North and Mezam in the East.[9] It equally shares an extensive border with the South West Region covering the Northwest, West and the Southwest of the study area. It has a total land surface of 1792 km^2 with a population of 138,693 inhabitants.[10] The Division is inhabited by the Widikum tribe. The area is equally characterized by highlands and lowlands. The highlands are predominantly mountains and hills with steep slopes, which often are less extensive due to the numerous accidental incised valley systems. Also, V-shaped valleys, deep and steep sided gorges and cascading waterfalls are common features. The occurrence of mountains is sporadic in the whole of the area extending from the Bamenda fault scarp, Tugie, Oshie and the Menka axis. This relief characteristic has impacted negatively on agricultural land and settlement expansion.

The methods employed in this study involved both data collection and analysis. The data collection process had two phases, which were the primary and secondary. Primary data was gained through field survey carried out in the study area in December 2017. Here, the various farming practices and their intensities were identified. The researcher equally embarked on evaluating the potentials of the different ecological niches as well as assessing the environmental peculiarities, investigating evidence of soil degradation and taking note of their spatial intensities. Also, interviews and focused group discussions were major sources of primary data. In total, fifty farmers were interviewed on the methods of farming adopted,

[8] Ibid: 7.

[9] Tawe and Neh (2018).

[10] BUCREP (2005).

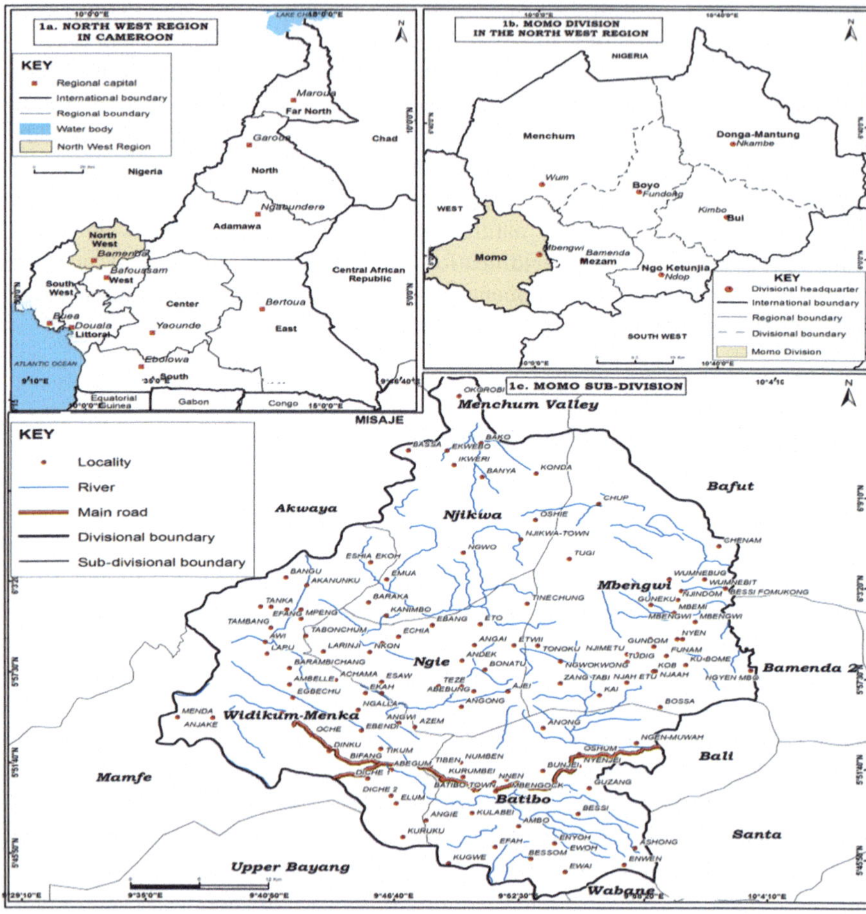

Fig. 1 Location of the Study Area. Source: Drawn by the author on the basis of Landsat images from the National Institute of Cartography, Yaoundé, Cameroon (2017)

agricultural productivity in terms of outputs and land management methods adopted. These, amongst others, were major sources of primary data collection.

In an attempt to justify soil degradation, a chemical test on soil samples collected from this sub region was performed. This was to ascertain the magnitude of change in the soil quality. Two sample sites were selected on the basis of their representation for the entire study area. Two soil samples were collected from two different plots, one in the highland savannah ecological zone and the other in the forested humid zone. The two soil samples were analyzed in a soil sample laboratory at the Agronomic Research Centre (IRAD) Bambui in the North West Region of Cameroon. The results obtained were compared to rate the magnitude of change in the soil constituents within the two zones due to the different agricultural intensification practices carried out in these areas. The secondary data involved the use of published and unpublished works, internet sources and journal articles. The results obtained were analyzed and presented in prose, tables, maps and figures.

3 Discussion of Results

The supposed stewardship of mankind over nature has been observed to have been wrongly instituted in most parts of the earth to levels where, if not checked, emerged consequences will continue to drag the human race into vulnerability. One of the core challenges humanity has to grapple with, is to sustainably managed land resources for agricultural production and other livelihood options, while maintaining the ecological integrity of the soil. Best practices of land use aimed at ensuring sustained food production however, bring to the fore the challenges of food production and soil degradation in Momo Division of the Western Highlands of Cameroon. These challenges focus on ensuring proper land use exploitation for agriculture, generating income to improve on livelihoods while maintaining the ecological integrity of the system and minimising the effects of climatic oscillations.

Agricultural production and food self-sufficiency in Momo Division of the Western Highlands of Cameroon has remained a challenge for the rural dwellers, as a significant proportion of the population depends absolutely on nature for survival. Coincidentally, within a few decades, the demographic increase in the area initiated a significant urge for increased food production so as to meet up with the demands of the growing population. This has resulted in the use of unsustainable methods in land exploitation, as nature has been forced to bear the heavy brunt.[11] This is evident in the decreasing trends observed in agricultural outputs, retreating arable lands and the frequent occurrence of soil erosion, landslides and other earth movement processes, which have become a daily routine in the area.

The scale of environmental hazards and frequency of occurrence initiated by human actions are enormous in this area. At present, more than three quarter of the landscape has been stripped off its natural vegetation, while there has been a drastic decrease in arable and rangelands. Areas such as Mbengwi, Njikwa, Batibo, Ngie and parts of the upper Menka highlands have completely lost their natural vegetation, while savannisation has been on the rise.[12] Human activities such as urbanisation, unsustainable farming practices like bush burning, the *"ankara"* farming system and overgrazing in this Western Highland zone in Cameroon are capable of altering the natural ecology due to conversion or transformation of the land from one use to another. Studies by Brassoulis[13] noted that some processes may be reversible, but others may be irreversible and can equally compromise the land use options and the agricultural productive capacity of the soil for future generations.

---

[11]Ibid: 7.

[12]Tawe and Neh (2018).

[13]Briassoulis (2000).

The desire and manner in which the local population exploits the land to make available their basic needs evidently compromises the possibility for future availability.[14] The quest for farmlands and settlements has left extensive surfaces striped off their natural vegetation. These have initiated higher rates of landslides, soil erosion, fertility decline and the loss of the biomass density of the earth. Some studies have reported that the spatial extent of land areas considered "lost" due to intense degradation increases by about 500,000 km^2 per year in developing countries. He further noted that the FAO reports show, that nature can take up to 400 years to produce one centimeter of soil, so soils' recovery far exceeds the span of multiple generations.[15] Therefore, though often perceived as a given resource, soil is a non-renewable natural resource and is increasingly threatened.

The local agricultural intensification practices carried out by the farming population of Momo Division do not incorporate soil rehabilitation measures. The quest for food and cash crop cultivation gives no meaningful consideration for the preservation of the soil, which is the fundamental livelihood asset to the population. This is in contravention to conventional standards by the United Nations Sustainable Development goal 15 and target 3, which seeks to address issues of achieving Land Degradation Neutral world (LDN) by 2030. In order to meet up with the growing food needs of the people, local agricultural intensification practices such as the slash and burn, bush burning, the "ankara" farming system among others, which do not usher well for soil rehabilitation, are utilised.

As the populations continue to grow, many locals have found agriculture as a source of employment since the secondary and tertiary sector in the area are still largely underdeveloped. The involvement of many in the sector in the midst of retreating arable and rangelands implies more land will be cleared, burnt and exposed, while settlements will continue to impact enormously on the forest and rangeland vegetation and further accelerate climate change in this area. Currently, the only measure to improve on soil fertility by the local farming population is allowing the farming plots to fallow. Unfortunately, increasing and uncontrolled population pressure is decreasing the fallow duration, which undermines the ability of the soil to gain resilience to sustain high crop yield. More still, due to the fact that land as a resource belongs either to a family or the community in Momo Division, there exists a tendency, where individuals, in the quest to make maximum profit, exploit unsustainably and degrade the resource base with impunity.

This section therefore focuses on examining the available land space for agriculture and other land use options in the study area.

[14]Tawe and Lambi (2014).

[15]Ibid: 7.

4 The Nature of the Landscape in Momo Division

The area is made up of mostly highlands, which constitute the most outstanding creation of forces of nature.[16] The altitudinal variation of the landscape ranges from about 185 m above sea level in the southwest section towards the Mamfe basin to about 2035 m in the northeast. Three relief zones are distinguishable based on altitudinal variations and relief configuration. These include: the more mountainous north-northeastern region with heights generally above 1000 m, the undulating central-southwestern section punctuated by troughs/hills of heights less than 1000 m above sea level and the lowland relief area, which is a narrow strip of plain in the southeast, through which the major river (Momo) drains.

From Fig. 2, the landscape is characterized by the outcrop of huge rock masses and hills especially in the northern zone, which portray different slope angles from the convex to the very steep, gentle and concave slopes at the river valleys. Approximately 78.78% of the landscape consists of highlands while the lowland relief covers 21.22% (Table 1)

From Table 1, the topographic contrast in the landscape has made the relief of the region generally difficult, both for human activities and agricultural production. Table 1 shows a general land cover of 17,9258 ha. From this total figure, 38,048 ha (21.22%) constitute the lowlands, 58,937 ha (32.89%) is made up of the intermediate highlands, which coincide to be areas of agriculture, cattle grazing and settlement expansion, while 82,274 ha (45.89%) constitute the extreme highlands. However, with a population density of 77.39 p/km^2, the pressure on the available habitable land is enormous, as the rush for economic fortunes to give life a meaning and the scarcity of livelihood assets is on the rise. In this case therefore, the pressure on the land for agricultural production and other land use options is equally gaining ground. The increasing demand for food crop production has accelerated pressure on the land for increased production.

As indicated in Table 1, the available low lying areas have been overexploited due to high pressure on scarce resources. It is worth noting, that a significant proportion of the land (45.89%) is made up of highlands, which are extremely hostile and regarded as "*no go zones*" for human activities or agriculture. The ecological niches are also very fragile since any imbalance on the slopes such as vegetation clearance for agriculture and mass movement processes make reconstitution difficult. The hilly nature of the landscape has also made agricultural development difficult and very unsustainable, initiating higher rates of surface wash processes and landslides. This is common especially in Menka, Ngie and parts of Njikwa.[17]

Equally, it has been observed that government policies on agricultural production in Cameroon are skewed towards the production of cash crop, since this is a major source of income for the government. In this light, very little attention is accorded to

[16] Ibid: 12.

[17] Ibid: 12.

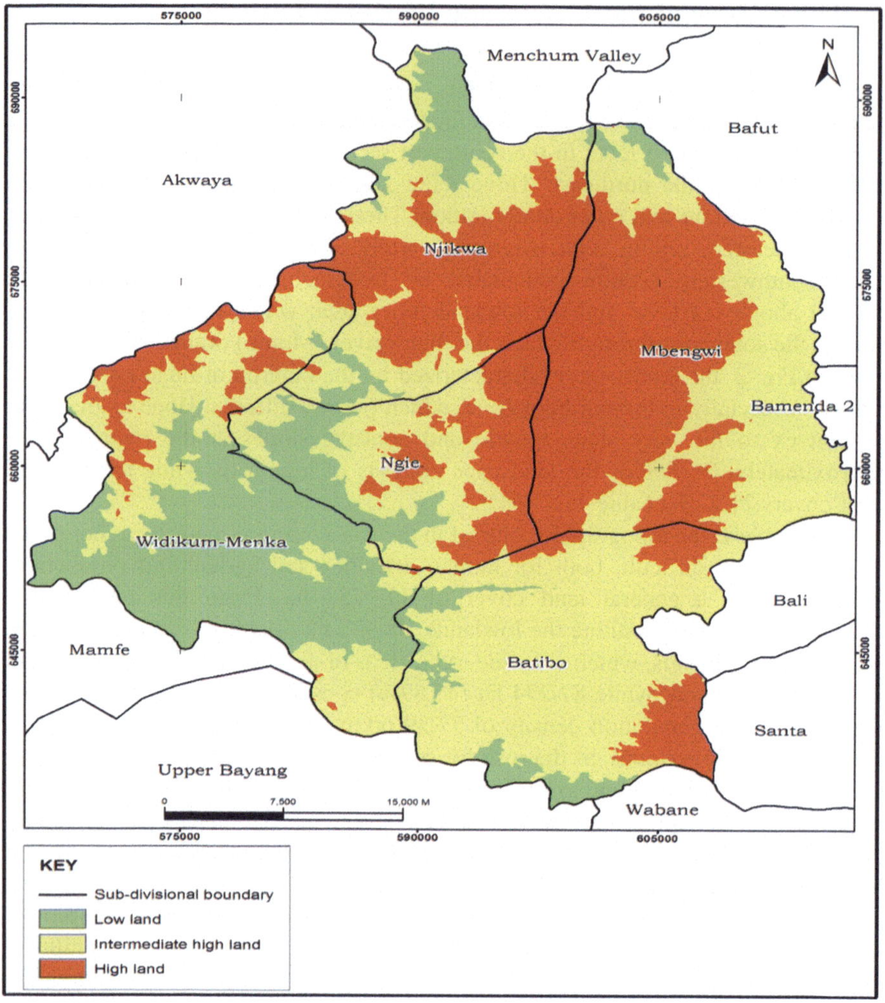

Fig. 2 The various Relief zones in Momo Division. Source: Drawn by the author on the basis of Landsat images from the National Institute of Cartography, Yaoundé, Cameroon (Ibid: 11)

Table 1 Proportions of lowlands to highlands in Momo Division

Nature of land	Area in hectares (ha)	Percentage (%)
Lowlands	38,,048	21.22
Intermediate highlands	58,937	32.89
Highlands	82,274	45.89

Source: Tawe and Neh (2018) (Ibid: 12)

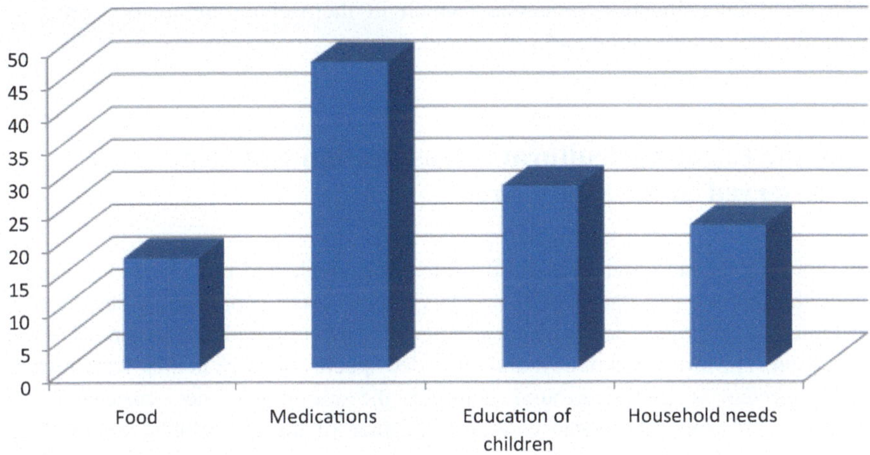

Fig. 3 Distribution of the uses of income from the sale of palm oil

food crop production. Since the rural farmers have to eke out a living from the land, they are placed in a dilemma whether to produce cash crops, that fetch higher incomes or food crops, which are readily consumed after production. This orientation in the Widikum-Menka production basin in Momo Division has pushed a significant proportion of the rural farmers to resort to palm oil production, with very little attention for food crops. The humid forest vegetation in the Widikum-Menka area has been heavily exploited for cash crops like oil palms and cocoa. In these circumstances, the available arable land for food crop production has been on a decrease over the years. The local farmers are therefore faced with the challenge of increasing food production to meet up with the mounting pressure of the growing population. With the increasing pressure for food production, the farmers have become *"environmental refugees"*, as they seek for promising agricultural havens across regional boundaries into the Mamfe basin of the South West Region of Cameroon. As observed during focused group discussions, the local farmers noted that conflicts over agricultural land are very common at the resource frontier zones in the Mamfe basin.

From the above observations, one could be tempted to say that a significant percentage of revenue generated from the sale of palm oil is used to buy food crops for most households. In an attempt to investigate how revenue generated from the sale of palm oil is used during focus group discussions, it was observed that about 80% of the revenues are used for healthcare (Fig. 3).

Figure 3 indicates that a significant proportion of income generated from the sale of palm oil is used to buy drugs for health related issues, followed by education of children and household needs, while very little is accorded for food. These observations explain why hunger was noted to be a common phenomenon in the whole of the study area. However, areas noted for food crop production were observed to be under local intensification processes, which in themselves are major agents of environmental deterioration and soil degradation.

5 Some Local Agricultural Intensification Practices Observed in the Study Area

The rapidly growing population amidst retreating arable and rangelands has necessitated the farming population of Momo Division to adopt coping strategies through local intensification practices (capital deficient), to meet up with the food need of the population. The utilisation of new and upgraded species of crops in their farming and grazing system is gaining ground. Through the adoption of new measures, the farming populations have been changing the face of the agricultural landmark in the study area. The local farming intensification measures observed range from farm preparation through soil fertility enrichment, irrigation, crop harvesting and processing techniques. The manner in which farming plots are prepared significantly determines the farm performance.[18] However, the increasing pressure on land for farming in the study area due to land scarcity has compelled the farmers to effect changes in their farming operations in order to render productivity sustained at acceptable levels to satisfy the ever growing need of the population. In doing this, local intensification practices, which are agents of soil degradation, are widely used.

6 Farm Clearing Intensification Techniques

The various farm clearing techniques identified in the field range from the old primitive methods of using fire as a means of clearing to the use of cutlasses and chemicals to kill grass, which have been noted as very unsustainable to soil productivity. Though fire as a means of farm preparation, as reported during field interviews, has long been losing its importance, it was observed that it has not been totally eliminated. However, the percentage of those using fire is significantly reducing in favour of the cutlass and chemicals. It was noticed that the use of fire in farm preparation varies across the study area, based on the peculiarities observed in the different ecological zones. As noted, the study area is made up of two ecological zones, which are the forest and the savannah. Field observations showed that fire as a means of farm preparation was more common in the savannah ecological zone than the forest zone. With this, the researcher resorted to finding out the reasons for the observed differences at the level of farm preparations in these different ecological zones. On this premise, a set of conditions were evaluated to

[18]Fondze (2010).

Fig. 4 Fire used in farm preparation in Menka (2017)

have necessitated the differences at the level of farm preparation in the two ecological zones. These include:

- The species of grass in the savannah ecological zone are different from those in the forest zone. From field interviews, it was reported that the cutlass is not efficient in eliminating some of the species of grass which are a deterrent to crop productivity. This therefore necessitated the use of fire as a means of farm preparation.
- Equally, due to the fact that the humus content in the savannah ecological zone is low, burning of the savannah becomes a means of increasing the soil fertility. Burning of the grass with fire has the ability to increase the phosphorus and potash content of the soils, which are excellent soil nourishment constituents.[19] It was reported during focused group discussions by farmers that areas burnt by fire in the process of farm preparation are associated with increase in output, but this only lasts for a year or two, after which the soil loses it fertility (Fig. 4).

Recent trends in the demand for more yields and the inability of the cutlass to ensure large scale clearing has resulted in farmers adjusting to current exigencies, where herbicides are used in the preparation of farming plots and in the process of removing weds from the farms. Herb selective herbicides are currently being used to clear large expanses within a limited period. This phenomenon is common in Widikum where herbicides are used in the preparation of cocoa farms, Batibo and Mbengwi.

[19]Lambi (2001).

Despite the above techniques, it was noticed that fire is still being used even in the forest ecological zone of the Widikum basin, Menka, Ngie and other areas, especially in newly opened farming plots in thick forest areas to ease access. This however, as observed in the field, does not discard the use of the cutlass in farm preparation, as it still remains the principal farm preparation tool. These techniques are still being used in association with each other by the same farmers or by different farmers in the same or different parts of the study area, due to the need for more farm yield.[20]

7 Soil Fertility Techniques

Significant changes have been observed in land cover mutations in Momo Division, which have necessitated improvements in farm management techniques in order to improve on crop yields. From field observations, most of the soil fertility intensification techniques identified were mostly common in the urbanising areas. These areas include Mbengwi, Batibo and Widikum, where there is the dire need for arable land in order to improve on food crop production. The soil fertility techniques identified include the use of green manure, animal wastes, compost, crop residues, chemical fertilizers, fallowing, slash and burn, soil burning or "ankara" among others. In an assessment of the various soil fertility intensification techniques identified, it was noted that the application of each of the techniques varies from one locality to another, based on the ecological peculiarities.

It was observed that green manure is generally used as a strategy to enhance soil fertility in the whole of the study area. Based on this, most farmers noted that the use of green manure is mostly common in farms which have not been burnt with fire in the process of land preparation. It was equally observed that farmers are increasingly seeing the essence of green manure as a soil fertility measure in order to improve on crop yields. The use of green manure was noted to be common in plantain and banana farms in the forest ecological zone of the Widikum production basin. In an attempt to evaluate output in farms using green manure, farmers indicated that it is an excellent method especially in the output of plantains and banana farms.

It was also noted that the use of animal manure as one of the soil enrichment intensification strategies is common in the savannah ecological zone of the study area. The droppings of animals such as pigs and fowl waste are used mostly in tomato and vegetable farms. This strategy, as noted, is to reduce expenditure on organic fertilizers. This observation was common with farmers who carry out market gardening in Widikum, Batibo and Mbengwi. Also, household waste, such as wood ash and crop residues, was observed to be one of the soil nourishment constituents in order to improve on farm output. The use of the above measures was common in the

[20]Ibid: 21.

Fig. 5 Wood ash on ridges for improved soil fertility and farm output (2017)

urbanising areas while elsewhere, bush burning and the "ankara" farming system are common and in some areas, nothing is applied to the soil for nourishment.

The use of wood ash in the farms, according to focused group discussions, is as a result of the gradual abandonment of the "ankara" farming system, which to the farmers is becoming more damaging to soil fertility and a drop in farm output.

The use of chemical fertilizers (NPK) is equally a common practice in soil fertility improvement as observed in the field. However, it was noted that the use of chemical fertilizer was more common in the savannah ecological zone than the forest zone (Fig. 5).

8 Land Exploitation Challenges in the Study Area

Generally, land exploitation practices for agricultural production in the study area are subsistent. The agricultural sector, which remains the only hope for the rural masses, is challenged with a series of physical and human constraints (Table 2)

Field observations indicate that three major problems are fast emerging in the study area as a result of unsustainable land exploitation practices for agricultural production. These are:

- Soil degradation.
- Increasing vulnerability of the population to livelihood assets and services.
- Increasing mobility of the rural masses within and out of the study area in search of agricultural havens.

In an attempt to justify land degradation, the researcher conducted a chemical test on soil samples collected in the area. This was to ascertain the magnitude of change in soil quality due to local agricultural intensification practices. Two sample sites

Table 2 An assessment of land exploitation challenges and soil degradation in Momo Division (2017)

Challenge	Impact on resources	Observation
The physical landscape	Foster slope instability, soil erosion and make ecosystem rehabilitation difficult	More than three quarter of the landscape is difficult for farmers to efficiently exploit
Soil infertility/ erosion	Deeply weathered laterites which are fast eroded and leached to render them poor to sustain crop productivity	A serious limitation to farmers. Without chemical inputs and innovations, yields will remain low
Climate	The impact of seasonality on land and water resource exploitation is high. Climatic uncertainty has altered the agricultural calenda	Agriculture is rain-fed. Low input of rain is liable to crop failures in the area since irrigation schemes are absent
Pests and diseases	There are crop and animal specific. Their prevalence is a huge damage to crops and animals in the area	Pests and disease control is insignificant. Seeds, growing crops and outputs are heavily damaged by pests and diseases
Land tenure systems	Tenure insecurity is high to guarantee efficient management of land resources	Ownership procedures are rigid but access to exploitation is very open.
Poverty	Foster the adoption of poor methods in the exploitation of resources. Resource rehabilitation and development is poor	Resource exploitation methods are unsustainable. They are liable to create livelihood problems and future poverty
Slash and burn agricultural practices	Leads to deforestation, soil infertility and the collapse of ecosystems and extinction of species	If not substituted with modern methods of cultivation, soil degradation will remain high in the area
Tilling systems	Cone-shape ridges result to high soil loss on hill slops	Contour ploughing, strip cultivation and terraces should be encourage
Lack of stakeholder support	Innovation diffusion in sustainable exploitation and management of natural resources is difficult	Urgent stakeholder intervention in all the land resource sectors is necessary
Inadequate farm to market roads	The economic value of resources (price) is low. Their conservation as wealth is not worthwhile	Roads/markets infrastructural development for the area is a prerequisite for poverty alleviation and development

Table 3 Summary of soil fertility parameters for the two samples (2017)

	Sample A (Savannah ecological zone)		Sample B (Forest ecological zone)	
Soil fertility parameters	Soil Depth 0–30 cm	Soil Depth 30–60 cm	Soil Depth 0–30 cm	Soil Depth 30–60 cm
Organic carbon %	3.92	1.98	2.30	1.21
Total Nitrogen %	0.38	0.05	0.12	0.09
Available phosphorous	11	6.3	1.0	1.0
K+ cmol (+)kg	0.38	0.09	0.22	0.00
Ca2+	2.01	0.02	1.88	0.57
Mg2+	1.40	0.01	1.48	1.30

were selected on the basis of their representation for the entire study area. Table 3 illustrates the results for sample "A", (soils within savannah ecological zone) and sample "B" (soils within the forest ecological zone).

From Table 3, the general evaluation shows that sample 'A' (Soil within the savannah ecological zone) has an average depth greater than 150 cm. The soils are well drained and dark reddish brown in colour. They range from clay loam to sandy clay, with low overall fertility.

The general characteristics of soils in sample 'B' (soils within the forest ecological zone) show that they are deep, well drained and dark brown in colour. The soils have higher clay content with moderate fertility.

The analysis was focused on evaluating the degree of change in the percentages of the different soil constituents at different depths. It however concentrates on the general soil fertility parameters, whose alterations have great implications on agricultural productivity and the soil health. These soil fertility parameters include organic carbon, total nitrogen, available phosphorus and bases (k+, mg+ and Ca+), which are crucial in sustaining agricultural productivity. The content of these soil parameters was observed to differ between 0–30 cm and 30–60 cm of soil depth. The results showed a decreasing trend in the above soil fertility components in the two samples.

Based on the results in Table 3, most of the soil fertility components have decreased for the two ecological sites. The decrease is significant for the top soil horizon (0–30 cm) than the horizon below (30–60 cm). The organic carbon percentage for the entire division has decreased severely, 3.92%–1.92% for the forest ecological zone and 2.30–1.21 for the savannah ecological zone in the top horizons. This decline is probably due to farming without or with inadequate soil amendments like organic matter, fertilizers and others. It is also because of the increasingly shorter fallow periods. The implications of the decrease in the percentage of organic carbon in the soils have been the observable drastic changes in the climatic cycles in this area, as noted in the disappearance of surface streams, the visible presence of dry valleys and the rise in the number of ephemeral streams, which in the past were rare to find.

The results equally indicate that the rate of decrease in the soil fertility components is higher for the savannah ecological zone than the forest ecological zone. As to why this disparity should occur in soil fertility decline between the two areas, the researcher raised the following justifications based on field observations:

- Topographically, the savannah is on a higher elevation (<800 m above sea level) with a general hilly and mountainous landscape, which fosters higher rates of soil wash than the forest area with a relatively undulating landscape.
- Another possible justification for this disparity between the two areas is the cattle grazing activity which is restricted to the savannah, and the excessive annual bush fires possibly account for its higher rates of soil degradation.
- The forest area is more accessible than the savannah. The accessibility of the forest has fostered innovation diffusion and modernisation in the agricultural

sector (the farmers easily access chemicals, seeds, field technicians, information); this cannot be as easily obtained in the rest of the division.

- The forest has a smaller surface area (21.22%) and a higher population concentration. This was observed to have caused such high population pressure on land in the forest area that most farmers are forced to modernise the sector. This is not the case in the rest of the sub region hence higher rates of land mismanagement and degradation.

In spite of the above spatial disparities in the rates of degradation, land resource degradation among other resources remains the major threat to rural livelihoods in the area. The rate of decrease in the parameters within the two decades is rated high. Despite this warning, the study disclosed that land degradation may persist for the rest of the century, since there are no blueprints for sustainable agriculture and land management for the division coupled with a broken public extension service in the area. However, rural vulnerability and population displacement to seek fertile soils elsewhere, though not a recent phenomenon in the study area, is at the present scale of mobility alarming. This is causing a majority of rural poor to become *"environmental refugees"*.

It has been observed that land within the study area has been highly exploited for the cultivation of crops. From field visits to some major crop zones and rangelands, it was observed that the various crops cultivated have a bearing with the soil types identified (Fig. 6). The hilly savannah areas with lateritic soils were noted for the cultivation of cereal crops such as maize, beans, groundnuts and equally cash crops like the Arabica coffee among others, while the humid forested zones were characterized with the cultivation of plantains, yams, cocoyams, cassava and cash crops like palm trees for the production of palm oil, cocoa and Robusta coffee. However, it was observed that the practices involved in the cultivation of these crops do not guarantee ecosystems productivity and soil sustainability, as these practices were noted to be primitive and subsistent. Crop productions per unit area were noted to be generally low. From field interviews, it was disclosed that the decline in crop yields increases with longer duration, such that most farms are abandoned by the third and fourth year of cultivation. Despite this, the local farming populations have persisted in crop production with the use of unsustainable practices as a means to raise income to meet up with their livelihood needs and further accelerating soil degradation and alterations in the climatic cycles in the area.

9 Conclusion

There is overwhelming evidence that the earth support system both locally and internationally is in crisis. The necessity to reduce the vulnerability of the increasing population for food and other material shortage in Momo Division in Cameroon and Africa at large has caused man to persist in his quest for *"hasty battles against nature"*. This chapter reveals that land resource exploitation and soil degradation in

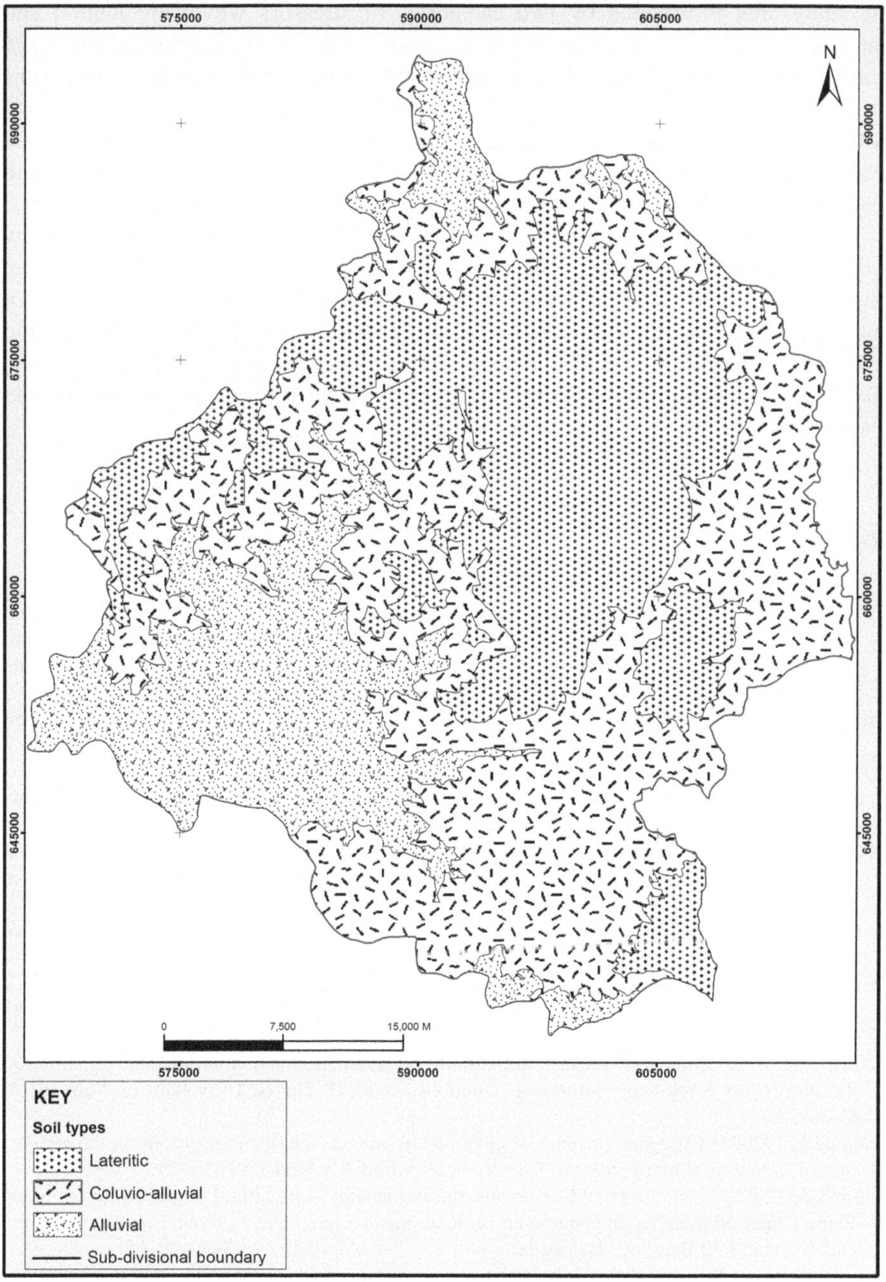

Fig. 6 Spatial distribution of soil types in Momo Division (2017)

the study area is affected by two categories of stressors which are natural and anthropogenic. In as much as the scale of the problem is expanding with large scale implications on rural livelihood sustainability and the environment, it remains grossly unaddressed, be it by the State or civil society organizations. However, it is not too late for actions to be taken, despite the effects of these stressors, as the resource base of the area, though greatly affected, still possesses the capacity and resilience to sustain human numbers and aspirations if properly managed.

It has been observed that the land use policies in the study area are governed by individual land use decisions. This is more so due to the near absence of the State in implementing land use decisions, coupled with a broken public extension service and institutional lapses. This chapter therefore suggests a legal framework in land resources exploitation with efficient enforcement mechanisms put in place that must take into consideration the usufruct right of the local farming population as land resource managers.

References

Birner R, Davis K, Pender J, Nkonya E, Anandayasekeram P, Janvier M, Cohen M (2006) From best practice to best fit: A framework for analyzing pluralistic agricultural advisory services worldwide (DSGD Discussion Paper No. 37). International Food Policy Research Institute. Retrieved from http://www.ifpri.org/sites/default/files/publications/dsgdp37.pdf

Briassoulis H (2000) Analysis of Land use Change: The theoretical and Modelling Approaches. http://www.wvu.edu/webbook/Braissoulis/chapter3(Theories).htm

Bureau Centrale des Recensement et des Etudes de Population (2005)

Central Intelligence Agency (2012) The world fact book: Cameroon. Retrieved from https://www.cia.gov/library/publications/the-worldfactbook/geos/cm.html

Chasek P, Safriel U, Shikongo S, Futran Fuhrman V (2014) Operationalizing zero net land degradation: the next stage in international efforts to combat desertification? J Arid Environ. https://doi.org/10.1016/j.jaridenv.2014.05.020

de Oliveira AC (2016) Audit of soil governance. In: Ginzky H (ed) The international yearbook of soil law and policy 2016. German Environment Agency Dessau, Dessau-Roßlau, Germany

Dewbre J, Borot de Battisti A (2008) Agricultural progress in Cameroon, Ghana and Mali: Why it happened and how to sustain it (OECD Food, Agriculture and Fisheries Working Papers, No. 9). https://doi.org/10.1787/241275631215

Fondze BB (2010) Population change, agricultural diversification and environmental dynamics in the Northwest Region of Cameroon. Unpublished Ph.D Thesis, University of Yaoundé 1, Cameroon

Fonjong LN (2004) Changing fortunes of government policies and its implications on the application of agricultural innovations of Cameroon. Nordic J Afr Stud 13(1):13–29

Lambi CM (2001) Environmental Constraints and Indigenous Agricultural Intensification in Ndop Plain, Upper Nun Valley of Cameroon. In: Readings in Geography. Unique Printers, Commercial Avenue, P.O Box 592, Bamenda

New King James Version of the Holy Bible

Tawe TI, Lambi MC (2014) Topographic constraints and agricultural production in the Widikum-Menka Highlands, Northwest Region of Cameroon. Afr J Soc Sci 5(3):16–29

Tawe TI, Neh AV (2018) The implications of unsustainable agricultural land resources exploitation in the Widikum-Menka Highlands, North West Cameroon. Int J Law Soc 1(1):34–45. https://doi.org/10.11648/j.ijls.20180101.15

Persistent Organic Pollutants and Soil Protection: National and Global Imperatives

Patricia Kameri-Mbote

1 Introduction

The unconsolidated mineral and organic material occurring at the earth's surface is what is meant by soil.[1] It refers to the various organized layers of the earth's surface formed by the mix of parent material, organisms and topography over a long period of time and acts as a natural medium for growth of plants.[2] The nature of the soil making process means that a single inch of soil takes very long periods of time, even centuries or millennia to be formed, based on the variances in location and geographical conditions.[3] It is apparent at once, that precisely because of the nature of the soil making process, in the event that soil degrades or erodes at a faster rate than it is formed, it is likely to become scarce with serious effects on the environment and humanity.[4] In such an event, the only option left would be an to attempt to create artificial soil which is problematic, not least because such process is expensive and difficult to scale up at a global level particularly in developing countries as reiterated by Locke.[5]

Soil presents a wide range of natural habitats and various environmental ecosystem services to humans, wildlife and supports diverse natural ecosystems.[6] Primarily soil is a source of livelihood for organisms but additionally for humans, it forms the

[1]Hannam and Boer (2002), p. 9.

[2]Richter (2007), pp. 961–962.

[3]White (2013), pp. 15–30.

[4]Marshall (2007), pp. 33–35.

[5]Locke (2007), p. 1.

[6]Howard and Lawson (2015), p. 1.

P. Kameri-Mbote (✉)
University of Nairobi, Nairobi, Kenya

H. Ginzky et al. (eds.), *International Yearbook of Soil Law and Policy 2019*,
International Yearbook of Soil Law and Policy 2019,
https://doi.org/10.1007/978-3-030-52317-6_3

bedrock of trade and development. Indeed soil is a critical component of biodiversity and provides a critical ecosystem for species, particularly as a "dynamic complex of plant, animal and micro-organism communities and their non-living environment interacting as a functional unit"[7] within terrestrial ecosystems. Due to the wide range of activities in the various environmental systems, soil is often subject to complex systems of property rights regimes, land uses, cultural, economic and political interactions.[8] The balancing of these complex regimes, competing soil uses, sustainable use of the soil and the uncertainties in the management requires development of suitable soil policies, governance mechanisms and monitoring of its utilization for sustenance of present and future generations. Principles for sustainable governance of soils are imperative for these activities if extraction, degradation through erosion and overuse are to be checked.

This Chapter examines soil governance and the impact of Persistent Organic Pollutants (PoPs) thereon globally. It uses Kenya to illustrate how nation states are dealing with the issues. It proceeds as follows: Section 1 is the introduction while Sect. 2 lays a background on soil and Persistent Organic Pollutants. Section 3 considers the elements of soil governance whereby it reviews the various policies and targeted strategies that have been initiated with a view to protecting soil as a natural resource. It also looks at PoPs underscoring their relevance for soil governance. Section 4 examines the international legal framework regulating PoPs. This leads to Sect. 5 that links law, soil governance and PoPs focusing on how international law has addressed these issues individually and the interfaces between them. Section 6 looks at soil governance and the regulation of Persistent Organic Pollutants in Kenya while Sect. 7 concludes.

2 Background on Soil and Persistent Organic Pollutants (PoPs)

2.1 Soil

Hanna and Boer broadly categorize soil as serving three main functions: ecological, cultural and land use functions.[9] Soil enables agriculture and thus ensures food security because humans and animals alike need plants to eat and these plants grow on soil. Soil also avails moisture, air, buffer against extreme temperatures, anchorage of roots and nutrients to plants thereby enabling their flourishing.[10] Efforts aimed at expanding agriculture, which degrade soil, are counterintuitive, as is protection of

[7]Assembly (1992a) Article 2.

[8]Allan (2008), p. 61.

[9]Hannam and Boer (2002), p. 10.

[10]Wild (2003), p. 21.

soil that unduly restricts or reduces agricultural production. The right balance is needed to ensure sustainability of soil for both the present and posterity.

As a host to biodiversity, soil contains one of the most diverse compositions of living organisms such as bacteria, fungi and other larger organisms.[11] So rich is the soil that studies indicate that a single gram of soil may contain millions of bacteria and thousands of species of living organisms.[12] Biodiversity is important and shold be protected protected because it provides stability and resilience in the ecosystem and the real economic benefits are of particular importance for the society. For instance, cultured soil bacteria has provided valuable new drugs.[13]

Soil also acts as a water purifier filtering the water it contains before such water reaches surface streams or groundwater, decomposing and absorbing harmful chemicals in the process.[14] Moreover, soil plays a direct role in regulating the global climate by being a critical component in the carbon cycle that enables sequestration (absorption) of carbon dioxide from the air, thus reducing global warming.[15] Indeed, soil is the largest surface carbon pool holding twice the carbon dioxide held by the atmosphere and thrice as much as that stored in terrestrial biomass.[16] Admittedly then, the potential of soil as a carbon sink has been underutilized and not much focus has been placed on it as has been on plants.[17]

Over and above the foregoing, soil plays a cultural role as it forms an essential part of landscapes holding crucial archaeological information relating to the history of humankind and the planet earth.[18] In essence then, soils are a cultural heritage that have similar significance to geographical origins of wines giving rise to geographical indications in intellectual property legal regime.

From the foregoing, it is evident that soil plays a principal role in the ecosystem, in enabling various land uses that are critical for survival of living organisms and as a cultural heritage. Yet, despite these invaluable benefits, soil has not been given as much attention in the broader corpus of international environmental law as has climate change or water pollution. This is partly the reason why soils have been degraded the world over, with deleterious spill over effects in all countries and particularly pernicious effects in developing economies.[19] For instance, a third of Sub Saharan Africa's population is undernourished, and this is likely to get worse with increasing populations.[20] This, and the relatively little attention accorded to the

[11]Brussaard (1997), pp. 563–564.

[12]Food and Agriculture Organisation of the United Nations, FAO (2015).

[13]Handelsman (2005), pp. 20–22.

[14]Wild (2003), p. 86.

[15]Solomon et al. (2007), p. 57.

[16]Roose et al. (2006), p. 13.

[17]Comis (2001), p. 4.

[18]Hannam and Boer (2002), p. 11.

[19]Winiwarter (2005), p. 490.

[20]Skoet and Stamoulis (2006), p. 5.

development of legal regimes to protect soil at the global level, makes soil governance an urgent need.[21]

2.2 Persistent Organic Pollutants (PoPs)

Population pressure and the need to boost production have resulted in innovations to satisfy human needs. This, particularly the quest to improve productivity, has had impacts on soil. Soil degradation may be attributed to various factors such as erosion caused by human activities and degradation by PoPs that are produced by chemicals and fertilisers applied to soil. The use of chemicals in industrial production, waste management and agricultural chemicals to improve soil viability or assist in mass crop production has grown rapidly over the years. These chemicals, though useful to human beings, have negatively affected the environment and soil is no exception.

POPs are organic carbon-based compounds that may encompass synthesized substances of agricultural and industrial usage or by-products resulting from both human and natural activity.[22] They resist natural biological degradation processes and hence persist in the environment. They are easily soluble in water and little soluble in fat which makes them bioaccumulate in fatty tissues of animals and human beings. This results in biomagnification of PoPs up the food chain with the predators having the highest concentration of these harmful chemicals.

Comprehensive studies have established the relationship between PoPs and decreased aquatic life, human diseases, animal and vegetation decline.[23] PoPs have adverse effects on human health such as neurophysiological effects including learning disorders, behavioural problems and poor motor coordination.[24] Global concern was particularly aroused due to PoPs capacity to move long distances due to their transportability. Despite the indigenous nature of the Arctic areas for instance, large deposits of PoPs have been noted having travelled long distances.[25] The cold temperatures are attractive to these chemicals as they are able to condense from their vapour form entering into water sources and soils.[26]

PoPs include pesticides, insecticides and transformers, while unintentionally developed ones occur through processes such as combustion.[27] The majority of the PoPs is intentionally produced for use in the industries and agricultural production, whereas others are unintentionally produced as by-products of other useful chemical processes. Some industrial processes that may result in the release of POPs

[21]Wyatt (2008), p. 206.

[22]Mansour (2009), p. 531.

[23]Bănăduc (2016), pp. 95–100.

[24]Resource Futures International for World Bank and CIDA (2001), p. 22.

[25]Van Oostdam et al. (1999), p. 22.

[26]Wenjie (2008), p. 9088.

[27]Wenjie (2008), p. 9087.

into the atmosphere and into the soil include: waste incinerators which may release hazardous wastes, waste oil refining, textile and leather dyeing, thermal processes such as in the metallurgical industry, industrial boilers and furnaces which use fossil fuels and manufacturing of chemicals such as chlorophenols among others.[28]

As pointed out above, increased demand for food and industrial development have led to the increased use of PoPs. Even in industrial production, these chemicals evaporate and once they condense end up in the soil or in water that later rains or ends up in soils as effluents. Soil is the biggest reservoir of PoPs due to their use in agriculture, the establishment of dumpsites for industrial waste and condensation that lands PoPs on soil.[29] Soil is therefore a critical factor in addressing PoPs. The ubiquity of PoPs calls for participatory, effective and efficient governance mechanisms.

3 Soil Governance and PoPs

3.1 Soil Governance

Soil governance incorporates the mechanisms developed through policies, rules, processes and strategies though which decisions on soil use are made, balancing the competing interests therein and their implementation by governments at international, national and local level. Essentially, at the various levels, different mechanisms are used including regulation, planning, economic and voluntary stakeholder participation.[30] It also incorporates soil regulation in order to avoid conflicts between stakeholders. Although soil governance has largely been limited to agricultural production and industrial use, there is increased concern on the soil ecological decisions that affect wildlife, human beings, plants and indigenous vegetation. Similarly, soil use allocations have largely concentrated on formal institutions such as legal prescriptions and legislations neglecting informal institutions of governance such as rules, norms, market considerations and attitudes towards soil uses.[31] In order to properly govern soil utilization among the competing interests, the informal and formal institutions are important.

Unlike water, bio-diversity and air, which are now universally accepted environmental public goods, soil continues to be a private good.[32] Accordingly, air and water are subject to concrete definitions of property rights, which influence the sustainable use of these environmental goods. Evidence shows that the more valued a resource becomes, there is a tendency to more aptly define the property rights in

[28]Wenjie (2008), p. 9090.

[29]Martina (2008), p. 3589.

[30]Juerges and Hansjürgens (2018), p. 1629.

[31]Juerges and Hansjürgens (2018), p. 1629.

[32]Juerges and Hansjürgens (2018), p. 1630.

such a resource. There is no clear delineation of property rights in the soil and soil has not yet been globally recognised as a public good. The conservation, governance and maintenance therefore continue to be at the whims of private land owners and communities.[33] These are largely voluntary and ensconced within the context of full property rights to the land where any interference with use of the soil attracts debates of compensation for programmes such as water purification by soils and carbon sequestration.[34]

Soil governance the world over has been through different policies and agreements at the global and regional levels, as well as national implementation plans and laws at the municipal level dealing with other matters. The momentum on soil governance has however, recently gained traction especially with the adoption of the Sustainable Development Goals (SDGs) in September 2015, which replaced the Millennium Development Goals.[35] SDG number 15 provides for protection, restoration and promotion of sustainable use of terrestrial ecosystems, sustainable management of forests, combating desertification, and halting and reversing land degradation and biodiversity loss. This necessarily encompasses prevention of soil degradation and makes proper soil governance an imperative for all nations that have adopted the SDGs. The SDGs were an outcome of the Rio+20 Conference held in 2012, which came up with a document entitled 'The Future We Want' which provided for measures of preventing desertification, degradation and drought as provided in paragraphs 205–207.

Essentially, the lack of a global concrete policy on soil may indicate that soil is yet to gain global recognition and nation states continue to neglect this important resource while private individuals see it as an economic resource only. This has led to a multi-layered soil regulation landscape where soil is haphazardly regulated only as a tangential issue in the regulation of other natural resources. This incoherent governance has created ineffective, inefficient and illegitimate soil governance.[36] The fragmented regulations address soil as an enabler of other productive resources such as water sustenance, industry development and human life support.

3.2 Impacts of PoPs on Soil Governance

Although cultivation in soils affects the balancing of flora and fauna and may be said to disturb the soil, it is reversible unlike PoPs, which last for a long time.[37] Their effect on soil, for instance killing some soil organisms are long term, this leaves the soil with few soil organisms that multiply in large numbers reducing the soil

[33]Weigelt et al. (2015), pp. 57–65.

[34]Kerekes (2011), pp. 315–338.

[35]Assembly (2015) Goal 3, 13 and 15.

[36]Juerges and Hansjürgens (2018), p. 1629.

[37]Edwards (1965), p. 330.

organism diversity. This imbalance negatively affects soil productivity. Harris in examining the effect of Dichlorodiphenyltrichloroethane (DDT) on soil noted the change in soil arthropods composition and population density thus creating mobilization and immobilization in the soil, which negatively affects plants growth.[38] Additionally, studies have shown that some PoPs kill enemies of harmful organisms leading to increased populations of harmful organisms on soils. For instance, some PoPs pesticides are known to kill surface beetles increasing aphids that eat the plants.[39]

Ordinarily, toxic chemicals alter the soil ecosystems fundamentally within a short time.[40] If the chemical is both toxic and persistent, the soil ecosystem is altered fundamentally and takes a long time to recover. Additionally, studies on the effect of DDT and aldrin on plants from soil indicate that they radically change the tryptophan nutrient content, which is important amino acids.[41]

The transportability and persistent nature of PoPs are important characteristics in looking at impacts on soil. More specifically, the chemicals can move from contaminated to healthy soil, making their effect expansive and capable of affecting large compartments of soil in different areas.[42] This reduces the productivity of soils in many areas.[43] Treatment of contaminated soils also becomes difficult due to its ability to move to faraway places and their insoluble nature including being in other ecosystems such as air and water.

As has been noted, soil remains a private good despite its global importance. This may be attributed to the linkage of soil to sovereignty and land. Further, soil unlike air and water, has trans-boundary effects.[44] Even at local levels, the government is mostly concerned with the productivity and economic viability of the soil. Soil governance therefore is also limited to the private, voluntary negotiations on soil utilizations.[45] However, like in the case of PoPs, use of the soil for agricultural produce, industrial practices and dumpsites has a trans boundary effect and calls for global alarm. Instead of soil serving human good, PoPs accumulation has resulted in it being a source of diseases and death to human beings.

Presence of PoPs contaminants and their effects on human health and the environment and particularly in the soil calls for a change in the soil governance. As has been indicated, policies, strategies, social and economic interactions and relationships in soil are not concretely interrelated and often are subsumed into land use and allocation. Although land use and planning allocates the various uses of the land, sustainability of the soil is rarely considered. PoPs call for a deeper

---

[38] Harris (1969), p. 20.

[39] Aktar (2009), p. 8.

[40] Pimentel and Edwards (1982), p. 597.

[41] Pimentel and Edwards (1982), p. 598.

[42] Pimentel and Edwards (1982), p. 599.

[43] Klánová et al. (2008), pp. 416–423.

[44] Juerges and Hansjürgens (2018), p. 1634.

[45] Juerges and Hansjürgens (2018), p. 1634.

interrogation of the land use and planning beyond land zoning to include possible effects of land use on the soil itself. This would go beyond economic aspects of land in determining use of the land and to PoPs regulation through multi-agency and interdisciplinary approaches. In this regard, allocation of rights and utilization and interactions of the stakeholders in soil governance, such as the government, farmers and land owners have to collaborate with soil scientists, health workers, industrialists and other professions. Such an approach will assist in preventing harmful effects of soil rights allocation and utilization. Traditional land use and planning approaches are inadequate in dealing with PoPs, hence the need for the engagement of more stakeholders.

PoPs countermeasures require a greater grip on land management, governance and sustainable use by relevant regulatory authorities. Control measures on contamination, use and regulation in public land may be easier than on private and community land. As noted above dealing with both soil organic matter and PoPs require targeted measures. Corrective measures to check the impacts of PoPs require large-scale involvement of the farmers and those who interact with soils on daily bases.

Industrial and domestic wastes are a big source of PoPs when deposited on the soil. The proper treatment of these wastes before disposal is therefore critical. This should be complemented by an integrated soil management policy that incorporates all stakeholders. Evidently, PoPs are a threat to human and animal survival and although some are prohibited and the use of others regulated, their continued use affects soil productivity and preservation. Unclear property rights and lack of a comprehensive soil governance regime makes the regulation of PoPs a difficult endeavour.[46] Policy and decision makers need view soil as a distinct natural resource and formulate policies to comprehensively address the challenges facing its productivity and sustainable use.

3.3 PoPs and Land Rights

Traditionally, Blackstonian absolute property rights' concept was prevalent with libertarian theorists.[47] To them, the owner of property is a sovereign in the dominion, unrestricted by both the government and adjacent owners. He has a right to exclusive possession, use, misuse and transfer. Proponents of absolute property rights view government as a facilitator of acquisition of dominions and protector of the rights. In their view, imposition of duties on land utilization is immoral, illegal and an encroachment of property rights.[48]

---

[46]Juerges and Hansjürgens (2018), p. 1634.

[47]Krier (2014), p. 589.

[48]Baron (2013), p. 57.

Technological developments and ecological concerns as public interest issues on the impact of the exercise of the sovereignty on the dominions have however, diluted the property absolutism concept over time. A.M Honoré demystifies the contents of the unity of ownership, defining it as a conglomeration of various rights, which collectively give the owner rights to effectively utilise the land.[49] He analyses eleven incidents of ownership and argues that an interest in land could be any of the incidents and such interests are subject to limitations and prohibitions. Ownership constitutes the right to possession, use, manage, income, capital, security, transmissibility, absence of term, execution, prohibition of harmful use and residuary character. In this sense, property is seen as a bundle of sticks tied together, which can be disentangled and different persons can hold each of the sticks. For instance, the government can hold some of the bundle of sticks for public good.

One area that Honoré's bundle of sticks has had an influence is the protection of the environment with the state playing an interventionist role by regulating property of rights. Environmentalists argue that although private property rights law provides remedies for misuse of rights through for instance, nuisance torts, there are some public goods such as air, oceans and mountains which are not allocated to individuals but whose protection is important.[50] In such instances, the state is required to regulate the use of the land by private individuals for the benefit of public goods and more importantly for future generations' enjoyment of both the private and public lands. In regulating pollution, which primarily results from the use of land, the exercise of property rights is restricted to avoid harm to others.

While states do not ordinarily limit the right to possession, use of land is regulated through planning laws to control developments, agricultural lands and industrial use among others.[51] Traditionally this regulation focuses on permitted and prohibited activities. The nature of PoPs however calls for focusing on both the activity and the modalities of the use. In agricultural fields, this would mean limiting the chemicals to be used and specifying allowed limits in agricultural production. Evidence shows that the level of soil organic matter also affects the fate of PoPs in soil and to this extent the government may require increase by farmers of the contents of the organic matter.[52] In industrial development, in addition to the restriction of the kind of chemicals used, measures on safe and solid waste disposal should also be put in place. Land owners could be required to adhere to safe disposal regulations, which entail curtailing the right to decide how the land would be utilized and managed.

[49]Kay (2016), p. 513.

[50]Meyer (2009), pp. 99–127.

[51]Juerges and Hansjürgens (2018), p. 1631.

[52]Hayat et al. (2010), pp. 217–222.

4 Legal Framework Governing PoPs

The period between the 1960s and 1980s marked an increase in the use of pesticides, which were considered more effective in increasing food production, particularly in the United States.[53] Additionally, during the Second World War, DDT remained a major malaria control drug for overseas soldiers.[54] It was the continued use of these PoPs that attracted scientists' attention to their effects on human and wildlife health. Rachel Carson in her book *Silent Springs* brought to the fore the effect of DDT on wildlife and human beings in places far away from their sources.[55] Over the years, the global community became cognizant of the effects of PoPs in chemicals deemed essential for everyday activities, including commerce. For most of these chemicals, there is neither sufficient information nor assessment of risks.

While dealing with chemicals is mainly a municipal law issue, the environmental risks posed by such chemicals may be of concern to international law where the chemicals can be widely dispersed throughout the environment and/or persist and accumulate in the environment over long timeframes, thus posing threats of trans boundary harm. International action to address the negative consequences of chemicals was signalled in Chapter 20 of Agenda 21.[56] Prior to this, chemical regulation was discussed within the broader rubric dealing with waste and hazardous substances, with the main focus placed on national action on treatment of wastes and illegal exportation of waste to developing countries.[57]

The need for specific regulation of chemicals was prompted by the increase in international trade of hazardous substances (toxic chemicals; pesticides; nuclear materials). International environmental law was slow in regulating them because of the limited scientific information and governments' failure to disclose properties. From the 1980s, the focus was on soft law Instruments[58] with the introduction of prior informed consent (PIC) requirements and the UNEP International Register of Potentially Toxic Chemicals.[59]

4.1 The Rotterdam Convention

The Rotterdam Convention on PIC Procedures for Certain Hazardous Chemicals and Pesticides in International Trade was concluded in 1998.[60] It applies to banned or

---

[53]Murphy (2001), pp. 892–893.

[54]Bishopp (1945), pp. 373–378.

[55]Carson (2002), p. 57.

[56]Assembly (1992b) Chap. 20.

[57]Assembly (1989), p. 1, African Union (1998), p. 1.

[58]FAO (1985), p. 1.

[59]UNEP (1976), p. 4.

[60]Assembly (1998), p. 1.

severely restricted chemicals and severely hazardous pesticides with parties sharing responsibility and engaged in cooperative efforts in international trade of hazardous chemicals to protect human health and environment.[61]

Its objectives are to promote shared responsibility and cooperative efforts among parties in international trade of certain hazardous chemicals in order to protect human health and the environment from potential harm and contribute to their environmentally sound use, facilitating information exchange about their characteristics, by providing for a national decision-making process on their import and export and by disseminating these decisions to Parties.[62]

The PIC procedure draws upon FAO and UNEP voluntary schemes in applying PIC for chemicals listed in Annex III to the Convention.[63] Currently there are forty chemical in the Annex with room for the Conference of Parties to add further chemicals.[64] Unlike its equivalent in the Stockholm Convention on Persistent Organic Pollutants,[65] the Chemicals Review Committee is not obliged to observe the precautionary principle in carrying out its decision-making tasks. The Convention also makes provision for general information exchange[66] and technical assistance,[67] though not for financial assistance to developing countries.

4.2 *The Stockholm Convention*

The Intergovernmental Forum on Chemical Safety (IFCS) was established in 1994 as a high-level environmental policy forum for policy makers from national government and interested members of the public.[68] The UNEP Governing Council in 1995 requested that an international assessment of an initial list of 12 PoPs (that eventually became known as the dirty dozen; see Table 1) be undertaken by the IFCS. Based on this assessment, IFCS was then to make recommendations for international action to deal with POPs for consideration by the Governing Council of UNEP.

In June 1996, based on its assessments IFCS concluded that there was sufficient information justifying the need for international action, including a global legally binding instrument, to minimize the risks from the 12 POPs through measures to reduce and/or eliminate their emissions or discharges. The Governing Council in its meeting in 1997 invited UNEP to prepare and convene an intergovernmental

---

[61] Assembly (1998) Preamble.

[62] Assembly (1998) Article 1.

[63] UNEP, (http://wedocs.unep.org/bitstream/handle/20.500.11822/26721/7749iPICMechan_EN.pdf?sequence=1&isAllowed=y) accessed 15th July 2019.

[64] Assembly (1998) Annex III to Rotterdam Convention.

[65] Assembly (2001), p. 1.

[66] Assembly (2001) Article 14.

[67] Assembly (2001) Aricle 16.

[68] IISD (2010).

Table 1 Dirty Dozen

1	Aldrin
2	Chlordane
3	Dichlorodiphenyltrichloroethane (DDT)
4	Dieldrin
5	Endrin
6	Heptachlor
7	Hexachlorobenzene
8	Mirex
9	Toxaphene
10	Polychlorinated biphenyls (PCBs)
11	Dioxins
12	Furans

negotiating committee (INC), so as to prepare an international legally binding instrument for implementing international action, initially beginning with the 12 PoPs. They also requested the INC to establish an expert group to develop criteria and a procedure for identifying additional POPs as candidates for future international action.

The Convention was adopted and opened for signature at a Conference of Plenipotentiaries held from 22 to 23 May 2001 in Stockholm, Sweden. It entered into force on 17 May 2004, ninety days after submission of the fiftieth instrument of ratification, acceptance, approval or accession in respect of the Convention. At its fourth meeting in Geneva, Switzerland, the Convention was amended by including additional chemicals to the list in Annexes A and B. The Convention, as noted above, initially listed twelve PoPs but the listing has now grown to twenty-two (15 pesticides and seven chemicals and byproducts). Many PoPs are unknown, with PCBs, DDT and Dioxin being the most widely known.

Coined in precautionary language,[69] it requires states to ensure close monitoring of these chemicals. Precaution requires that action be taken to avoid harm without waiting for conclusive scientific evidence. This provides an opportunity for action to prevent degradation of soils. Other relevant principles guiding the Convention and relevant for protection of soils are: Polluter pays principle,[70] Common but differentiated responsibilities[71] and responsibility not to cause harm.[72] The Convention also acknowledges the right of states to utilise resources within their jurisdictions and their obligation to control activities within their jurisdiction to ensure they do not

[69]Article 2 of the Convention provides that "Where there are threats of serious or irreversible damage, lack of full scientific certainty shall not be used as a reason for postponing cost-effective measures to prevent environmental degradation." This resonates with Principle 15 of the Rio Declaration on Environment and Development, 1992.

[70]Assembly (2001) the Preamble.

[71]Assembly (2001) the Preamble.

[72]Perrez (1996).

cause trans-boundary harm.[73] Its main objective is to protect the human health and environment against PoPs.[74]

The Convention recognizes that no government can respond to PoPs threats alone, hence the call for international cooperation. It also acknowledges that PoPs continue to be used and encourages members to control them as they aspire to completely phase them out. In this regard, state parties are required to undertake several actions. The first is the duty to adopt measures to reduce or eliminate the release of POPs into the environment from intentional production and use.[75] Other measures include safe handling of existing stockpiles,[76] prohibition of the importation or exportation of PoPs except for environmentally sound disposals,[77] stringent restriction and elimination of byproducts of combustion processes, some of which are very ordinary[78] and reduction of emissions of unintentionally produced PoPs.[79]

To actualize these obligations, state parties are required to prepare National Implementation Plans (NIPs).[80] In implementing the Convention and the NIPs, a critical challenge is lack of data and information about sources, releases and environmental levels of PoPs in developing countries. Parties are therefore called upon to cooperate and exchange information on elimination, reduction of PoPs and the discovery of alternatives.[81] The parties are also required to share information with the public and carry out public awareness campaigns and engage the public in tackling the PoPs.[82] The issue of information is particularly pertinent for developing countries that have limited funding for research. This affects screening, finding alternatives and creating public awareness. Kenya's National Action Plan for the Stockholm Convention on Persistent Organic Pollutants for instance notes that in addition to information availability challenges, synthesizing the data to levels where it is consumable and usable is still a big challenge.[83]

---

[73] Assembly (2001) Preamble.

[74] Assembly (2001) Article 1.

[75] Assembly (2001) Article 3.

[76] Assembly (2001) Article 6.

[77] Assembly (2001) Article 3.

[78] Assembly (2001) Article 5.

[79] Assembly (2001) Article 5.

[80] Assembly (2001) Article 7.

[81] Assembly (2001) Article 9.

[82] Assembly (2001) Article 10.

[83] Ministry of Environment, Water and Natural Resources (2014).

5 Law, Soil Governance & PoPs

5.1 General Linkages

Soil governance within a context of PoPs must be considered among broader issues that lead to soil degradation. These include firstly climate change and poverty, over-concentration of people or livestock in highly exposed areas and interrelationships that make impoverished persons more likely to be over-concentrated in an area and more likely to be occupying poor quality soils that easily degrade.[84]

Secondly, there are linkages between soil management and human rights. Proper soil management leads to the fulfillment of human rights. As a consequence, and as noted by Klaus Töpfer, soil protection policies should be predicated on a human rights framework, giving special focus to the rights of the marginalized and vulnerable persons in society.[85] Moreover, gender issues are still relevant for soil governance, given that women particularly in developing economies including Kenya are the main stakeholders in agriculture. Women play a crucial role in agriculture and natural resource management and they should be involved in the governance of and decision making over soil.

Thirdly, considering the climate change, poverty and human rights interrelatedness, soil governance needs to be integrated into related programmes and processes from the local to the international levels. It should also cut across sectors and take a long timeframe perspective, if they are to be comprehensive and effective.

5.2 Soil Governance Under International Law

As noted above, the soil governance framework is not contained in a specific set of laws and policies but rather is dispersed in many international, regional, national and local documents. For instance, multilateral environmental agreements relating to climate change, biodiversity, hazardous waste or wetlands have soil governance tenets.[86] Global treaties such as the Convention on Biological Diversity (CBD),[87] the United Nations Convention on Climate Change (UNFCCC)[88] and the United Nations Convention to Combat Desertification in those Countries experiencing Serious Drought and/or Desertification, Particularly in Africa (CCD)[89] have elements of soil governance, but they are not adequate *per se*, to found an effective

[84]UNCCD (2008), p. 7.

[85]Töpfer (2015), p. 2.

[86]Wyatt (2008), p. 166.

[87]Assembly (1992c).

[88]Assembly (1992c).

[89]Assembly (1994).

international agreement relating to soil management.[90] They are however very important, as they signal global agreement on issues related to soil governance.

The CBD's objective is to promote conservation of biological diversity, its sustainable use and fair and equitable benefit sharing agreements from the utilization of the biodiversity. Although this convention does not directly address soil governance, conservation and preservation of the biological biodiversity are land uses that impact on soil usage, policy directions and decisions a country makes. Similarly, the Kyoto Protocol[91] to the UNFCCC requires states to reduce their greenhouse gases emissions to a level that would "prevent dangerous anthropogenic interference with the climate system".[92] It recognizes the impact of land uses and soil on climate change.[93] Given that soil supports various soil utilizations and land uses, the stakeholders have to bear the climate change angle in mind.

The CCD's objective is to combat desertification and mitigate the effects of drought in countries that experience desertification and drought, particularly in Africa.[94] It requires states to have long-term strategies on land productivity, rehabilitation and sustainable management of land.[95] The convention obligates the states to involve local communities in decisions on combating drought and desertification.[96] It also requires the cooperation of the states at the international and regional level and with non-governmental organizations, landowners, women and the youth.

Both the Basel[97] and Bamako[98] conventions that deal with hazardous chemicals are relevant for soil governance, because these chemicals can degrade the quality of the soil. Their control of the transboundary movement of hazardous wastes contributes to soil governance, as hazardous wastes find their way into soils. The same can be said of the Stockholm and Rotterdam conventions on PoPs and PIC discussed above. The fact that many chemicals they address have direct or indirect interaction with soil as discussed above makes them a critical plank of soil governance.

In addition to the above is the European Charter for the Protection and Sustainable Management of Soil,[99] the Thematic Strategy for Soil Protection,[100] European Soil Charter,[101] Alpine Convention Soil Conservation Protocol[102] and the Association of Southeast Asian Nations Agreement on the Conservation of Nature and

---

[90]Boer and Hannam (2015), p. 7.

[91]Assembly (1998).

[92]Assembly (1998) Article 5.

[93]Assembly (1998) Article 3.

[94]Assembly (1994) Article 2.

[95]Assembly (1994) Article 2(2).

[96]Assembly (1994) Article 3.

[97]Assembly (1989).

[98]African Union (1998).

[99]European Council (2003).

[100]Commission of the European Communities (2006).

[101]Council of Europe (2003).

[102]European Union (2005).

Natural Resources.[103] Other governing instruments include the World Soils Agenda,[104] Tutzing Initiative for a Soil Convention[105] and the Soil World Charter developed by the *Food and Agriculture Organization of the United Nations (FAO).*[106] *The Charter establishes a set of principles on the maximum use of soil resources both for increased productivity and sustainable use. Decision makers on land use allocations are encouraged to adhere to the principles for stewardship of the land.*[107] It is important to point out, that the Protocol on the Implementation of the Convention Concerning the Protection of the Alps of 1991 in the Area of Soil Protection[108] is the only legally binding convention that is focused on the protection of soil at the moment.

In Africa, the African Convention on the Conservation of Nature and Natural Resources[109] seeks to harmonize and coordinate policies in the conservation and preservation of the natural resources including land and soil. In fulfilling this objective, the Convention requires respect for the right to development of the people. It requires states to develop long-term strategies for conservation and sustainable management of soils, including the adoption of land use plans based on land capability, sustainable farming, to control soil erosion and to enact laws on land tenure geared towards achievement of the objectives of the Convention.[110] To this end, the Convention requires sustainable use of the soil depending on soil capability based on scientific investigations, traditional knowledge and experience. Further, under the Nairobi Declaration on the African Process for Combating Climate Change,[111] governments commit themselves to integrate climate change adaptation measures into national and regional development plans, policies and strategies, including land uses and agriculture. The adaptation methods impact on land uses and hence the soil utilization.

---

[103] Association of Southeast Asian Nations (1985).

[104] Hurni and Meyer (2002), p. 67.

[105] Schellnhuber (2001), p. 1.

[106] FAO (2015), p. 1.

[107] FAO (2015), p. 2.

[108] Approved by The Council of the European Union, Decision 2006/516/EC, 27 June 2006 which stated thus in its decision: "One of the main goals of the Protocol on Soil Protection is the safeguarding of the multifunctional role of soil based on the concept of development. Sustainable productivity of soil must be ensured in its natural function, as an archive of natural and cultural history and in order to guarantee its use for agriculture and forestry, urbanism and tourism, other economic uses, transport and infrastructure, and as a source of raw materials."

[109] African Union (1968) Ar. 3.

[110] African Union (1968) Ar. 3.

[111] African Ministerial Conference on the Environment (2009).

6 Soil Governance and PoPs at the National Level: Kenya

At the national level, different countries have adopted different approaches to soil management generally and in mitigating the impact of PoPs. These approaches are contained in laws and policies that constitute the soil governance mechanisms of each country. It has been observed that various countries have been innovative in framing their approaches to soil governance and in dealing with the resulting land management problems. We will examine the legal and institutional framework relating to PoPs in Kenya, as an example of a national approach to soil governance.

Kenya has a comprehensive legal and institutional framework relating to the environment, though it does not comprehensively cover PoPs or soil governance specifically. Nonetheless, the laws and policies in place may arguably be employed in dealing with the impact of PoPs on soil and thus to promote sustainable soil governance. Our assessment however is that the framework can be improved with a little tweaking and rearrangement as detailed in the substantive discussion below. It is worth pointing out here that Kenya continues to rely largely on chemicals for agricultural and industrial production. Some of these chemicals contain PoPs and continue to pollute the environment both in air as vapour, in soil killing microorganisms and up the food chain affecting human beings. Kenya's National Implementation Plan[112] for the Stockholm Convention for the period between 2014 and 2019 presents the current state of PoPs in Kenya. As of 2014, there were several sites contaminated with pesticides and the plan notes that there is still high concentration of contaminated soils in the country. It calls for strategies to decontaminate and restore the soils. The main sources of contamination are pesticides and waste management dumpsites in Dandora and Industrial area sites having the high concentration of PoPs.[113]

6.1 The Constitution

The Constitution of Kenya 2010 provides for the right of every citizen to a clean and healthy environment.[114] To further actualise this provision, Article 69 of the Constitution obligates the State to protect the environment, while imposing a corresponding duty on citizens to protect and conserve the environment.[115] In addition, by virtue of the decentralized form of governance established under the Constitution and the further sharing of functions as stipulated in the Fourth Schedule of the Constitution, the devolved units (47 counties) are under an obligation to enact

---

[112]Ministry of Environment, Water and Natural Resources (2014).

[113]Kidaha (2016), p. vi.

[114]Republic of Kenya (2010) Article 42.

[115]Republic of Kenya (2010) Article 70.

and design their own laws, policies and practices to accord with the constitutional obligations.

Besides the Constitution, the Ministry charged with protection of environment drafted and adopted the National Environment Policy 2014,[116] which aims at availing a holistic framework on environmental protection and to guide the enactment of implementing laws. Some of the provisions in the Policy that are relevant to PoPs, at least as far as they concern the use and production of chemicals and waste management include: right to clean and healthy environment, sustainable resource use, public participation, subsidiarity, precautionary and polluter pay principle, and right to development.[117] Even more significant is the recognition by the Policy itself that there is no policy guiding the management of chemicals.[118] In response, the Policy provides that the government will integrate the chemicals management into the development plans; implement the Strategic Approach to International Chemicals Management (SAICM),[119] which was approved in the first chemicals management conference in Dubai in 2006 and develop and implement a chemicals management policy.[120] Similarly, the Policy also provides for waste management committing the government to: integrated national waste management strategy, promoting economic incentives for managing waste and promotion of cleaner production, waste recovery, recycling and re-use.[121] The provision in the Policy that is however directly relevant to the use of PoPs and their impact on soil management is on toxic and hazardous substances, considering that PoPs may be aptly classified under this head. The Policy notes that toxic and hazardous substances are extensively used in agricultural and industrial sectors, and exposure to these substances has chronic and acute health effects.[122] As a countermeasure, the Policy commits the government to: developing a national policy on toxic and hazardous substances, establishing appropriate toxic and hazardous substances disposal facilities and collaborating with the international community in curbing dumping of obsolete or banned toxic substances.[123]

Besides the Constitution, soil governance tenets can be gleaned from environmental, land and agriculture laws and policies. We will look at a snapshot of these and the way they address or could address PoPs.

[116]Ministry of Environment, Water and Natural Resources (2013), p. 1.

[117]Ministry of Environment, Water and Natural Resources (2013), pp. 9–13.

[118]Ministry of Environment, Water and Natural Resources (2013), p. 30.

[119]Ministry of Environment, Water and Natural Resources (2013), p. 30.

[120]Ministry of Environment, Water and Natural Resources (2013), p. 30.

[121]Ministry of Environment, Water and Natural Resources (2013), p. 39.

[122]Ministry of Environment, Water and Natural Resources (2013), p. 40.

[123]Ministry of Environment, Water and Natural Resources (2013), p. 40.

6.2 Environmental Law

The main statute relating to environmental protection is the Environment Management Coordination Act (EMCA), No. 8 of 1999, as amended in 2015. As the framework environmental law in Kenya, EMCA provides for the establishment of the National Environment Management Authority (NEMA) which is the body charged with implementing virtually all the statutory obligations relating to environmental management. NEMA conducts environmental impact assessments for all projects, including the setting up of chemical industries that may manufacture, distribute or deal with PoPs. In addition, there is a requirement under the Act for NEMA to conduct regular environmental audits[124] and thereafter prepare a State of the Environment Report on an annual basis.[125] More specifically however, provisions relating to PoPs may be found in sections 91–100 of EMCA. Section 91 provides that the Standards and Enforcement Review Committee[126] shall, in consultation with other lead agencies, recommend to NEMA the standard criteria for classifying the various hazardous wastes in accordance with their characteristic/ quality, which may be: hazardous, corrosive, carcinogenic, flammable, persistent, toxic, explosive, radioactive or any other category of waste. In addition, section 92 proceeds to empower the Minister in charge of the environment, in consultation with NEMA, to promulgate regulations to classify, label, register, package, advertise, control importation and exportation, monitoring, disposal, restriction and banning and distribution and storage of chemicals and materials in accordance with their toxicity.

Section 93 creates offences for persons who discharge or act contrary to the provisions of the Act with regard to the handling of these toxic and hazardous substances, which in our view include the POPs. It is further provided under section 94 of EMCA that the Standards Committee shall in consultation with lead agencies prepare and submit to NEMA draft standards on the concentration of pesticide residues in raw agricultural commodities and also establish procedures for labelling, registering, advertising, packaging and importation and exportation of pesticides and toxic substances. It is an offence to deal with pesticides and toxic substances in a manner contrary to the provisions of the statute.[127] To operationalize these provisions regarding the impact of POPs on the environment including soil, there are draft Chemical Regulations 2017, which are yet to be adopted. When adopted, they will avail a mechanism for registration of all hazardous chemicals by NEMA and provide for labelling, packaging, distribution and all other processes involved. There are however other operative regulations that fulfil some of the objectives such as the Waste Management Regulations 2006, Water Quality Regulations and the Air Quality Regulations 2012.

[124]Government of Kenya (1999) s. 68.

[125]Government of Kenya (1999) s. 9.

[126]Government of Kenya (1999) s. 70.

[127]Government of Kenya (1999) s. 98, 99.

There are other sectoral statutory enactments that play some of the roles and fill the gaps that may not necessarily be covered by the main framework law, EMCA, and the yet to be adopted regulations. For instance, the Occupational Health and Safety Act (OSHA) 2007 provides for the management of any place inhabited by people to ensure the safety and health of the people. With regard to safety from toxic chemicals, OSHA obliges the safe handling and transportation of chemicals, provision of material safety data sheets about all toxic chemicals, labelling and marking of all containers containing the chemicals and classification of all chemicals according to their dangerous characteristic or properties and their effects on humans.

In addition, there is also the Pest Control Products Act,[128] which regulates *inter alia*, the importation, exportation, manufacture, distribution and use of products used for the control of pests. It proscribes the manufacture, storage, distribution, use or advertisement of pesticide products otherwise than in accordance with the Act[129] and the regulations or that which is false, deceptive or misleading as regards its character, value, quality, composition or safety.[130] Further, the Act prohibits the importation or export of pesticide products that are contrary to the stipulations or do not conform to the set standards. This Act supplements the provisions of the Standards Act,[131] which empowers the Kenya Bureau of Standards to ensure that all products, including PoPs that are brought into Kenya as insecticides or other chemicals conform to the laid down standards for safety.

The East Africa Community Customs Management Act 2004 also prohibits entry of prohibited goods into the East African Community region including Kenya. These goods include hazardous wastes as defined under the Basel Convention, as well as various agricultural and industrial chemicals such as Caplafol, Chlordirneform, Chlorobenxilate, DDT, some of which include POPs. Further, the Malaria Prevention Act,[132] though directly dealing with the prevention and control of malaria is relevant in as far as DDT, a PoP, is used in the control of the disease. DDT is cited as a PoP and its use is restricted and only allowed in disease-vector control under Annex B of the Stockholm Convention. An assessment of the framework relating to PoPs in Kenya[133] has observed that it is critical that a review of legislation be done to align with the SAICM Implementation Plan and the National Implementation Plan for the Stockholm Convention. It also points to the need to finalize the draft Chemical Regulations 2017.

With respect to the institutional framework, the Ministry of Environment and Natural Resources is the main institution charged with the formulation of policies, coordinating and driving the policy agenda of creating awareness and training in chemical and waste management. According to the assessment report above, the

---

[128]Government of Kenya (1982).

[129]Government of Kenya (1982) s. 3.

[130]Government of Kenya (1982) s. 3(2).

[131]Government of Kenya (1973).

[132]Government of Kenya (1929).

[133]Nyakang'o (2017), p. 1.

Ministry suffers from a lack of coordinated awareness and training mechanisms at the national level, lack of a database on chemical information and similar awareness programs for various stakeholders and the general public, lack of regular training on chemical management for key personnel at the Ministry and the transient nature of the chemicals unit which is anchored in the Directorate of Multilateral Environment Agreements as a project management unit rendering it unsustainable in the long run.

With regard to NEMA, the chemicals management unit is currently under the Directorate of Compliance and Enforcement, at least until the draft Chemical Regulations come into effect. The assessment report indicates that NEMA suffers from shortage of staff with inadequate capacity to monitor PoPs from both open burning and processing in industry.[134] The devolvement of the waste management function to the counties, which lack capacity at the moment, also complicates efforts towards promoting management of POPs. Other challenges facing NEMA with respect to chemicals management are: lack of data on chemical information and laboratory equipment and lack of a Pollutant Release and Transfer Register for chemicals, despite proposals for its development under the Water Quality, Waste and Air Regulations.

There is also the Pest Control Products Board, which is charged with ensuring that all pest control products are registered for use, whereby the Board considers the efficacy, quality, safety and economic value.[135] It also ensures proper labelling, maintaining and updating a list of registered pesticides that are restricted, allowed and banned. It is also mandated with inspecting licensed premises that deal with pesticides and in collaboration with other agencies impounds illegal and counterfeit products, smuggled, unlabelled, expired and repacked products. In addition, the Board has developed a draft GHS Manual for Classification and Labelling of Pest Control Products in Kenya[136] to accord with the Globally Harmonized System of Classification and Labeling of Chemicals (GHS) of the UN. The assessment report indicates that the Board however lacks the capacity to analyse the various POPs in pesticides.

Besides the various state actors, the non-governmental organisations such as the Centre for Environmental Justice and Development and the Greenbelt Movement also form part of the institutional framework for dealing with PoPs. These NGOs also suffer from inadequate financial and human capacity as well as coordination with the Ministry. Industry associations such as the Agrochemicals Association of Kenya and Kenya Association of Manufacturers are also critical, as they constitute the industry, which prepares or imports the products that may well contain PoPs. The assessment report however notes that these industry associations need to partner with government and other players so as raise awareness as well as being in need of government financial support and institutional restructuring to create a section

[134]Nyakang'o (2017) para. 4.1.2.

[135]Government of Kenya (1984), s. 4.

[136]Pest Control Products Board (2016), p. 7.

dealing with chemical and waste management.[137] The other key institutions are the county governments, given that waste management is a devolved function[138] and universities and research institutions such as the International Centre for Insect Physiology and Ecology (ICIPE), Kenya Medical Research Institute (KEMRI) and Kenya Agricultural and Livestock Research Organization (KALRO).

6.3 Land and Agriculture

The way land is managed has a direct impact on soil. The Constitution has elaborate land governance principles, which include sustainable and productive management of land resources[139] and sound conservation of ecologically sensitive areas.[140] Article 66, which provides for the regulation of land use and property for the purpose of land use planning among others, provides an opportunity to ingrain tenets or soil governance and PoPs control. Additionally, the Land Act[141] and the Community Land Act[142] contain mechanisms for ensuring sustainable management of public, private and community land. This is further bolstered by the inclusion of sustainable development as one of the national values and principles of governance in Article 2 of the Constitution.

Kenya's first ever-National Land Policy identified soil degradation as one of the grave issues facing the Kenyan land sector.[143] Additionally, it provides for need to combat soil degradation. The National Land Use Policy[144] contains principles that are important in soil governance for maintenance of sustainable soils for various uses. It recognizes the needs for sustainable land usage, combating environmental degradation and engagement of the public in the preparation and approval of development plans.

The Agriculture, Fisheries and Food Authority Act No. 13 of 2013 provides that the Cabinet Secretary in consultation with the National Land Commission should make guidelines for purposes of soil conservation. The regulations should guide agricultural activities through prohibitions and regulation. To this end, the utilization of soil may be governed within some of the regulated or prohibited uses.

---

[137]Nyakang'o (2017) para 4.3.2.

[138]Government of Kenya (2010) Fourth Schedule Part 2.

[139]Government of Kenya (2010) Article 60 (1) (c).

[140]Government of Kenya (2010) Article 60 (1) (e).

[141]Government of Kenya (2012) s. 6.

[142]Government of Kenya (2016), s. 20.

[143]Government of Kenya (2009), p. 30.

[144]Government of Kenya (2017).

7 Conclusion

PoPs are some of the most dangerous pollutants released into the atmosphere through human activities. As noted above, they comprise toxic pesticides, industrial chemicals and other industrial highly toxic waste. Their effect is amplified by their long life, ability to travel long distances and their accumulation in fatty tissues of living organism's body, causing adverse health effects.

That PoPs need to be addressed is not arguable, considering the problems associated with them. How to deal with them comprehensively is the issue at hand. Given that most of the POPs enter the environment and human body through activities on the soil, it is critical that all relevant bodies manage the soil in a manner that takes care of or mitigates the possibility of contamination by these PoPs. In addition, owing to the transnational and intergenerational nature of the effects of PoPs, a robust soil governance framework is an imperative. Soil governance and mitigating the impacts of PoPs are inexorably linked. Soil occurs at the lowest level and should be addressed through mechanisms at that level. Land, agricultural and environmental policies at the national level present a critical entry point for both governing soil and regulating PoPs. However, a combination of mechanisms is called for when considering PoPs because of the trade connection and their movement across boundaries.

We have noted that PoPs continue to be used despite their being hazardous to human health and the environment in general. Agricultural production and industrialization are also on the rise. Both have implications for soil governance and PoPs. With regard to PoPs, the adoption of a multiple track to deal with their use and proliferation is as pragmatic as it is realistic. Soil, on the other hand, is yet to gain global and national recognition as an important natural resource and a public good that requires protective mechanisms such as those advanced for water and air.

Given the primacy of soil, a soil regime defining entitlements, permissible uses, public and private good elements in soil needs to be developed. Failure to do this will relegate soil utilization to the periphery, thereby causing that no due consideration will be given to the need to sustain and preserve it for future generations. PoPs impact on soil and addressing them should be an integral part of a soil governance regime at the international, national and local levels. It is heartening to note that soil governance and PoPs regulatory regimes have gained traction. The elements already developed should form the foundation for more concerted efforts towards the development of a global soil charter with principles that inform national laws. The soil governance regime should borrow from both formal and informal norms and institutions. Inclusion of all sectors and actors is imperative for an effective regime, where an integrated approach, ecological concern for soil, participatory governance including women and more inclusive decision-making are taken onboard. To ensure success, the regime should be anchored with a monitoring and evaluation framework that tracks specific components.

References

African Ministerial Conference on the Environment (2009) Nairobi declaration on the African process for combating climate. African Union, Nairobi
Aktar W, Sengupta D, Chowdhury A (2009) Impact of pesticides use in agriculture: their benefits and hazards. Interdisciplinary Toxicol 2(1):1–12
Allan C (2008) Can adaptive management help us embrace the Murray-Darling Basin's wicked problems?. In: Adaptive and integrated water management. Springer, Berlin, pp 61–73
Association of Southeast Asian Nations (1985) Association of Southeast Asian Nations' Agreement on the Conservation of Nature and Natural Resources. s.n. Kuala Lumpur
Bănăduc AC (2016) The impact of persistent organic pollutants on freshwater ecosystems and human health, 1st edn. Lucian Blaga University of Sibiu, Sibiu
Baron JB (2013) Rescuing the bundle-of-rights metaphor in property law. Univ Cincinnati Law Rev 82:57
Bishopp FC (1945) Insect problems in world war II with special references to the insecticide ddt. Am J Public Health Nations Health 35(4):373–378
Boer B, Hannam I (2015) Developing a global soil regime. Int J Rural Law Policy 2015(1):1–13
Brussaard L (1997) Biodiversity and ecosystem functioning in soil. Ambio:563–570
Carson R (2002) Silent spring. Houghton Mifflin Harcourt
Comis D, Becker H, Stelljes KB (2001) Depositing carbon in the bank. Agric Res 49(2):4
Commission of the European Communities (2006) Thematic strategy for soil protection. s.n., Brussels
Edwards CA (1965, January) SOME SIDE-EFFECTS RESULTING FROM USE OF PERSISTENT INSECTICIDES. In Annals of Applied Biology (Vol. 55, No. 2, p. 329). Natl Vegetable Res Station Wellsbourne, Warwick Cv35 9ef. Assoc Applied Biologists, England
European Council (2003) European Charter for the protection and sustainable management of soil. European Union, Brussels
European Union (2005) Protocol on the implementation of the Alpine Convention of 1991 in the field of soil conservation Soil Conservation Protocol. Official Journal of the European Union, Brussels
Food and Agriculture Organization (FAO) (2015) Soils and Biodiversity. [Online] Available at: http://www.fao.org/fileadmin/user_upload/soils-2015/images/EN/WSDPosters_Promotional_Material/En_IYS_food_Print.pdf
Food and Agriculture Organization of the United Nations (FAO) (1985) International code of conduct on the distribution and use of pesticides, s.l.: FAO
Government of Kenya (1929) Malaria Prevention Act 2016. Kenya Gazette, Nairobi
Government of Kenya (1973) Standards Act 1973. Kenya Gazette, Nairobi
Government of Kenya (1982) Pest Control Products Act 1982. Kenya Gazette, Nairobi
Government of Kenya (1999) Environment Management and Coordination Act 1999. Kenya Gazette, Nairobi
Government of Kenya (2009) National Land Policy 2009, Sessional Paper No. 3. Government Printers, Nairobi
Government of Kenya (2010) Constitution of Kenya 2010. Kenya Gazette, Nairobi
Government of Kenya (2012) Land Act 2012. Government Printers, Nairobi
Government of Kenya (2016) *Community Land Act 2016*. Kenya Gazette, Nairobi
Government of Kenya (2017) National Land Use Policy 2017. Government Printers, Nairobi
Handelsman J (2005) How to find new antibiotics: metagenomics could be the way to mine the soil beneath our feet. The Scientist 19(19):20–22
Hannam I, Boer B (2002) Legal and institutional frameworks for sustainable soils: a preliminary report (No. 45). IUCN
Harris C (1969) Insecticide Pollution and Soil Organism. 100:14–29

Hayat MT, Xu J, Ding N, Mahmood T (2010) Dynamic behavior of persistent organic pollutants in soil and their interaction with organic matter. In: Molecular environmental soil science at the interfaces in the earth's critical zone. Springer, Berlin, pp 217–222

Howard TM, Lawson A (2015) Soil governance: accessing cross-disciplinary perspectives. Int J Rural Law Policy 1:1–8

Hurni H, Meyer K (2002) A world soils agenda: discussing international actions for the sustainable use of soils. Geographica Bernensia

IISD (2010) IISD Reporting Services. [Online] Available at: http://enb.iisd.org/process/chemical_management-ifcsintro.html

Juerges N, Hansjürgens B (2018) Soil governance in the transition towards a sustainable bioeconomy–a review. J Clean Prod 170:1628–1639

Kay K (2016) Breaking the bundle of rights: conservation easements and the legal geographies of individuating nature. Environ Plan Econ Space 48(3):504–522

Kerekes CB (2011) Property rights and environmental quality; a cross-county study. Cato J 31:315

Kidaha J (2016) Assessment of new persistent organic pollutants-pesticides in air and soil samples from sites in Nairobi and Mount Kenya. University of Nairobi, Nairobi

Klánová J, Matykiewiczová N, Máčka Z, Prošek P, Láska K, Klán P (2008) Persistent organic pollutants in soils and sediments from James Ross Island, Antarctica. Environ Pollut 152 (2):416–423

Krier JE (2014) Of property rights and rights to property. Ohio NUL Rev 41:589

Locke S (2007) "Artificial soil". ABC Net. Available http://www.abc.net.au/site-archive/rural/nt/content/2006/s1924714.htm

Mansour SA (2009) Persistent organic pollutants (POPs) in Africa: Egyptian scenario. Human Exp Toxicol 28(9):531–566

Marshall J (2007) Artificial soil: quick and dirty. New Sci 195(2616):33–35

Meyer JM (2009) The concept of private property and the limits of the environmental imagination. Polit Theory 37(1):99–127

Ministry of Environment, Water and Natural Resources (2013) National Environment Policy, 2013. Ministry of Environment and Natural Resources, Nairobi

Ministry of Environment, Water and Natural Resources (2014) Kenya National Implementation Plan for Stockholm Convention on Persistent Organic Pollutants 2014-2019. Ministry of Environment, Water and Natural Resources, Nairobi

Murphy SD (2001) US signing of persistent organic pollutants treaty. Am J Int Law 95(4):892

Nyakang'o JB (2017) Institutional needs analysis for chemicals and waste management in Kenya. GEF & UNDP, Nairobi

Perrez FX (1996) The relationship between "permanent sovereignty" and the obligation not to cause transboundary environmental damage. Environ Law:1187–1212

Pest Control Products Board (2016) Agrochemicals Association of Kenya. [Online] Available at: http://new.site.agrochem.co.ke/wp-content/uploads/2017/11/2016-KENYA-GHS-MANUAL_15_07_2016.pdf

Pimentel D, Edwards CA (1982) Pesticides and ecosystems [Pollutants, food chains, cycling, effects]. BioScience American Institute of Biological Sciences

Resource Futures International for the World Bank and CIDA (2001) Persistent organic pollutants and the Stockholm convention: a resource guide. s.l.:s.n

Richter DD Jr (2007) Humanity's transformation of Earth's soil: pedology's new frontier. Soil Sci 172(12):957–967

Roose E, Lal R, Feller C, Barthès B, Stewart BA (eds) (2006) Soil erosion and carbon dynamics, vol 376. CRC Press, Boca Raton

Schellnhuber HJ (2001) World in transition: conservation and sustainable use of the Biosphere, Vol 1. Earthscan

Skoet J, Stamoulis KG (2006) The state of food insecurity in the world 2006: Eradicating world hunger-taking stock ten years after the world food summit. Food & Agriculture Org

Solomon S, Qin D, Manning M, Averyt K, Marquis M (eds) (2007) Climate change 2007-the physical science basis: Working group I contribution to the fourth assessment report of the IPCC, Vol. 4. Cambridge University Press, Cambridge
Töpfer K (2015) 'Transformation is Needed – Transformation is Possible' (Chairman's Conclusions, Global Soil Week). [Online] Available at: http://globalsoilweek.org/
UN General Assembly (1989) Basel Convention on the control of transboundary movements of hazardous wastes and their disposal. United Nations, Basel
UN General Assembly (1992a) Agenda 21. United Nations, Rio de Janeiro
UN General Assembly (1992b) Convention on biological diversity. United Nations, Rio de Janeiro
UN General Assembly (1992c) United Nations framework convention on climate change. United Nations, Rio de Janeiro
UN General Assembly (1994) United Nations Convention to combat desertification in those countries experiencing serious drought and/or desertification, particularly in Africa. United Nations, Paris
UN General Assembly (1998) Kyoto protocol to the United Nations framework convention on climate change. United Nations, Kyoto
UN General Assembly (2001) Convention on persistent organic pollutants. UNEP, Stockholm
UN General Assembly (2015) Transforming our world: the 2030 agenda for Sustainable Development
UNCCD (2008) Human Rights and Desertification: Exploring the Complementarity of International Human Rights Law and the United Nations Convention to Combat Desertification: Desertification, Land Degradation and Drought. Secretariat of the United Nations Convention to Combat Desertification in cooperation with the Swiss Agency for Development and Cooperation, Druck Center Meckenheim GmbH 7(1)
UNEP (1976) The International Register of Potentially Toxic Chemicals. s.l.: UNEP
UNEP. The Informal Prior-Informed Consent (iPIC) Mechanism. s.l.:s.n. available at http://wedocs.unep.org/bitstream/handle/20.500.11822/26721/7749iPICMechan_EN.pdf?sequence=1&isAllowed=y
Van Oostdam J, Gilman A, Dewailly E, Usher P, Wheatley B, Kuhnlein H et al (1999) Human health implications of environmental contaminants in Arctic Canada: a review. Sci Total Environ 230(1–3):1–82
Weigelt J, Müller A, Janetschek H, Töpfer K (2015) Land and soil governance towards a transformational post-2015 Development Agenda: an overview. Curr Opin Environ Sustain 15:57–65
Wenjie L (2008) Cold-trapping of persistent organic pollutants in the mountain soils of Western Sichuan, China. Environ Sci Technol 42(24):9086–9091
White RE (2013) Principles and practice of soil science: the soil as a natural resource. John Wiley & Sons
Wild A (2003) Soils, land and food: managing the land during the twenty-first century. Cambridge University Press
Winiwarter V (2005) Jared Diamond. Collapse: how societies choose to fail or survive. Environ Hist 10(3):538
Wyatt AM (2008) The dirt on international environmental law regarding soils: is the existing regime adequate. Duke Environ Law Policy Forum 19:165

Prospects and Challenges for Soil Protection Law in Zambia

Pamela Towela Sambo

1 Introduction

According to the World Bank, Zambia is a low-middle income country (LMIC) in sub-Saharan Africa with a population of approximately 17 million people. Soil is a significant resource for Zambia. This chapter starts with Sect. 2 below analysing scientific approaches to defining the term 'soil'. Section 3 relates sustainable soil management to mining, agriculture and manufacturing, being major industries that support the economy in Zambia. As part of the global community, every country has an international legal obligation to secure the protection of local environments. To this end, Sect. 4 of this chapter will critically analyse selected international legal initiatives that effect sustainable soil management and to which Zambia is a party. Section 5 evaluates soil related law and policy in Zambia.

2 What Is Soil?

Legal and policy documents in Zambia show that soil is not defined in precise terms. Section 2 (1) of the *Plant Pests and Diseases Act*, however, provides that soil means "a growing medium which is neither sterilised nor inert." The literature defines soil in various ways, including: as a three- dimensional natural and dynamic thin layer on the surface;[1] or a product of weathered and weathering mineral matter, mixed with

[1]Osman (2013), pp. 1–6.

P. T. Sambo (✉)
University of Zambia, School of Law, Lusaka, Zambia

H. Ginzky et al. (eds.), *International Yearbook of Soil Law and Policy 2019*,
International Yearbook of Soil Law and Policy 2019,
https://doi.org/10.1007/978-3-030-52317-6_4

decomposed and decomposing organic matter;[2] or the foundation in and on which plants grow if water and plant food are available.[3] In a nutshell, soil is the basis of several life-giving necessities such as clean water, food and development opportunities.[4]

Although there is no coordinated approach towards sustainable soil management in Zambia, there are some constitutional and other legislative provisions that make provisions related to soil management. These are considered in detail in Sect. 5 of this chapter.

2.1 Soil Classification in Zambia

Understanding soil classification in the geographical context of Zambia as a land-locked country is important because it underscores sustainable use and management of the resource. Historically, soil classification has facilitated research and awareness raising among the general public, stakeholders and policy makers. This is usually made against the background of the major global environmental challenges such as food security and loss of biodiversity.[5] By way of background to soil classification, Africa, with a total land mass of about 30.7 million square kilometres and a population exceeding 746 million, has generally lagged behind in agricultural development. Sub-Saharan Africa to which Zambia belongs, with the exclusion of South Africa is ranked the poorest developing region.[6] With this background, it means that there is immense pressure on the soils of this region making it even more urgent to intensify sustainable soil management initiatives. Soil classification together with other data, is used to make assessments of land productivity and sustainability. Impliedly, this brings soil fertility and sustainability into the picture.[7]

The Soils Map of Zambia was first published in 1983 and it was correlated to the Soil Taxonomy (Soil Survey Staff, 1975), revealing over 350 soil series.[8] According to the 2016 National Agricultural Policy, Zambia is divided into three major agro-ecological regions (I, II, and III). Rainfall performance and the quality of soils differ across these regions. Roughly, Region I covers parts of Southern, Eastern and Western Provinces; receives less than 800 mm of rainfall annually and constitutes 12% of Zambia's total land area. It consists of loamy to clayey soils on the valley floor and course to fine loamy shallow soils on the escarpment suitable for the

[2]Merrill (1896)

[3]Houghton and Charman (1986), Gericke (1936) and FAO (2015).

[4]Ludwig (2018), pp. 1485–1489; LAL (2015).

[5]Bouma (2012), pp. 554–555.

[6]Jirstrom (2018), pp. 17–53; Kasimbazi (2019).

[7]Dewitte (2013); Eswaran (1997), pp. 1–18.

[8]https://esdac.jrc.ec.europa.eu/content/soils-map-republic-zambia-republic-zambia-atlas-sheet-no-12.

production of crops such as cotton, sesame, sorghum, groundnuts, beans, sweet potatoes, cassava, rice and millet and has potential for the production of various irrigated crops including fruits and vegetables and is also suitable for extensive cattle production. Region II covers the Central, Lusaka, parts of Southern, Eastern and Western provinces and is home to inherently fertile soils over a total of 42% of the country's land surface.[9] Region III constitutes 46% of the country's total land area comprising the Copperbelt, Luapula, Northern, Muchinga and North-western Provinces. With the exception of the Copperbelt Province, this Region is characterised by highly leached, acidic soils. The agricultural potential of the Region is enhanced by the application of agricultural lime and organic matter.

3 Sustainable Soil Management and Economic Outlook

It is indisputable that soil is an important natural resource, to the extent that every country's economy rests upon it. The previous section has discussed what soil is, while this section will evaluate its relevance to the country's economic activities, the predominant ones being mining and agriculture. A snapshot of the economic outlook is that agricultural production declined in 2018 due to poor rainfall distribution in most of southern Africa, but is expected to rebound in 2019. In addition, mining output is expected to increase by 4%–5% in 2019.[10] In view of these economic circumstances, the question to be addressed is whether the current legal and policy framework is capable of supporting a sustainable soil management regime which consequently contributes to the economic well-being of the country.

3.1 Sustainable Soil Management

Sustainable soil management according to the Food and Agriculture Organisation (FAO) is defined on the basis of Principle 3 of the revised World Soil Charter:

> Soil management is sustainable if the supporting, provisioning, regulating, and cultural services provided by soil are maintained or enhanced without significantly impairing either the soil functions that enable those services or biodiversity. The balance between the supporting and provisioning services for plant production and the regulating services the soil provides for water quality and

[9]National Agricultural Policy (2016), p. 17.

[10]https://www.afdb.org/en/countries/southern-africa/zambia/zambia-economic-outlook/; Africa's Pulse (2019).

availability and for atmospheric greenhouse gas composition is a particular concern".[11]

This definition shows that when the numerous soil functions are maintained or enhanced and this does not distort biodiversity, the resource has been sustainably managed. Included in this definition is the relationship between water quality, soil health and a balanced atmospheric greenhouse composition.[12]

In Sect. 2, it has been shown that soil is essentially a life giving and life supporting resource. Sustainable soil management is therefore an important concept that must be met with legislative support. There are many facets of soil sustainability that need to have some coercive force only possible through legislation.

3.2 *Economic Outlook in Relation to Soil*

In order to understand how soil impacts the economic outlook, it is important to briefly discuss the core drivers of soil degradation in Zambia. According to the 2017 Zambia Environment Outlook Report, human induced activities are responsible for most land and soil degradation as a result of mining, construction and quarrying, especially in the Copperbelt, Lusaka and North- Western Provinces.[13] The evidence of environmental degradation from these activities is seen in deforestation, unfilled mine pits and trenches, uncontrolled disposal of mine wastes and tailings and soil erosion. Despite the economic benefits of mining to the Zambian economy, it brings numerous negative soil- related impacts such as land degradation, ground and surface water pollution, formation of sinkholes, erosion, loss of biodiversity and soil contamination.[14]

Distinct from mining, deforestation is fuelled by clearing forests for agricultural purposes, charcoal and fuelwood production.[15] In addition, the employment of unsustainable agricultural practices has also been identified as a key driver of soil degradation in Zambia, especially in the northern part of the country where shifting cultivation is mostly practiced.[16] This challenge in the agricultural sector is gradually being overcome by the implementation of more sustainable and climate-smart agricultural methods, such as conservation farming. Conservation farming aims at improved soil and water conservation combined with reduced dependence on mineral fertilizer resulting in increased and more stable yields.[17]

[11]FAO (2017), p. 3.

[12]Kamunde-Aquino (2018) and Vargas-Rojas (2019).

[13]ZEMA (2017), p. 51.

[14]ZEMA (2017), p. 84.

[15]MLNR (2017), pp. i–iii, 1–5.

[16]Chidumayo (1987), pp. 15–18.

[17]Engel et al. (2017), p. 13.

The core drivers of soil degradation identified above show the need for law and policy to be strengthened, if sustainable soil management is to be achieved in Zambia. This will be discussed in Sect. 5 of the chapter.

The economic development strategy in Zambia is framed in two major plans; the National Long-Term Vision 2030 (Vision 2030), and the Seventh National Development Plan (SNDP). Vision 2030 is the country's first written long-term plan, expressing citizens' aspirations to become a prosperous, dynamic industrial middle-income nation.[18] The Vision 2030 is also a long-term strategy that identify objectives and policy instruments to achieve an integrated approach. Secondly, the SNDP adopts the Sustainable Development Goals (SDGs), the African Union (AU) Agenda 2063, as well as the Regional Indicative Strategic Development Plan.

Vision 2030 is poised to be operationalised through 5-year development plans which started with the Fifth National Development Plan (2006–2010), Sixth National Development Plan (2011–2016) and the current Seventh National Development Plan (SNDP) (2017–2021). There are five (5) pillars or strategic development areas contained within the SNDP:

(i) economic diversification and job creation;
(ii) poverty and vulnerability reduction;
(iii) reducing developmental inequalities;
(iv) enhancing human development; and
(v) creating a conducive governance environment for a diversified economy.[19]

The first two of these five pillars are directly relevant to sustainable soil management. The first pillar, 'economic diversification and job creation', presupposes that the country's economy will grow, reduce dependence on the extractive industries, especially copper mining and as a result reduce unemployment. In similar vein, the second pillar, 'poverty and vulnerability reduction' is aimed at reducing poverty and vulnerability in Zambia. These two pillars are related to the discussion in the latter part of this chapter. Further, these pillars are all buttressed by development outcomes, strategies, and programmes and key performance indicators, linked to the SDGs, which are themselves the foundation of global sustainable soil management initiatives.

The economic profile of Zambia consists of mining, agriculture, manufacturing and tourism, accounting for approximately 70%, 23%, 6% and 1% respectively towards the National Gross Domestic Product.[20] This is in line with observations that some of the factors that drive economic growth in least-developed countries (LDCs) in Sub-Saharan Africa include "availability of natural resources, poor

[18] http://www.mndp.gov.zm/wp-content/uploads/2018/05/Vision-2030.pdf accessed on 1st May, 2019.

[19] http://www.mndp.gov.zm/wp-content/uploads/2018/05/7NDP.pdf accessed on 10th May, 2019.

[20] Mulungu and Ng'ombe (2017), p. 15.

economic policies, access to the sea, tropical climate, volume of exports, a longer life expectancy and increased investment rates."[21]

The medium-term outlook in the Seventh National Development Plan is that GDP growth is forecast to rise by about 5% and will be largely attributable to agriculture, energy and mining.[22] This projected growth of the GDP will undoubtedly place enormous strain on soil health. In recent research specific to Zambia, it was found that "although most newly cultivated soils in sub- Saharan Africa may be rich in nutrients and organic matter, they would lose their productivity within 25 years under continuous cultivation in the absence of organic/inorganic fertiliser application..."[23] This is another timely 'wake- up call' to double efforts at creating a coordinated and sustainable soil management regime in Zambia, and more broadly in sub Saharan Africa.

In order for agricultural returns to have a more meaningful impact on the country's GDP, it has been suggested that:

> There is need to attract capital in agriculture. Deliberate efforts must be made to allow farmers adopt new technologies and encourage the use of machinery. Even though there is zero tax on agricultural equipment, credit constraints have left most farmers unable to take advantage of this measure... Policies to educate the population and reduce illiteracy levels, and encourage labour reallocation to more efficient sectors like services would help improve resource use in the economy overall.[24]

This suggestion is plausible because it points out that education and awareness raising are critical aspects that need to be addressed. More focus should be channelled to education on the importance of soil to agriculture and ultimately the Zambian economy. More concerted effort should also go towards ensuring that soil health is preserved over a longer period if agriculture is to emerge as the economic mainstay in Zambia.

4 International Law and Sustainable Soil Management

The international law perspective of sustainable soil management consists mostly of domestic implementation of international environmental agreements. In Zambia, the *Ratification of International Agreements Act* No. 34 of 2016 provides for the ratification of international agreements and sets out their domestication process. This legislation is relevant to sustainable soil management in Zambia because it facilitates the implementation of soil related international agreements, through legislation or other enforceable means.

---

[21]Ibid, 19.

[22]Seventh National Development Plan.

[23]Hamamoto (2018).

[24]Mulungu and Ng'ombe (2017), pp. 15–16.

Montanarella and Lobos Alva provide a historical description of the way in which soils have been discussed in UN literature in recent decades. They demonstrate that, even though soils are essential to sustainable development, they have never been the specific focus of any multilateral environmental agreement (MEA).[25] However, as a crosscutting natural resource in the environmental sphere, soils are considered within the three "Rio Conventions" negotiated at the United Nations Conference on Environment and Development (UNCED) in Rio de Janeiro in 1992 as follows:

(i) United Nations Frame- work Convention on Climate Change (UNFCCC);
(ii) United Nations Convention on Biological Diversity (CBD); and
(iii) United Nations Convention to Combat Desertification (UNCCD).

These global environmental agreements are considered as the framework conventions which individual countries can use to implement sustainable development initiatives. In relation to sustainable soil management, these three conventions create a basis for managing the effects of climate change. In this case, soil carbon is important, along with the conservation of biological diversity. As argued in this chapter, biological diversity is sustained by healthy soils, which are in turn, enhanced by the prevention of desertification and land degradation.

Zambia is a state party to the UNFCCC, CBD and UNCCD. The process of domesticating these conventions has been on going. In relation to the UNFCCC, Zambia is a signatory and has also ratified the convention. In furtherance of the country's commitment towards domesticating the UNFCCC, the National Policy on Climate Change (NPCC) was adopted in 2016. The objective of the NPCC, is "supporting and facilitating a coordinated response to climate change issues, enabling the country to realign its climate sensitive sectors of the economy and contributing towards achieving the objective of the UNFCCC."[26] The NPCC further provides for the creation of administrative structures and enactment of legislation and to support the policy's aspirations. With these positive strides, the process of domesticating the provisions of the UNFCCC is well on course in Zambia.

The objectives of the CBD have, to a large extent, been domesticated in the *Forests Act* (2015) as well as the *Zambia Wildlife Act* (2015). Both pieces of legislation are fairly recent and also specifically refer to the implementation of CITES and CBD in their preambular sections. The *Forests Act* (2015) goes a step further to specifically domesticate provisions of the UNFCCC and UNCCD. Overall, the provisions of Zambia's framework environmental legislation, the *Environmental Management Act* (2011) are focused on giving effect to all international environmental agreements to which Zambia is a party.[27]

Regionally, the Revised African Convention for the Conservation of Nature and Natural Resources, 2017, negotiated under the auspices of the African Union makes clear and deliberate provision for soil management. Specifically, Article VI relates to

---

[25] Montanarella and Lobos (2015), pp. 41–48.

[26] NPCC (2016), p. 2.

[27] Sambo (2019), p. 651.

land and soil. The provision obliges members to take "effective measures to prevent land degradation. . .using long term integrated strategies". It can be argued here that the long- term integrated strategies could be affected through appropriate legislative means in the African Union member- countries. Further, Article VI implores members to improve soil health in their jurisdictions through "scientific and local knowledge mechanisms."

In a research that focused on content analysis of sectoral policies in the agriculture, energy and forestry sectors in Zambia, it emerged that "although developing countries like Zambia are ratifying international environmental conventions, measures are often not drafted into national policies and linkages remain largely superficial."[28] This finding is cause for concern given that the global environment and its constituent elements such as soil, water, air and biological diversity are largely regulated by international law.

This section has underscored the interaction of international, national and local legislative mechanisms towards sustainable soil management. In the context of the Revised African Convention for the Conservation of Nature and Natural Resources, it is clear that the use of local or traditional knowledge in relation to the preservation and sustainability of the soil resources is encouraged.

There are also international non-binding instruments which are relevant to soil management. These include the World Soils Policy, 1982, Revised World Soil Charter, 2015, Voluntary Guidelines on Sustainable Soil Management and the Sustainable Development Goals (SDGs), 2015. These initiatives have stemmed from the global desire to improve agriculture and enhance food security. Accordingly, the three main international organisations that work around agriculture and food security are the Food and Agricultural Organisation, World Food Programme and International Agricultural Fund. These instruments, read together, recognise the need for good soil governance coupled with local initiatives informed by traditional and indigenous knowledge. For instance, the World Soil Charter recognises international actors, individuals and the private sector alike as 'stewards of the soil' and implores intergenerational uses towards soil. Further, the use of incentives to farmers for sustainable agricultural practices is another idea employed towards sustainable soil management.

Healthy soil and land resources are crucial to the achievement of the objectives and targets of many of the SDGs set out in Agenda 2030. More specifically, the SDGs relating to food security health, water security/resources, climate change, biodiversity, and land management are directly based on global soil sustainability. Even though each of the SDGs can be summarised in terms of food, health, water, climate and ecosystems, the descriptions cover a wide variety of subtopics that require input by many different disciplines. Healthy soils contribute to general ecosystem services, defined as "services to society that ecosystems provide", which requires cooperation between and among different disciplines.[29] As shown

[28]Kalaba (2014), p. 181.

[29]Keesstra (2016).

in the discussions in Sects. 2 and 3 above, understanding the importance of soil requires a good understanding of the scientific processes that support ecosystems and their constituent elements. Despite not being binding, the SDGs have since 2015 been well- received in the international community, with a good number of countries devising national plans and strategies to implement them.

The last two decades have seen spirited focus on the international regulation of agricultural soils to enhance food security. Overall these initiatives have lacked coordination at the global level. For instance, none of these initiatives refers to the other and there is no commonly identified target on which to anchor the campaign for soil sustainability. These shortcomings have led to ideas mooting an international convention dedicated to soil protection.[30] Such a soil-specific convention would then integrate these pre-existing initiatives, for instance, a lot of international support has been garnered around climate change. This support can be used to centralise measures for sustainable soil management. At international level, this would somewhat compel concerted national efforts for sustainable soil management.[31]

5 National Law and Policy for Sustainable Soil Management

This section critically evaluates the legislative framework in Zambia in order to understand the legal and policy perspectives of sustainable soil governance. It will consider property rights in soil, and some sectoral laws and policies which are relevant to sustainable soil management, such as environment, land, agriculture, water and forestry.

5.1 Constitutional Framework on Soil Governance

The Constitution is the supreme law in Zambia. According to Article 1(1), all written laws, must be consistent with the provisions of the Constitution. In relation to the sustainable management and use of soil, Article 43 stipulates the responsibilities of every Zambian citizen, and provides that every citizen has the mandate to protect and conserve the environment and to sustainably utilise natural resources.[32] The Constitution however falls short of creating an environmental right to a clean, healthy and sustainable environment. The right to a clean, healthy and sustainable environment is the bedrock of enforcing many other rights which hinge on the right to life. No doubt

[30] Kibugi (2018).

[31] Alori and Nwapi (2014), pp. 98–114; Hannam and Boer (2004); Ginzky (2018).

[32] Article 43 (1) (c).

that the argument for sustainable soil management can be included within this constitutional environmental right frame.

In addition, Article 253 makes provision for the principles of land policy and that land must be used sustainably.[33] Article 255 further outlines the principles upon which the environment must be managed and developed, as well as natural resources governance. The Constitution echoes the polluter pays principle, provides for the sustainable conservation and protection of ecologically sensitive areas, habitats, species and other aspects of the environment. Furthermore, the Constitution provides for the concurrent and exclusive functions of national and provincial government, which include soil conservation.

These Constitutional provisions enable: the enactment of law by Parliament; the creation of subsidiary laws such as statutory instruments and regulations; and, bylaws through local government for the protection, conservation and sustainable management of soil. This is an opportunity that can be utilised to develop specific initiatives for a coordinated soil management framework in Zambia.

5.2 Environmental Legislation and Soil Governance

Section 4 of this chapter has identified that the *Environmental Management Act*, EMA (2011) is the framework legislation for the environment in Zambia. In respect of all environmental matters, the provisions of the EMA (2011) are only secondary to the Constitution.[34] The EMA (2011) has robust provisions for the protection of all aspects of the environment, including soil.[35]

Section 2 of the EMA (2011) defines a number of terms that are used in the legislation. There are two (2) terms that are defined with direct relevance to soils and these are "element" and "sustainable management". The first term "element" in relation to the environment means "any of the principal constituent parts of the environment including water, atmosphere, *soil,* vegetation, climate, sound, odour, aesthetics, fish and wildlife". The second term "sustainable management" means the "protection and management of use of the environment, in a manner that, while enabling human beings to provide for their health, safety, social, cultural and economic well-being, *(a)* safeguards the life-supporting capacity of air, water, *soil* and ecosystems; *(b)* maintains the life-supporting capacity and quality of air, water, *soil* and ecosystems, including living organisms, to enable future generations to meet their reasonably foreseeable needs."

The other references to soil in the EMA (2011) are in sections 75 and 76 of the EMA. Section 75 provides that the Zambia Environmental Management Agency (ZEMA) should work in consultation with local authorities and other appropriate

[33] Article 253 (1) (d).

[34] Section 5.

[35] Sambo (2019), pp. 647–664.

authorities, within 5 years of the commencement of the Act, which lapsed in 2016, to identify hilly areas at risk from environmental degradation. The EMA (2011) further identifies hilly areas as at risk from environmental degradation by being prone to soil erosion, landslides occurring or being likely to occur, vegetation cover being removed or being likely to be removed from the area at a rate faster than it is being replaced; or any other land use activity in that area being likely to lead to environmental degradation. The Minister of Environment is accordingly empowered to issue a list of landscapes and hilly areas that are regarded as being at risk from environmental degradation, and the ZEMA must maintain a record of all these matters and devise strategies to ameliorate these soil- related problems.

The totality of the provisions under section 75 of the EMA are very progressive and proactive. A search in the ZEMA registry, however, revealed that there is no record of hilly areas that are at risk of environmental degradation, nor are there strategies devised to address the causes of degradation. ZEMA are also supposed to facilitate public awareness of how to address the problem of environmental degradation, in collaboration with the local authorities and Minister responsible for the environment. The EMA (2011) creates a sound basis upon which the main drivers of land degradation in Zambia can be addressed, in the quest for general environmental well-being. It is, however, yet to be seen how these provisions will be implemented.[36]

Section 76 (1) of EMA (2011) relates to the management of fisheries, water, forestry and wildlife resources, and provides that they shall be managed in accordance with the provisions of the respective pieces of sectoral legislation. It can be inferred from this that due regard must be hard to soil sustainability in relation to the protection of natural resources such as water, fish and other aquatic life, forests and wildlife. Provisions like this need to be clearly made in the legislation so that there is coordination of policies and regulations that can foster soil sustainability in Zambia.

Section 105 is also relevant to sustainable soil management in Zambia as it makes provision for the enforcement of environmental restoration orders. Here, there is a discharge of a contaminant or pollutant into the environment in an amount, concentration or manner that constitutes a risk to human health or property, or that causes or has the potential to cause adverse effects, serve an environmental restoration order take any measures that will assist in reducing or eliminating the risk or harm and to take any measures to *(a)* take such action as will prevent the continuation or cause of pollution; *(b)* restore land, including the replacement of soil, the replanting of trees and other flora and the restoration as far as may be, of outstanding geological, archaeological or historical features of the land or the area contiguous to the land.

[36] Sambo (2019), pp. 662–663.

5.2.1 Soil Governance and EIA

The EMA (2011) makes provision for environmental impact assessments (EIAs) as systematic examinations which must be conducted to determine whether or not an activity or a project has or will have any adverse impacts on the environment. As shown in the previous section, the EMA (2011) provides that soil is an integral component of the environment. In order to support the enforcement on the carrying out of EIAs, EIA regulations exist under the Act.[37] Regrettably, these "current" EIA regulations are dated 1997, having been issued pursuant to the repealed *Environmental Protection and Pollution Control Act* (EPPCA 1999). From a strict environmental rationale, these regulations should not be in force, having been technically repealed together with the outdated parent legislation in 2011. The ZEMA, however continues to enforce the 1997 EIA regulations based on the provisions of the law that where any Act of Parliament is repealed, any Statutory Instrument issued thereunder remains in force as long as it is not inconsistent with the repealing written law.[38]

The Environmental Protection and Pollution Control (EIA) Regulations, 1997 are relevant to sustainable soil management in Zambia because they make provision to consider soil fertility as a possible adverse environmental effect. Regulation 8 (3) refers to ecological considerations which must be taken into consideration as follows:

(a) Biological diversity such as effect on number, diversity, breeding sites, etc. of flora and fauna, breeding populations of fish and game; and effects on the gene pools of domesticated and wild sustainable yield.
(b) Sustainable use including the effects on *soil fertility*, nutrient cycles, aquifer recharge, water run-off rates, aerial extent of habitats; and bio-geographical processes.

This provision shows that EIAs are purposed to protect the totality of biological diversity and ensure sustainability. It has been noted in the earlier discussions that soil is important to the ecosystem- balance equation. The provisions of regulation 8 (3) outline the components of ecosystems that need to be protected. The regulations, however under-estimate the value of soils by only considering their fertility, and not necessarily their sustainability. As pointed out, these regulations were made pursuant to a repealed piece of legislation, which was repealed to pave way for the enactment of a more forward- looking framework legislation, encompassing current global trends in environmental well- being. Against this background, it can be argued that the 1997 regulations must be immediately replaced by more robust and modern provisions which can be used to anchor the sustainable soil campaign in Zambia.

[37]Statutory Instrument No. 28 of 1997, the Environmental Protection and Pollution Control (Environmental Impact Assessment) Regulations, 1997.

[38]Section 15 of the Interpretation and General Provisions Act, chapter 2 of the laws of Zambia.

5.2.2 Soil Governance and Physical Planning

The *Regional and Urban Planning Act* Number 3 of 2015 is the current legislation that governs physical planning in Zambia. The Act exists to "ensure sustainable urban and rural development by promoting environmental, social and economic sustainability in development initiatives and controls at all levels of urban and regional planning". This means that the implementation of the Act must take into consideration all facets of the environment.

Section 3 (1) (c), (d) and (g) of the Act provide that planning procedures shall incorporate environmental standards and requirements specified in any law relating to the environment and natural resources as well as social and economic demands impacting on an area. Further, section 3 (2) (l) refers to the protection of natural surroundings and landscapes, including water bodies and forests. Furthermore, it is provided in various isolated sections of the Act that the planning function includes enhancing the beauty and environmental sustainability of the areas under planning. The provisions also emphasize that baseline strategic EIAs need to be conducted prior to the implementation of development works. The appropriate Minister is required to issue regulations "guiding all aspects of environmentally sensitive areas, protection and care of environmentally, culturally and historically significant places or buildings." The Act also includes provisions for public notification, consultation and participation.

With this analysis of the principal planning legislation, there is no specific reference to soil sustainability, other than soil being incidental to the environment. In fact, most provisions that refer to the environment are not even structured in such a way that it can be inferred that soil and its importance on balancing ecosystem were in contemplation when the legislation was enacted. It is worth mentioning that a number of provisions in the legislation can be relied upon to introduce soil protection and sustainability.

5.2.3 Public Participation and Environmental Awareness

Levels of environmental awareness in Zambia are relatively low.[39] A much more concerted effort is required to make the connections between the 'dirt' and environmental well- being if awareness is to be raised about soil as a critical component of the environment. The importance of soil and consequences of soil erosion are likely to be more appreciated in relation to agriculture. The real significance of soil, however is that is supports all life- giving organisms through stable ecosystems. It is therefore imperative to design an awareness strategy that resonates with every day activities. The levels of public awareness of environment- related issues can also be enhanced by implementing the State's constitutional mandate to encourage public

[39] Sambo (2019), p. 653.

participation across all spheres of governance, particularly in relation to natural resources use and environmental management.

The relevant public participation provisions are found in the Constitution and the EMA (2011).[40] Article 255 of the Constitution provides for principles of environmental and natural resources management and development. One of them is that there must be "effective participation of people in the development of relevant policies, plans and programmes."[41] Further, Article 257 (d) of the Constitution mandates the State to encourage public participation in the utilisation of natural resources as well as environmental management. These constitutional provisions on public participation can be extended to sustainable soil management initiatives, even though this is not expressly stated.

Part VII of the EMA (2011) covers sections 91 to 94, the totality of which, gives the public the right to be informed of any intentions by public officers to make any decision affecting the environment, which arguably includes soil management. The public is also empowered to participate in the formulation of policies, strategies and any plans related to the environment.

Other means of public participation provided in the EMA (2011) include the right to access information and inspect environmental documents, conduct awareness campaigns and debates in the environmental interest. In this regard, there are a wide range of soil- specific activities and strategies that can be initiated by the public. The EMA (2011) incorporates relatively good environmental law provisions that can be exploited to enhance soil sustainability in Zambia. However, the environmental awareness levels of the Zambian populace are generally low.

5.3 Mining Law and Soil Governance

Mining is a complex array of activities that are one of the highest causes of global soil pollution. Some of the activities involved in the process include emissions, waste disposal, combustion, mineralization and many others.[42] This situation is also evident in Zambia, where mining activities account for over 90% of water, air and land pollution in Zambia.[43] The levels of soil contamination attributed to mining activities is high and evidence abounds in parts of the country used for mining activities.[44] Given this background, it is important for this section to analyse the provisions of the mining regulatory framework in order to ascertain the existence of measures to contain soil degradation.

---

[40] Sambo (2019), pp. 655–656.

[41] Article 255 (l), Constitution (2016).

[42] Asami (1988), p. 143; Glazewski (2005), p. 456.

[43] Lindahl (2014).

[44] Kříbek (2019), pp. 159–173.

Most of the challenges in addressing pollution and soil contamination in the Zambian mining sector are historical.[45] One of the major historical challenges is the laxity of the Development Agreements that were signed between the Government of the Republic of Zambia and mining investors. For instance, the Development Agreements had weak environmental governance mechanisms and environmental remediation was never prioritised.[46] This was coupled with a relatively weak policy and legislative framework in the pre- 2008 period. The enactment of the *Mines and Minerals Development Act*, 2008, alongside the EMA (2011), saw the gradual strengthening of the mining and environmental legal framework. In 2015, the *Mines and Minerals Development Act,* MMDA (2015) was enacted with a view to further strengthen mining and environmental regulation in Zambia.

5.3.1 Mining Regulatory Framework

The *Mines and Minerals Development Act*, MMDA (2015) and the EMA (2011) form the primary regulatory framework for the mining industry in Zambia. Other pieces of legislation such as the *Forest Act* (2015)[47] and the *Zambia Wildlife Act* (2015)[48] make provision for mining rights to subsist in national forests, national parks, game management areas, community partnership parks or birds and wildlife sanctuaries, subject to EIAs being carried out, in line with the provisions of the EMA (2011).

In addition, the provisions of the *Zambia Development Agency Act*, ZDA (2006) are vitally important. The objective of the ZDA Act (2006) is to "foster economic growth and development by promoting trade and investment in Zambia through an efficient, effective and coordinated private sector led economic development strategy."[49] As with sectoral legal frameworks, the ZDA Act (2006) does not address sustainable land-use investments. This is a significant omission because sustainability or environmental considerations are central to any investment,[50] especially mining and others that are land based. The ZDA Act, however, requires that in considering an application for a license or permit, regard must be had to "the impact the proposed investment is likely to have on the environment."[51] In light of this, it is open to argue that the need to achieve sustainable soil management is incorporated in the ambit of monitoring environmental impacts of investments, including mining. The environmental and social safeguards inherent in environmental assessments has been more fully discussed in Sect. 5.2.1.

---

[45]Lindahl (2014).

[46]Lungu (2008) and Parliamentary Report, Zambia (2014).

[47]Section 15, Forests Act (2015).

[48]Section 16, Zambia Wildlife Act (2015).

[49]Preamble, ZDA Act (2006).

[50]Sambo (2015), p. 13.

[51]Section 69, ZDA Act (2006).

The MMDA (2015) contains provisions that regulate mines and minerals operations and environmental protection in relation to mining activities. The purpose of the MMDA (2015) is to revise the law relating to mining exploration, mining and mineral processing as well as ensuring safety, health and environmental protection in mining operations and other matters incidental to mining in Zambia.[52]

Section 4 of the MMDA makes provision for the general principles that need to be taken into consideration for mining and minerals development in Zambia. All of the principles are relevant in the quest for sustainable soil management. Generally, the principles assert the non-renewability of mineral resources that they must be conserved and used prudently in order to secure the needs of the present and future generations. Other principles are utilising mineral resources in order to ensure socio-economic development and in safety, health and environmental protection, avoidance of wasteful mining practices and prevention of adverse environmental effects.

Section 6 of MMDA establishes the Mining Licensing Committee whose function it is to consider applications for mining rights and determine their subsistence. The composition of this Committee is drawn from the mining industry, Geological Survey department, and representatives from the ministries of environment, land, finance and labour. Provision is also made for the incorporation of representation from the office of the Attorney-General, the Zambia Development Agency, and Engineering Institute of Zambia. This wide- ranging composition ensures that all sectors of the economy are taken into consideration. Of particular interest, the Geological Survey department and ministries of land and environment are incorporated because they are critical in advancing initiatives for sustainable soil management.

Of further importance to this discussion, section 12 of MMDA prohibits exploration, mining or mineral processing without licence or environmental impact assessment granted by the Zambia Environmental Management Agency (ZEMA). Any person who contravenes this requirement commits an offence and is liable, upon conviction to payment of a fine or imprisonment for a term not exceeding 7 years, or to both.[53] This provision is important for averting future environmental damage, but it cannot forestall historical contraventions that have yet to be remediated.

Section 36 covers the suspension of production or closure of a mine for failure of the mining rights holder to comply with the conditions of the mining right or mineral processing licence. This includes the creation of an unsafe environment or causing uncontrollable pollution from mining operations, amongst other reasons. It is clear that if such a provision is to be enforced, soil sustainability can be enhanced in mining and surrounding areas. Since the enactment of the MMDA (2015), a number of mining companies have been held accountable for on- going pollution, through court prosecutions.[54]

---

[52]Preamble to the Mines and Minerals Development Act, 2015.

[53]Section 29 (3), EMA (2011).

[54]Mulenga (2019).

Part V of the MMDA covering sections 80 to 87 provides for safety, health and environmental protection. Specifically, section 80 provides for the consideration of the environment and human health when granting mining rights or mineral processing licences. When processing an application for mining rights, the Committee is mandated to ensure that the applicant will be able to discharge the duty to "conserve and protect the air, water, soil, flora, fauna, fish, fisheries and scenic attractions; and the features of cultural, architectural, archaeological, historical or geological interests."[55] Generally, a mining licence will not be granted where the applicant is unable to show evidence of measures that will be taken to guarantee environmental well- being.

Section 87 (1) provides for making mining rights holders strictly liable for any harm or damage caused by mining operations or mineral processing operations. Such liability is attached to the person who directly contributes to the act or omission which results in the harm or damage. Where there is more than one person responsible for the harm or damage, the liability shall be joint and several. Further, in the event of harm or damage being caused to the environment or biological diversity, compensation shall include the cost of reinstatement, rehabilitation or clean-up measures which are incurred and where applicable, the costs of preventive measures.

Furthermore, the MMDA (2015) extends liability to the following:

(a) harm or damage caused directly or indirectly by the mining or mineral processing operations to the economy or social cultural conditions;
(b) negative impact on the livelihood or indigenous knowledge system or technologies of any community;
(c) disruption or damage to any production or agricultural system;
(d) reduction in yields of the of the local community;
(e) air, water or soil contamination or damage to biological diversity;
(f) damage to the economy of an area or community; or
(g) any other consequential disorder.[56]

This provision is important because it covers all aspects of the environment, including soil that are prone to pollution or destruction from mining operations and which are rarely recognised in practice. The examples of social cultural conditions, community livelihoods, agricultural systems are all anchored on soil sustainability.

Section 87 (7) widens *locus standi* (the right to bring an action or to be heard) by moving beyond the directly affected person(s) to include a wide range of interested and affected persons, including environmental Non-Governmental Organisations. Any of these parties can bring an action in respect of breach or threatened breach of any provision relating to damage to the environmental, biological diversity, human and animal health or to socio–economic conditions.

The net effect of these progressive legal provisions can only be seen once the implementation and enforcement mechanisms are improved. The lack of efficient

[55]Section 80 (1) MMDA (2015).

[56]Section 87 (5), MMDA (2015).

coordination between and amongst the different line ministries has also contributed to failure to effectively enforce the law. As the regulatory framework standards, it is feasible to seek enforcement of certain provisions in the interest of achieving sustainable soil management in Zambia

5.4 Agricultural Laws and Soil Governance

The Zambian agriculture sector is classified into two main groups; (a) traditional agriculture which is mainly for producing subsistence food; and (b) commercial agriculture, which is basically for commercial production.

The Second National Agricultural Policy (NAP) 2016–2020 is operationalised through the Ministry of Agriculture and local authorities. Amongst other objectives, the NAP promotes the sustainable management and use of natural resources through sustainable land management technologies such as conservation agriculture and appropriate stock densities. These measures are important for facilitating sustainable soil management. Some of the advantages of conservation farming that enhance sustainable soil management are that water infiltration and root development of plants are improved.[57] Conservation farming also ensures the precise application of fertilizer and other inputs next to the plants where it is most likely to do good to the plant and surrounding soils. Further, conservation farming is renowned by agricultural experts for holding the potential to restore soil fertility to land damaged by years of excessive ploughing and heavy applications of chemical fertilizer. Furthermore, it improves farm yields and incomes with moderate input use, pointing out that conventional hand-hoe technologies that are replaced, are responsible for unnecessary soil removal.[58]

Healthy soils are critical to agriculture, and both are essential to enabling food security. In this regard, focus must be had on the intersection of agriculture, sustainable development and other topical issues such as climate change, health, environmental degradation, pollution control, biotechnology and human rights.[59] It is important to understand how agricultural legislation addresses soil-related challenges, such as combating land and soil erosion and degradation. Legal and institutional frameworks, policies and technologies are important mechanisms through which agricultural sectors and systems may be reformed and strengthened to attain soil sustainability.

The policy and legislative framework for agriculture in Zambia includes the NAP and a number of pieces of legislation, some of which are analysed in this section. The NAP has ten (10) objectives and the formulation of four of these is directly relevant to soil sustainability; namely:

---

[57]Baudron (2007).

[58]http://www.fao.org/conservation-agriculture/en/.

[59]Alori and Nwapi (2014).

(i) increase agricultural production and productivity by promoting improved seed, efficient use of fertilisers and agro-chemicals, efficient water resources usage and promotion of mechanisation in farming areas;
(ii) ensure the sustainable use and management of natural resources across the agricultural sector;
(iii) improve food and nutrition security through the promotion and diversification of agricultural production and utilisation;
(iv) increase private sector participation in agricultural development by strengthening the legal and regulatory framework which includes the formulation of new legislation.

With this policy background in mind, this section discusses the current legislative framework with its many short-comings vis a vis sustainable soil management. The Agricultural Lands Act provides for and regulates tenant farming schemes, such as the alienation of land and the issuance of leases and grants for farming purposes. This piece of legislation has no provisions relating to soil sustainability. The Agriculture (Fertiliser and Feed) Act is another piece of legislation applicable to the agriculture sector, the definitional section of which defines "fertiliser", as any substance or compound of substances used in the improvement or maintenance of the growth of plants or of the productivity of the soil. There are no other provisions explaining soil productivity or whether fertilisers and other chemicals used in the agricultural sector are tested or licensed.

This short review shows that there is need to review agricultural legislation as explained in the Agricultural Policy. Effective regulatory and monitoring capacity in institutions such as the Zambia Compulsory Standards Agency, Zambia Environmental Management Agency, the Ministry of Agriculture amongst others is required. In order to improve agricultural output, it is important to take sustainable soil management into consideration. A lot of legislative gaps exist in the agricultural legislative framework and the process of enacting new legislation has started. It is hoped that this window of opportunity can be utilized to infuse progressive ideas on how best to legislate in the interest of soil sustainability in Zambia.

5.5 Soil Governance, Land Tenure and Property Rights

Land is not only a factor of production, it is also a source of agricultural output and income, an asset, and therefore the origin of wealth, prestige, and power.[60] Land is also important because through its constituent elements, such as soil, it contributes to the survival of other terrestrial forms of life. Land use concerns the products and/or benefits obtained from land. In Zambia, these are mainly divided into agriculture, wetlands, wildlife and forest; and urban development. There are several factors that

[60] De Janvy and Sadoulet (2016).

affect land use in Zambia, including population growth, encroachment on protected areas, climatic hazards, agricultural development and demand for construction materials amongst others.[61] The use to which land can be put is mainly influenced by land tenure system and its management.

5.5.1 Land Tenure Systems

In Zambia and many other African countries, there are generally two types of land tenure systems; leasehold and customary. According to section 2 of the Lands Act, all land is held by the President in trust for and on behalf of the people of Zambia. All land is vested in the President who holds it in perpetuity for and on behalf of the people of Zambia. Customary land is estimated to be 94% of the total land area in Zambia, while 6% is under leasehold. These statistics appear out of date and can only be confirmed once the ongoing land audit under the Ministry of Lands and Natural Resources is completed.[62]

Under customary tenure, land is administered by chiefs using practices based on customs and beliefs among a particular community. Accessibility of land to the citizenry is determined by the cultural set up of the particular chiefdom that is either the matrilineal or patrilineal systems.[63] Customary land tenure generally lacks security of tenure because the rights of landholders are not clear until land is fully surveyed and converted to leasehold tenure. Research has shown that insecure land tenure can lead to the adoption of unsustainable land management practices and inability to make investments for sustainable land management.[64] In such an eventuality, sustainable soil management cannot be guaranteed.

Leasehold tenure is regulated by statutory law and the Ministry of Lands is principally responsible for land administration and management. The day to day administration of land is delegated to the Commissioner of Lands who is legally mandated to make grants or dispositions of land to any person who qualifies to hold land within the provisions of the Lands Act. The country's local authorities are also empowered to administer land within their districts and have responsibility for land-use planning.[65]

In recent years, there has been an increase in demand for conversion of land from customary to leasehold tenure. The two major factors influencing this increase are enhanced security of tenure and the greater security offered to lenders through the mortgaging of the land title. Land on title is key for enhancing economic empowerment as it can be used as collateral.

---

[61]Mususa (2012).

[62]ZEMA (2017).

[63]ZEMA (2017).

[64]Way (2006).

[65]Section 6, Lands Act (1995).

5.5.2 Land Property Rights

There are a number of existing laws with a focus on land. For purposes of understanding land in the context of being property, this section focusses on the provisions of the Constitution and the Lands Act. The Constitution provides for equitable access to land and ownership of land, security of land rights as well as sustainable and productive management of land resources. Further, the Constitution specifies the principles of land policy, environmental and natural resources management and establishes a Land Commission purposed to "administer, manage and alienate land, on behalf of the President."[66] Apart from the provisions on sustainable use of land, Article 16 (1) of the Bill of Rights under the Constitution provides for protection from deprivation of land in the following terms:

> Except as provided in this Article, no property of any description shall be compulsorily taken possession of and no interest in or right over property of any description shall be compulsorily acquired, unless by or under the authority of an Act of Parliament which provides for payment of adequate compensation for the property or interest or right to be taken possession of or acquired.

The reference to "no property of any description" includes both movable and immovable property, including land. Property rights in land are acquired through processing title which is recognised as proof of ownership.

From the Kenyan perspective, Kibugi argues that property rights provide "the most direct legal entitlement and decision-making authority to make choices of utilisation."[67] In the case of Zambia, it can be seen that security of tenure under customary tenure is not guaranteed. This means that even the choice of utilisation over customary land is hampered, for instance, access to credit may be limited.

The Lands Act of 1995 provides for the continuation of leaseholds and leasehold tenure, continued vesting of land in the President and alienation by the President. It also provides for statutory recognition and continuation of customary tenure as well as the conversion of customary tenure into leasehold tenure.[68] Customary tenure can be converted into private leasehold tenure, however, once converted, customary land rights are extinguished and the land cannot be converted back to customary tenure.

[66] Article 253, Constitution of Zambia, Amendment No. 2 of 2016.

[67] Kibugi (2016), p. 223.

[68] Section 8, Lands Act (1995).

5.6 Other National Concerns Related to Soil Governance

There are other issues that are relevant to the creation of a sustainable soil management system in any country which have been identified from the literature. These are explained below:

5.6.1 Urbanisation and Poverty

Zambia is one of the most urbanised sub-Saharan African countries, with 42.1% of its population living in urban areas.[69] Historically, this was attributed to mining and other economic activities in urban Zambia that were not extended to the rural and peri urban areas.[70]

High urbanisation begets numerous environmental problems such as immense strain on natural resources such as water and soil. From the scientific perspective, efficient water reticulation is dependent on sustainable soil health.[71] It therefore follows that where soil health is compromised, so is the efficiency of water reticulation. For instance, a research conducted in New Zealand investigated the effects of high urbanisation on soil and established that proper land-use planning is required for urban expansion and its absence would gradually erode New Zealand's food production potential.[72] More specifically, the research showed that urbanisation differentially affected most of New Zealand's versatile soils and if such trends were to continue, a large percentage of fertile lands could be lost over the next 50–100 years. Further, it was found that during the same period, impacts of climate change, population growth, and resource limitations would make agricultural access to versatile soils even more important. It was also pointed out that the available land-use data was inconsistent and incomplete thereby making it difficult to analyse and interpret the results. Furthermore, the research found that soils are "non-replaceable national assets that require long- term protection, appropriate policies and land-use management planning supported by robust land- use data, in order to ensure that future generations enjoy the same range of options for their food production as are presently enjoyed.[73]

Consequently, high urbanisation rates can be linked to high levels of poverty. According to research, high urbanisation, poverty and environmental degradation are intricately connected. The majority of poverty- stricken populations engage in unsustainable land use practices in order to sustain their livelihoods.[74] It can be

[69]Mususa (2012), pp. 571–5873.

[70]Osei-Hwedie (1996); Huth (1984), pp. 1–16.

[71]Haase (2009), pp. 211–219.

[72]Rutledge (2010).

[73]Rutledge (2010) and Grinlinton (2019).

[74]Nkonya (2008), pp. 1–5; 7–9.

argued that land use planning, in terms of changing land use can address both already existing and future poverty.

The reality of poverty in Zambia and anywhere else is that it is an identifiable further strain on natural resources already under severe exertion. This chapter agrees with the view that poverty is a difficult concept to define, and as such has no universally accepted definition.[75] Research has been undertaken in the Zambian context pointing out that "lack of access to basic needs – primarily food, and other necessities such as housing, clean water, clothing, basic infrastructure, and transportation and importantly, food insecurity" are indicators, or definitions of poverty.[76] At this juncture, it is important to interrogate the interplay of poverty with sustainable management of natural resources, particularly soil. As with high urbanisation, poverty puts immense pressure on natural resources and leads to other offshoot problems such as conflicts which also exacerbate the quality of soil.[77] In the most urbanised areas of Zambia, like the Copperbelt, Central and Lusaka Provinces, and other towns along the line of rail, the weight exerted on natural resources and soil in particular is immense. It must be mentioned these are also the same areas under strain with mining effects. It is therefore important that particular attention be had to soil sustainability in these areas.

The applicable law and policy in respect of natural resources usage, mining and poverty are the EMA (2011), MMDA (2015), Forests Act (2015) and the Zambia Wildlife Act (2015) which have already been discussed in this chapter. The point to emphasize here, however, is that the EIAs must be carried out in accordance with the legal provisions. This would ensure that all adverse impacts of any development project, whether social, economic, environmental or cultural are factored in.

5.6.2 Deforestation

Zambia is home to huge natural forest reserves which are threatened by deforestation, whose rate is well above the global and regional average. Deforestation continues to threaten biodiversity and undermines key ecosystem services such as climate and water regulation. The forestry sector is regulated by the Forests Act (2015) which declares, establishes and ensures sustainable management of all forests in Zambia.

Section 2 of the Act defines "forest produce" as including soil, algae, bamboos, bark, bedding, bees, honey and a lot of other forest-based products, while the term "forest resources" refers to vegetation, wood and non-wood products and forest ecological services and "includes the maintenance of soil quality, control of erosion" provision of organic materials and modulating climate. This piece of legislation does

[75]Bulawayo (2019), pp. 3–4.

[76]Bulawayo (2019).

[77]Linking Gender, Environment, and Poverty for Sustainable Development: A Synthesis Report on Ethiopia and Ghana (2012), p. 59.

not necessarily provide for soil protection. In Section 5, a duty is placed on the Department of Forestry to take "all necessary measures to ensure the realisation of the exploitation of forest resources, the maintenance of sustainable forest management and the conservation and sustainable use of forest resources". This includes the maintenance of soil quality and control of erosion as shown in the interpretation clause to the Act.

Section 16 (2) (a) of the Forests Act provides that a mining right shall not be granted in a National Park, Community Partnership Park or bird or wildlife sanctuary without an EIA conducted in accordance with procedures specified by the Environmental Management Act (2011). These procedures are supposed take into consideration, the need to conserve and protect the air, water, soil, flora, fauna, fish, fisheries and scenic attractions in or on the land over which the right is sought. The totality of these provisions show that soil health is a key consideration in the Forest Act.

5.6.3 National Disaster Management and Climate Change

Zambia is prone to extreme meteorological events such as droughts and floods, with major impacts on the environment, food security[78] and public health. Climate change adds to existing environmental stress and is likely to negatively affect Zambia's natural resource base and exacerbate existing problems of land degradation, flood regulation, water purification and droughts. Undoubtedly, the strain on soil health would worse the overall picture.

The Disaster Management Act (2010) is purposed to *inter alia* "... establish and provide for the maintenance and operation of a system for the anticipation, preparedness, prevention, coordination, mitigation and management of disaster situations..."[79] The definitional section refers to related terms such as disaster, hazard and emergency to mean more or less the same event, which interacts with all forms of life is potentially harmful on the environment. To start with, a 'disaster' means:

> An event that is associated with the impact of a human induced or natural hazard, which causes a serious disruption in the functioning of a community or society, causing widespread human, material or environmental losses which exceeds the ability of the affected community or society to cope with the hazard using its own resources.

Further, a 'hazard' is taken to be:

> A potentially damaging physical event such as an earthquake, a hurricane, flood, drought, fire, epidemic, phenomenon or human activity, which may cause injury or the loss of life, damage to property, social and economic disruption or environmental degradation, and includes latent conditions that may represent future threats and can have different origins, natural and human induced.

[78] Brown and Funk (2008), pp. 580–581.

[79] Preamble, Disaster Management Act No. 13 of 2010.

Furthermore, an 'emergency' is taken to mean:

> An event, actual or imminent, which endangers or threatens to endanger life, property or the environment and which requires a significant and coordinated response.

It can be seen that the definitions of disaster, hazard and emergency necessarily impact on soil as a foundational element of the environment. Despite this scenario, the legislation falls short of specific reference to soil preservation. Further, in the absence of climate change- specific legislation in Zambia, it is arguable that the Disaster Management Act (2010) is the closest in filling that gap. In the midst of overwhelming scientific evidence that soils contain a huge carbon stock which is capable of preventing carbon dioxide building up in the atmosphere and thereby compounding the problem of climate change,[80] this omission is a missed opportunity in achieving the objectives of the Disaster Management Act (2010). This clear relationship between sustainable soil management and mitigation of climate effects needs to be harnessed in the legislation.

To facilitate implementation of the Disaster Management Act (2010), section 4 creates the Disaster Management and Mitigation Unit (DMMU) as being responsible generally, to "take all necessary measures in order to prevent, alleviate, contain and minimize the effects of disasters." Implicitly, soil sustainability would form a part of the environment that would require remediation or restoration after a disaster. The Act is however silent on this. At best, there ought to be cross reference to the provisions of the EMA (2011), the framework environmental law. In the absence of a clear reference to soil restoration in such a situation, it means that it is open to argue that the mandate of the DMMU ends at carrying out the functions that have been specifically itemised in the legislation.

6 Conclusion

This chapter has identified the core drivers of soil degradation and further shown that there is no coordinated legislative approach to sustainable soil management in Zambia. Apart from the provisions of the Environmental Management Act (2011) that have been discussed in this chapter, there is no dedicated focus on sustainable soil management in any of the other pieces of legislation and policy documents. It has been noted that Zambia is a vast country with the abundance of natural resources that feed on healthy soils. Therefore, the interconnectedness of ecosystems necessitates that soil and soil health must be prioritised in the legislative framework. An analysis of literature from the scientific perspective has revealed that there are concerted global efforts aimed at enhancing sustainable soil management. Given the uniqueness of the core drivers of soil degradation in different parts of the world, it is important to also consider targeted legislative measures. Whilst it is acknowledged

[80] Lorenz and Lal (2015), pp. 35–37; Janzen (2015), pp. 19–21.

that the law is only but one mechanism amongst others for resolving societal challenges, a coordinated legislative framework for enhancing sustainable soil management would be a good starting point for harnessing favourable results. In the absence of a clearly mapped out sustainable soil governance regime in Zambia, the country will continue losing biodiversity and encountering negative environmental effects.

References

Africa's Pulse (2019) An analysis of issues shaping Africa's economic future. https://doi.org/10.1596/978-1-4648-1421-1. Accessed 1st May 2019

Alori ET, Nwapi C (2014) The international legal regime for sustainable soil. In: Huks R, Ako T, Olawuyi D (eds) Food and agricultural law: readings on sustainable agriculture and the law in Nigeria. Afebabalola University, Ado Ekiti, pp 98–114

Asami T (1988) Soil pollution by metals from mining and smelting activities. In: Salomons W, Förstner U (eds) Chemistry and biology of solid waste. Springer, Berlin. https://doi.org/10.1007/978-3-642-72924_7. Accessed 20th July 2019

Baudron F et al (2007) Conservation agriculture in Zambia: a case study of Southern Province

Bouma J et al (2012) Soil information in support of policy making and awareness raising. Curr Opinion Environ Sustain 4(5):552–558

Brown ME, Funk CC (2008) Food security under climate change. Science 319:580–581

Bulawayo M, Ndulo M, Sichone J (2019) Socio- economic determinants of food insecurity among Zambian households: evidence from a national household survey. J Asian Afr Stud 54 (6):800–818. 1–19

Chidumayo E (1987) A shifting cultivation land use system under population pressure in Zambia. Agroforest Syst 5:15–25. https://doi.org/10.1007/BF00046411. Accessed 17th July 2019

De Janvy A, Sadoulet E (2016) Access to land and development. https://link.springer.com/content/pdf/10.1057/978-1-349-95121-5_2126-1.pdf. Accessed on 12th July 2019

Dewitte O et al (2013) Harmonisation of the soil map of Africa at the continental scale. Geoderma 211:138–153. https://doi.org/10.1016/j.geoderma.2013.07.007, 138–153, Accessed 20th Apr 2019

Engel E et al (2017) Towards inclusive and sustainable rural transformation in Sub-Sahara Africa. Centre for Rural Development, Berlin. https://www.sle-berlin.de/files/sle/publikationen/Final%20Report_Towards%20Inclusive%20and%20Sustainable%20Rural%20Transformation%20in%20Sub-Saharan%20Africa.pdf. Accessed 19th July 2019

Eswaran H (1997) An assessment of the soil resources of Africa in relation to productivity. Geoderma 77(1):1–18. https://doi.org/10.1016/S0016-7061(97)00007-4. Accessed on 19th Apr 2019

FAO (2015) What is soil? FAO, Rome. http://www.fao.org/soils-portal/about/all-definitions/en/. Accessed on 1st May 2019

FAO (2017) Voluntary guidelines for sustainable soil management Food and Agriculture Organization of the United Nations. FAO, Rome, p 3. http://www.fao.org/3/a-bl813e.pdf. Accessed on 19th Apr 2019

Gericke WF (1936) https://science.sciencemag.org/content/88/2294/568.pdf-extract. Accessed on 1st Apr 2019

Ginzky H (2018) The sustainable management of soils as a common concern of humankind: how to implement it? In: Ginzky H, Dooley E, Heuser I, Kasimbazi E, Markus T, Qin T (eds) International yearbook of soil law and policy 2017. International yearbook of soil law and policy, vol 2017. Springer, Cham

Glazewski J (2005) Environmental law in South Africa, 2nd edn. Lexis Nexis, Durban

Grinlinton D (2019) Sustainable management of urban soils: the New Zealand approach. In: Ginzky H, Dooley E, Heuser I, Kasimbazi E, Markus T, Qin T (eds) International yearbook of soil law and policy 2018. International yearbook of soil law and policy, vol 2018. Springer, Berlin

Haase D (2009) Effects of urbanisation on the water balance – a long-term trajectory. Environ Impact Assess Rev 29(4):211–219

Hamamoto T et al (2018) Small- scale variability in the soil microbial community structure in a semi- developed farm in Zambia. Appl Environ Soil Sci 2018:1–6. https://doi.org/10.1155/2018/7939123. Accessed on 19th Apr 2019

Hannam I, Boer B (2004) Drafting legislation for sustainable soils: a guide. IUCN, Gland

Houghton PD, Charman PEV (1986) Glossary of terms used in soil conservation. Soil conservation service, Bathurst. Available at http://www.scs.nsw.gov.au/__data/assets/pdf_file/0004/49441 0/Glossary_of_Terms_Used_In_Soil_Conservation.pdf. Accessed on 14th Apr 2019

Huth MJ (1984) The impact of rapid population growth, expanding urbanisation, and other factors on development in Sub Saharan Africa: the contrasting responses of Tanzania and Kenya. Int J Sociol Soc Policy 4(2):1–16. https://doi.org/10.1108/eb012963. Accessed on 1st May 2019

Janzen HH (2015) Beyond carbon sequestration: soil as a conduit of solar energy. Eur J Soil Sci 66 (1):19–32. https://doi.org/10.1111/ejss.12194

Jirstrom M et al (2018) African smallholder farmers on the move: farm and non- farm trends for six African Countries 2002–2015. In: Djurfeldt A, Dzanku F, Isinika A (eds) Agriculture, diversification, and gender in rural Africa- longitudinal perspectives from six countries. Oxford University Press, Oxford, pp 17–53

Kalaba FK, Quinn CH, Dougill AJ (2014) Policy coherence and interplay between Zambia's forest, energy, agricultural and climate change policies and multilateral environmental agreements. Int Environ Agreements Polit Law Econ 14:181. https://doi.org/10.1007/s10784-013-9236-z. Accessed on 22nd Apr 2019

Kamunde-Aquino N (2018) Who owns soil carbon in communal lands? An assessment of a unique property right in Kenya. In: Ginzky H, Dooley E, Heuser I, Kasimbazi E, Markus T, Qin T (eds) International yearbook of soil law and policy 2017. International yearbook of soil law and policy, vol 2017. Springer, Cham

Kasimbazi E (2019) Soil protection law in Uganda. In: Ginzky H, Dooley E, Heuser I, Kasimbazi E, Markus T, Qin T (eds) International yearbook of soil law and policy 2018. International yearbook of soil law and policy, vol 2018. Springer, Cham

Keesstra SD et al (2016) The significance of soils and soil science towards realization of the United Nations Sustainable Development Goals. www.soil-journal.net/2/111/2016/. Accessed 14th Apr 2019

Kibugi R (2016) Evaluating the role of private land tenure rights in sustainable land management for agriculture in Kenya. In: Ginzky H, Dooley E, Heuser I, Kasimbazi E, Markus T, Qin T (eds) International yearbook of soil law and policy 2017. International yearbook of soil law and policy, vol 2017. Springer, Cham

Kibugi R (2018) Soil health, sustainable land management and land degradation in Africa: legal options on the need for a specific African soil convention or protocol. In: Ginzky H, Dooley E, Heuser I, Kasimbazi E, Markus T, Qin T (eds) International yearbook of soil law and policy 2017. International yearbook of soil law and policy, vol 2017. Springer, Cham

Kříbek B (2019) Soil contamination near the Kabwe Pb-Zn smelter in Zambia: Environmental impacts and remediation measures proposal'. J Geochem Explor 197:159–173. https://doi.org/10.1016/j.gexplo.2018.11.018. Accessed 10th May 2019

Lal R (2015) The soil–peace nexus: our common future. Soil Sci Plant Nutr 61(4):566–578. https://doi.org/10.1080/00380768.2015.1065166. Accessed 11th May 2019

Lindahl J (2014) Environmental impacts of mining in Zambia- Towards better environmental management and sustainable exploitation of mineral resources. http://resource.sgu.se/produkter/sgurapp/s1422-rapport.pdf. Accessed 11th May 2019

Linking Gender, Environment, and Poverty for Sustainable Development: A Synthesis Report on Ethiopia and Ghana (2012) Report No. P125713, 59 http://siteresources.worldbank.org/EXTSOCIALDEVELOPMENT/Resources/244362-1265299949041/6766328-1270752196897/ESW-Gender-Poverty-Development.pdf Accessed 8th May 2019

Lorenz K, Lal R (2015) Managing soil carbon stocks to enhance the resilience of urban ecosystems. Carbon Manage 6:35–50. https://doi.org/10.1080/17583004.2015.1071182. Accessed 19th July 2019

Ludwig M et al (2018) Measuring soil sustainability via soil resilience. Sci Total Environ 626:1484–1493. https://www.sciencedirect.com/science/article/pii/S0048969717327468 Accessed 10th May 2019

Lungu J (2008) Copper mining agreements in Zambia: renegotiation or law reform? Rev Afr Polit Econ 35(117):403–415. http://www.jstor.org/stable/20406529. Accessed 19th July 2019

Merrill G (1896) The principles of rock weathering. J Geol 4(7):850–871. http://www.jstor.org/stable/30054908

Ministry of Lands and Natural Resources, MLNR (2017) National investment plan to reduce deforestation and forest degradation (2018–2022), www.climateinvestmentfunds.org. Accessed 19th July 2019

Ministry of National Development Planning, MNDP (2016) National policy on climate change

Ministry of Agriculture (2016) National agricultural policy

Montanarella L, Lobos A (2015) Putting soils on the agenda: the three rio conventions and the post-2015 development agenda. Curr Opinion Environ Sustain 15:41–48

Mulenga C (2019) Judicial mandate in safeguarding environmental rights from the adverse effects of mining activities in Zambia. Potchefstroom Electron Law J 22:1–33, 2019. https://ssrn.com/abstract=3377116. Accessed on 10th July 2019

Mulungu K, Ng'ombe J (2017) Sources of economic growth in Zambia, 1970–2013: a growth accounting approach. Economies 5:15. https://www.mdpi.com/2227-7099/5/2/15. Accessed 11th May 2019

Mususa P (2012) Mining, welfare and urbanisation: the wavering urban character of Zambia's Copperbelt. J Contemp Afr Stud 30(4):571–587. https://doi.org/10.1080/02589001.2012.724873

Nkonya E et al (2008) Linkages between land management, land degradation and poverty in Sub-Saharan Africa: the case of Uganda, IFPRI Research Report 159, Available at http://www.ifpri.org/publication/linkages-between-land-management-land-degradation-and-poverty-sub-saharan-africa. Accessed 22nd July 2019

Osei-Hwedie BZ (1996) Environmental protection and economic development in Zambia, Available at http://sanweb.lib.msu.edu/DMC/African%20Journals/pdfs/social%20development/vol11no2/jsda011002007.pdf

Osman KT (2013) Soils: principles, properties and management. Springer, London

Parliamentary Report, Zambia (2014) Committee on lands, environment and tourism on the Auditor Generaal's report on environmental degradation caused by mining activities for the fourth session of the eleventh national assembly appointed on 25th September, 2014. http://www.parliament.gov.zm/sites/default/files/documents/committee_reports/Report%20-%20Auditor%20General%27s%20Report%20-%20Environmental%20degradation.pdf

Rutledge DT et al (2010) Thought for food: impacts of urbanisation trends on soil resource availability in New Zealand. https://www.grassland.org.nz/publications/nzgrassland_publication_49.pdf. Accessed 9th Apr 2019

Sambo PT (2019) The Environmental Management Act (2011): a basis for the growth of an environmental ethos and good environmental governance in Zambia? In: Kameri-Mbote et al (eds) Environmental law in Africa. Nomos Publishers, Baden- Baden, pp 647–664

Sambo PT et al (2015) Enabling legal frameworks for sustainable land- use investments in Zambia – legal assessment report. https://www.idlo.int/sites/default/files/pdfs/publications/Enabling%20legal%20frameworks%20for%20sustainable%20land%20use%20investments%20in%

20Zambia-%20Legal%20assessment%20report.pdf and https://doi.org/10.17528/cifor/005777. Accessed 19th July 2019

Vargas-Rojas R et al (2019) Unlocking the potential of soil organic carbon: a feasible way forward. In: Ginzky H, Dooley E, Heuser I, Kasimbazi E, Markus T, Qin T (eds) International yearbook of soil law and policy 2018. International yearbook of soil law and policy, vol 2018. Springer, Cham

Way SA (2006) Examining the links between poverty and land degradation: from blaming the poor toward recognising the rights of the poor. In: Johnson P et al (eds) Governing global desertification: linking environmental degradation, poverty and participation. Ashgate, Burlington, pp 27–41

Zambia Environmental Management Agency, ZEMA (2017) Zambia environment outlook report, www.zema.org.zm/index.php/publications/environmental-outlook-reprts/. Accessed 18th July 2019

Combatting Desertification Through Soil and Water Conservation and Environmental Rehabilitation Measures: Experiences from the Tigray Region, Ethiopia

Kelemework Tafere Reda and Desta Gebremichael Gidey

1 Background

Over 65% of the population in sub-Saharan Africa depends on agriculture for their livelihood.[1] In areas like Ethiopia, traditional plough-based farming systems and the overuse of land for centuries have led to soil degradation and infertility.[2] Northern Ethiopia (particularly Tigray and Amhara regions), have historically been under extreme demographic influences (i.e. population pressure leading to intensive subsistence farming), civil war, poverty and climate change that has increased the occurrence of drought and extreme weather variability and concomitant environmental degradation.[3] The ever-increasing population size coupled with the long standing tradition of breaking apart land holdings to successive generations, has led to problems of land fragmentation with diminishing average household land holdings ranging from 0.5 ha in the highlands to 1.5 ha in the lowlands.[4] Intensive use of arable, pastoral and forest land has resulted in low productivity, which in turn significantly challenges the country's quest for economic progress and poverty alleviation.[5]

[1]Scoones (2009).

[2]Nyssen et al. (2006, 2009).

[3]Balehegn et al. (2019).

[4]Munro et al. (2008).

[5]Corbeels et al. (2000).

K. T. Reda
Mekelle University, Mekelle, Ethiopia

D. G. Gidey (✉)
Mekelle University, School of Law, Mekelle, Ethiopia

H. Ginzky et al. (eds.), *International Yearbook of Soil Law and Policy 2019*,
International Yearbook of Soil Law and Policy 2019,
https://doi.org/10.1007/978-3-030-52317-6_5

According to various authors,[6] aerial photographs taken in many parts of the Tigray region at different intervals clearly reveal that environmental degradation, particularly in the highlands of northern Ethiopia (i.e. Tigray and parts of Amhara regions), is not a recent phenomenon but rather existed as early as the eighth Century and has to date, remained cyclic in nature. As a result, Tigray is identified as a highly degraded region within Ethiopia.[7]

Several measures were put in place, in various countries, to address environmental degradation.[8] Soil and water conservation (SWC) and biological rehabilitation (BR) are the most commonly implemented and studied interventions in Tigray.[9] One of the measures falling under BR is 'ex-closure' also known as 'enclosure',[10] which allows for degraded land to rest for several years before it can be used for productive activities. In the past, ex-closures were exercised to avoid the degradation of communal pastoral lands, particularly in poor quality areas. This was done through rotational grazing during the different seasons in countries like; Ethiopia, Somalia, Tunisia and Algeria. Ethiopia also boasts a good track record of conserving forests around religious centres, such as churches. Until the development of the system of enclosures by local communities, the best-protected areas of natural vegetation were the small patches of sacred forests around churches and monasteries, in Muslim burial grounds and around holy springs.[11] The fact that religious centers look less degraded can be attributed to spirituality in which people feel uncomfortable to encroach into holy lands and destroy the resources surrounding them.

In the Tigray region, a series of stringent measures were taken to reverse depletion of forest resources which were caused by; agricultural encroachment, overgrazing and injudicious utilization of woodland; one such measure is building ex-closures.[12] This entails establishing a 'physical boundary' or a 'social fence' (meaning communal restriction to access) to promote natural regeneration of plants and thus reduce further degradation of formerly degraded lands"[13] Ex-closures can thus be effective in restoring degraded soils as well as increasing soil carbon in the highlands of Tigray.[14]

Local communities initiated another measure of allocating degraded communal hillsides to the unemployed landless youth, so that they could plant trees on such unproductive land to help them earn incomes. The intervention thereby had double advantages. On one hand, what was categorised as wasteland became productive; on the other hand, people's livelihoods were transformed. Such a community initiative

---

[6]See for example Butzer (1981).

[7]Edwards et al. (2010).

[8]Balehegn et al. (2019).

[9]Balehegn et al. (2019), p. 6.

[10]See, for example, Edwards et al. (2010).

[11]Edwards et al. (2010), p. 43.

[12]Mekuria et al. (2011).

[13]Birhane et al. (2017).

[14]Mekuria et al. (2011).

was supported by the local administration and later became part of the regional policy on land use, particularly after 1997. Subsequent measures were also taken by the Bureau of Agriculture and Rural Development (BoARD) and a local NGO called Relief Society of Tigray (ReST), including providing seedlings and technical support and scaling it up to other places as best practice.[15]

As a result of such significant public mobilisation of activities by the government, the percentage of infertile land went down from 28.2% in 1997 to just 13% by 2012. Similarly, the extent of poverty was reduced from 41% to 26% during the indicated time. The Tigray regional state is now home to big protected woodland parks, including; Waldiba, Hirmi, Desia and Gra-kahsu, covering over 265,000 ha of the land mass in total.[16] As the recent literature indicates, nearly half of the three million hectares of land under enclosure management in Ethiopia are found in Tigray.[17]

2 Problem Statement

Soil and water conservation activities under the food for work scheme that were implemented in Ethiopia during the 1970s and 1980s were criticized for their top-down trajectories.[18] During the following decades, the Ethiopian government adopted an integrated participatory watershed management for poverty alleviation and sustainable livelihoods, particularly in rural area of the country. In order to ensure this, the country introduced what is commonly known as Accelerated and Sustainable Development to End Poverty (PASDEP). The growth and transformation plan for the period of, 2006–2011, was aimed to ensure food security through improved natural resource management practices mainly through the erection of physical SWC measures such as; soil and stone bunds, and the planting of numerous seedlings every year throughout the country.[19]

Natural resource conservation and rehabilitation practiced in Tigray, Ethiopia, have of late captured the attention of the international community and UN agencies owing to their effective and participatory nature. In 2012, the UNDP recognised the MERET project at Abreha We-Atsbeha, which was among the 25 initiatives to qualify for the Equator prize. More recently in 2017, the Tigray region of Ethiopia was awarded the World Future Council Gold Award for effectively tackling environmental and desertification, thereby reducing climate change-related risks.[20] Despite these and other specific success stories, however, a recent study indicates that Ethiopia has already experienced and is still suffering from a lot of natural

[15]Meaza et al. (2016).

[16]Hagos (2017).

[17]Balehegn et al. (2019), p. 6.

[18]Bewket (2007).

[19]MoFED (2006).

[20]All Africa 2019.

resource degradation due to the absence of a national land use policy and a nationally integrated land use plan.

In recent years, there have been global efforts to promote and stimulate country-specific initiatives for sustainable eco-systems development and climate resilient green economy, particularly in developing countries. While it is worthwhile to recognize significant achievements in specific countries through the assignment of incentives so that best practices can be replicated, the policy framework, modes operandi and institutional setups that bear successful results, as well as the challenges encountered must also be duly documented. The current study was aimed at identifying the specific natural resource conservation and management activities that led to such awards and recognition by UN agencies so that best practices can be scaled up in other parts of Ethiopia and beyond.

3 Objectives

The overall objective of the study is to analyse the activities that led to the Gold Award and the policy implications for Ethiopia and beyond.

Specific objectives are to:

1. Outline the relevant national (Ethiopian) and regional (Tigray) policies, strategies and legislation with regards to natural resources conservation and rehabilitation;
2. Examine the role played by various stakeholders in the process of rejuvenating natural resources to fight desertification and ensure food security in the area;
3. Assess the level of participation of people at the grassroots level during project planning and implementation as well as monitoring and evaluation;
4. Examine the achievements made and challenges encountered;
5. Identify lessons learnt and their policy implications for possible scaling up of best practices in future;
6. Make possible recommendations towards sustainable resource management and conservation in Ethiopia and beyond.

4 Methods

This chapter is the result of desktop research with all the relevant data collected from secondary sources such as reports and web materials. A limited number of interviews were conducted with key informants belonging to various socio-economic brackets, particularly focusing on those who played crucial roles in designing and implementing the initiatives in the Tigray region. Data was qualitatively analysed and presented in descriptive narrative form.

5 Description of Study Site

The Tigray regional state, also called Region 1, is one of the 9 ethnic-based regional states under Federal Democratic of Ethiopia. Bordered by Eritrea to the north, Sudan to the west, the Afar Region to the east, and the Amhara Region to the south and southwest, it is Ethiopia's 5th most densely populated region and is located at the northern tip of the country. Its capital city is called Mekelle and it is home to Tigrayasn, Irob and Kunama people. According to the Central Statistics Agency, it has a total population of 4,316,988, the bulk majority of whom live in the rural areas.

With an estimated area of 41,409.95 km^2, this region has an estimated density of 100 people per square kilometer. For the entire Tigray region, 992,635 households were counted, which results in an average of 4.4 persons per household, with urban households having on average 3.4 and rural households 4.6 people. The population was projected to be 5,247,005 in 2017.

Agro-ecologically, it is characterised by arid and semi-arid climate with high to moderate temperatures and an annual precipitation of 450 to 980 mm. Tigray is generally divided into three different Agro-ecological zones: Kola (lowland), which is below 1500 m above sea level; Woina Degua (mid-altitude) with altitudes ranging between 1500 and 2200 m above sea level; and Degua (Highland), which is above 2300 m above sea level. The regional economy is dependent on agriculture, which is largely rain-fed.[21]

6 The Policy Framework

6.1 The 1995 Constitution

The 1995 Constitution of the Federal Democratic of Ethiopia (FDRE) laid down the policy and legal framework for land administration in particular, natural resources and environmental governance in general. As per Article 40 (3) of the Constitution, it states that "The right to ownership of rural and urban land, as well as of all natural resources, is exclusively vested in the State and in the peoples of Ethiopia." As stipulated in the same provision:

> Land is a common property of the Nations, Nationalities and Peoples of Ethiopia and shall not be subject to sale or to other means of exchange." Regarding land, the Constitution confers only usufructuary rights to Ethiopian farmers (Art 40(4)), Ethiopian pastoralists (Art 40(5)), and private investors (Art 40(6)) in accordance with conditions to be set by law. Citizens are given "the full right to the immovable property" they build and to the permanent improvements they bring about on the land by their labour or capital, including the right to alienate, to bequeath, and when the right of use expires, to remove the property, transfer title, or claim compensation for it (Art 40(7)). The Constitution provides protection against eviction to Ethiopian farmers (Art 40(4)) and the right not to be displaced from their own

[21]Reda (2007).

lands to Ethiopian pastoralists (Art 40(5)), respectively. However, the Constitution does not provide an absolute protection from both eviction and displacement as the State is mandated to "expropriate private property for public purposes subject to payment in advance of compensation commensurate to the value of the property" (Art 40(8)).

The Constitution also mandated that the federal government, amongst others to, "enact laws for the utilization and conservation of land and other natural resource" (Art 51(5)). Members of the federation, i.e. the regional states, may "administer land and other natural resources in accordance with Federal laws" (Art. 52 (2) (d)), but the formulation of broad policy and laws on utilization, conservation of land and other natural resources rests with the federal government. In the chapter prescribing the national policy principles and objectives (Chapter 10), the Constitution reiterated the duty of the government to hold land and other natural resources on behalf of the people and to deploy them for the common benefit and development of the people (Art 89(5)).

The Constitution devoted a couple of provisions to the right to sustainable development, as well as the right to a clean and healthy environment. These rights are recognised among the list of fundamental rights and freedoms enshrined in Chapter 3 of the Constitution. With regard to the right to (sustainable) development, Article 43 of the Constitution provides substantive and procedural guarantees as follows:

1. The Peoples of Ethiopia as a whole, and each Nation, Nationality and People in Ethiopia in particular, have the right to improved living standards and to sustainable development.
2. Nationals have the right to participate in national development and, in particular, to be consulted with respect to policies and projects affecting their community.
3. All international agreements and relations concluded, established or conducted by the State shall protect and ensure Ethiopia's right to sustainable development.
4. The basic aim of development activities shall be to enhance the capacity of citizens for development and to meet their basic needs.

With regard to environmental rights, the Constitution stipulates that "All persons have the right to a clean and healthy environment" (Art 44(1)). In the chapter dealing with national policy principles and objectives (Chapter 10), the Constitution provides further policy prescription for "Environmental Objectives" (Art 93):

1. Government shall endeavour to ensure that all Ethiopians live in a clean and healthy environment.
2. The design and implementation of programmes and projects of development shall not damage or destroy the environment.
3. People have the right to full consultation and to the expression of views in the planning and implementation of environmental policies and projects that affect them directly.
4. Government and citizens shall have a duty to protect the environment.

With these stipulations, the Constitution has laid down the general policy framework for further policy and legislative actions on land, natural resources and environmental governance.

6.2 *Strategies and Policies*

Building upon the Constitutional stipulations, Ethiopia adopted various sectoral and cross-sectoral strategies, policies focusing on development, natural resource conservation and environmental protection. Almost a year after the adoption of the FDRE Constitution in 1995, Ethiopia adopted a national conservation strategy, which was under formulation a few years before the adoption of the Constitution. The conservation strategy of Ethiopia claims to take a holistic view of natural, human-made and cultural resources and their use and abuse.[22] It posits that the "strategy for conservation should concern itself with both economic development and care for the natural resources and for the environment.[23] It outlines as its purpose the assessment of the status and trends in the use and management of the resource base of Ethiopia, the formulation of a policy and strategy framework addressing the key issues identified, and the development of a Federal Action Plan and Investment Programme, including legislative measures and management and operational arrangements for implementation.[24]

Building upon the notion of sustainable development enshrined in the Constitution and the strategic direction of the CSE, Ethiopia adopted a comprehensive Environmental Policy (EP) in 1997, which covers a variety of sectoral and cross-sectoral environmental issues. In fact, the EP was extracted from the CSE.[25]

The overall policy goal of the EP is reproduced as follows:

> The overall policy goal is to improve and enhance the health and quality of life of all Ethiopians and to promote sustainable social and economic development through the sound management and use of natural, human-made and cultural resources and the environment as a whole so as to meet the needs of the present generation without compromising the ability of future generations to meet their own needs.[26]

This policy goal captures the essence of sustainable development in its broad sense, as provided in the widely accepted Report of the World Commission on Environment and Development (commonly referred as 'Brundtland Report'),

[22] CSE, Para. 52.

[23] CSE, Volume 1, para. 239.

[24] Ibid, para. 52.

[25] Sustainable Development and Poverty Reduction Program of Ethiopia (SDPRP), Ministry of Finance and Economic Development (MOFED), 2002, Addis Ababa, 121.

[26] Environmental Policy of Ethiopia, section 2.1.

namely: development that meets the needs of the present without compromising the ability of future generations to meet their own needs.[27]

Consistent with the above quoted overall policy goal, the EP outlines a set of specific policy objectives covering a wide range of issues related to conservation, sustainable use of both renewable and non-renewable resources, integrated cost-benefit analysis of development interventions, pollution prevention, empowerment and participation of people and their organisations, and public awareness raising.[28] Moreover, the EP provides a number of guiding principles to help shape the formulation of all subsequent policies, strategies and programmes at sectoral and cross-sectoral level and to assess their consistency and implementation.[29] The right to live in a healthy environment, the principles of sustainable development and public participation, integration, prevention, precaution, equity, conservation, EIA, and the polluter-pays principle are included among the key guiding principles. Furthermore, the EP outlines a wide range of cross-sectoral and sectoral environmental policies.

The sectoral environmental policies (Sect. 3) cover a wide range of issues, including; Soil Husbandry and Sustainable Agriculture; Forest, Woodland and Tree Resources; Genetic, Species and Ecosystem Biodiversity, Water Resources; Energy Resources, Mineral Resources, Human Settlements, Urban Environment and Environmental Health, Control of Hazardous Materials and Pollution from Industrial Waste; Atmospheric Pollution and Climate Change and Cultural and Natural Heritage (Sects. 3.1–3.10). The cross-sectoral environmental policies (Sect. 4), on the other hand, deal with different issues around Population and the Environment; Community Participation and the Environment; Tenure and Access Rights to Land and Natural Resources, Land Use Plan; Social and Gender Issues, Environmental Economics;, Environmental Information System, Environmental Research, Environmental Impact Assessment and Environmental Education and Awareness (Sects. 4.1–4.10).

Both the sectoral and cross-sectoral environmental policies give due attention to issues of sustainable land and natural resources governance, in general, and sustainable soil governance, in particular. For example, the sectoral environmental policy on 'Soil Husbandries and Sustainable Agriculture' (Sect. 3.1) aims to: promote the use of appropriate organic matter and nutrient management for improving soil structure, nutrient status and microbiology in improving soil conservation and land husbandry (3.1(c)); safeguard the integrity of the soil and to protect its physical and biological properties, through management practices for the production of crops and livestock which pay particular attention to the proper balance in amounts of chemical and organic fertilizers, including green manures, farm yard manures and compost (3.1(d)); promote effective ground cover as one of the most important factors in soil

[27]Report of the World Commission on Environment and Development (hereafter 'Brundtland Report'), Chapter 2, para. 1.

[28]Environmental Policy, section 2.2(a-i).

[29]Environmental Policy, at 4.

erosion control, taking advantage of the wide range of sustainable agronomic, pastoral and silvicultural approaches used in the various areas of Ethiopia as potentially flexible alternatives to mechanical soil conservation systems (3.1(e)); ensure that planning for agricultural development incorporates in its economic cost-benefit analysis the potential costs of soil degradation through erosion and salinisation as well as soil and water pollution (3.1(j)); ensure that inputs shall be as diverse and complementing [sic] as the physical, chemical and biological components of the soil require, and shall not focus solely on a quick and transitory increase in plant nutrients to the long-term detriment of soil structure and microbiology (3.1(k)); institute the stall feeding of domesticated animals through a combination of providing agricultural residues, on-farm produced forage and fodder as well as the cutting and carrying of grasses from meadows and hillsides in order to encourage revegetation of grazing lands and the reduction of soil erosion (3.1(l)); use biological and cultural methods as well as resistant or tolerant varieties or breeds, pheromones or sterile male techniques in an integrated manner as a pest and disease management method in preference to chemical controls (3.1(o)); safeguard human and environmental health by producing adequate regulation of agricultural (crop and livestock) chemicals (3.1(p)).

The sectoral environmental policy on ‘Forest, Woodland and Tree Resources’ (Sect. 3.2) was established with multiple similar aims, such as to: ensure that forestry development strategies integrate the development, management and conservation of forest resources with those of land and water resources, energy resources, ecosystems and genetic resources, as well as with crop and livestock production (3.2(c)); assist the natural process of afforestation of uncultivable areas by controlling felling and grazing and by planting judiciously selected local species, as well as by other affordable interventions (3.2(e)); promote changes in agricultural and natural resource management systems which will limit the need for free grazing of animals in protected forest areas (3.2(h)). Similarly, the sectoral environmental policy on ‘Water Resources’ (Sect. 3.4) recognises the interaction between the ecosystem and water quality and quantity (3.4(b)) on one hand, and on the other, the interface between water bodies and land (3.4(d)) and the need for their integration.

The policy attention given to land utilisation and conservation and other related issues seems unsurprising in the Ethiopian context, given the country’s strategic focus on Agricultural Development-Led Industrialisation (ADLI), which prioritises agriculture as a key driver of economic development. The idea in Ethiopia’s ADLI is that the growth in agriculture will induce overall economic growth through structural transformation to industrialisation. Ethiopia has been using ADLI since 1994 as an evolving strategy and implementing it under different succeeding programmes and plans, such as the; Sustainable Development and Poverty Reduction Program (SDPRP), 2002/03-2004/05; A Plan for Accelerated and Sustained Development to End Poverty (PASDEP), 2005/06-2009/10; Growth and Transformation Plan I (GTP I), 2010/11–2014/15; and Growth and Transformation Plan II (GTP II), 2015/16-2019/20. All these policy instruments treat poverty reduction, sustainable development and natural resources conservation as inseparable parts of the national development endeavour.

With specific regard to natural resources conservation and development, GTP II provided an assessment of the previous 5 years performance of the First Growth and Transformation Plan (GTP I) and expressed its commitment to scale up the success stories, stating as follows:

> . . .the accomplishments with regards to natural resources conservation and development activities are identified as one of the success stories in agriculture and environment sectors during the last five years. Performance has surpassed targets by a significant margin in the area of natural resource conservation and management. Community[-]based watershed developments that improved water, soil and biological resources surpassed the target set in the Plan. Through area closure, degraded lands have been rehabilitated; community based watersheds infrastructures that prevent soil loss and enhance water percolation have been constructed over and above the targets set. Accordingly, the area of land closed for rehabilitation increased from 3.2 million hectares in 2009/10 to 10.9 million hectares by 2014/15. Community[-]based watershed infrastructure development expanded to 20.2 million hectares during the plan period. Besides, area of denuded lands rehabilitated being covered by multipurpose trees increased from 6.1 million hectares in 2009/10 to 16.2 million hectares by 2014/15. Along with the natural resources conservation, an estimated 2.34 million hectares of land were developed through small scale irrigation schemes during the plan period. These good lessons could be scaled up and replicated to other sectors in the upcoming Plan (GTP II).[30]

Like the other policy instruments mentioned above, the Ethiopian Climate-Resilient Green Economy Strategy (CRGE), which was formulated in 2011 with the vision to achieve middle-income status by 2025 in a climate-resilient green economy, appreciates both the importance and the challenge to achieve sustainable economic development (CRGE, 1). The traditional economic development path, as indicated in the CRGE, could deliver the required growth, but at the cost of significant agricultural land expansion (inducing and accelerating deforestation), soil erosion, and higher emissions as well as at the risk of reaching the limits to further development (CRGE, 23). "In order to offer a viable alternative to the conventional development path without foregoing growth in the short term and significant advantages thereafter," the Strategy goes on "a set of initiatives has been identified that can provide the required increase in agricultural productivity and resource efficiency". (CRGE, 23). These initiatives include the efficient utilisation, conservation and rehabilitation of natural resources such as land and forestry, among others (See CRGE, 23–24).

However, despite the existence of these and other policy and strategic initiatives regarding natural resources conservation, in general, and sustainable land utilisation, in particular, the absence of a nationally integrated land use policy is considered a major challenge to sustainability in Ethiopia.

---

[30]FDRE Growth and Transformation Plan II (GTP II), 2015/16-2019/20, Volume I: Main Text (National Planning Commission, May 2016 Addis Ababa) 27.

7 Combining NRM with Food Security: Achievements Made

Environmental rehabilitation projects in Tigray aim not only at rehabilitating forest and vegetation cover, but also boosting agricultural productivity with an ultimate target of improving the quality of life of people in the region.

One of the most pronounced achievements ever recorded in Tigray relates to the issue of youth empowerment through agricultural and reforestation-based employment opportunities; with people of varying ages and genders voluntarily participating in community development projects. The build-up of social capital led to strong support networks, which, in turn, helped to establish a strong institutional framework to build resilient communities in the fight against environmental degradation, poverty and economic deprivation. It is astonishing to witness mothers carrying their little ones on their backs to join their fellow male compatriots in the field. The collective determination and commitment for change has resulted in the rehabilitation of over 1.2 million hectares of land with a dramatic increase in forest coverage from just 2% a few decades ago to about 13% today. Priority is given to the development of the upper basin with the aim of improving the moisture retention capabilities of the land.[31] This has been a significant leap forward in a village where over 42 metric tons of soil was being washed away by erosion annually in a given hectare of land.[32]

Specific measures taken included access to credit, diffusion of new ideas and technologies in agricultural development combined with environmental rehabilitation activities. The bulk majority of farm households were advised to dig shallow wells and ponds to harvest rainwater during the wet seasons, most of which were prepared in 2005 alone. Farmers were encouraged to cultivate high value crops and produce vegetables and fruits in their backyard as a way of generating extra income. This was in no way limited to rural areas. In urban and semi-urban areas, rooftop and roadside water harvesting techniques were introduced to raise the water table.[33] A piecemeal approach was put in place in which about 37 model farmers implemented the programme first and they later reported their incomes significantly increased as a result.[34]

[31]Hagos (2017).

[32]Ibid, 2017.

[33]Reda (2007).

[34]Meaza et al. (2016).

8 Restoring Degraded Land: Experiences from Rural Tigray

Historically, conservation measures were put into place in a limited number of places in the Tigray region. For example, during the 1970s USAID-supported terracing programmes were initiated around Mekelle and Adigrat, also stimulating the Institute of Agricultural Research (IAR) based in Mekelle, which played a role in the terracing and reforestation of some 4700 ha of steep lands in over 13 areas of the region. Similarly, between 1982 and 1988, the Ministry of Agriculture together with the World Food Programme (WFP) constructed over 94,000 ha of hillside terracing, 18,000 km of bunds and 500 km of check dams under the Food for Work Programme. In the Tigrean People's Liberation Front (TPLF) occupied areas, attempts were made to establish local NGOs like the Relief Society of Tigray (ReST) and mobilise people for environmental rehabilitation activities.[35]

The northern Ethiopian highlands generally and rural Tigray in particular have suffered from recurrent droughts and crop failure for the last several decades. Following repeated incidents of food shortages that necessitated a series of relief aid, particularly in the 1990s, the government provided rural communities with an option to relocate to other fertile lands to make them economically self-sufficient. While some decided to do so, the proposal was rejected by a large proportion of the people because of the rural communities' belongingness and strong attachment to their land. Instead, many decided to remain rooted in their respective villages and reclaim their land through a serious of resource management and rehabilitation schemes, which included; soil and water conservation through water harvesting and hillside land management, bans on encroachment of livestock and humans into forest areas for grazing and harvesting of wood, respectively. Soil and stone bunds were erected along vast tracts of land to circumvent run-off and prevent subsequent soil erosion. Each household prepared a 4m by 4 m pond to harvest rainwater. Check dams[36] were constructed in the rivers and hand dug wells were established for dry-season irrigation projects, which benefited many households.[37] Through successive years of awareness campaigns, it was possible to establish contour bunds to enable terrace farming on slopes in a way that prevents erosion; while the human pressure on hillside resources was also minimised by enforcing rules against free grazing and the cutting of trees for fuel. Tigray presents a unique experience of environmental rehabilitation because human-friendly conservation practices were put in place. For example, the need for firewood for immediate fuel consumption was sufficiently addressed by allowing households to grow eucalyptus and grevillea (Silky Oak) and introducing energy saving technologies. New breeds of dairy cows were introduced to emphasise the shift in focus to quality of milk

[35]ibid, 2019.

[36]Check dams are small structures often erected by communities, constructed using rocks aim at regulating the flow of water and sediment as part of a strategy to control soil erosion.

[37]ibid, 2002.

production from the different animals rather than just mere livestock numbers. Similarly, organic honey production was encouraged through the introduction of modern beehives and by establishing viable markets both within and outside the country.

Government-led afforestation programmes were also launched from several nursery sites in which millions of seedlings were planted around homesteads (and mountainous areas where trees are needed to cover barren, non-agricultural land) over the past couple of decades. Community and state forest areas were well protected through effective formal and informal resource management rules.[38] On one hand, there are salaried personnel regularly stationed in particular forest areas to prevent intruders from cutting trees; on the other hand, in some areas, vigilante groups are mandated to conduct surveillances, and bring to justice those that violate the rules.

9 Factors for Success

Success in natural resource management and conservation in Tigray can be attributed to a multitude of factors, including the commitment of the community to change and transformation, strength and resilience in the formal and informal leadership structures,[39] sound government policies and synergy with local NGOs and microfinance institutions, and strong social capital and networking at the local community level. The overarching Agricultural Development Led Industrialization (ADLI) at the national level, which was translated into mass mobilisation campaigns and youth responsive land policy in the Tigray region (especially since 2008) created a conducive environment for uptake. Government-led social protection measures and other rural development interventions such as the agricultural extension system (distribution of agricultural inputs, i.e., seed varieties, inorganic fertilisers and other chemicals) are instrumental in creating viable livelihoods in the villages. Under the Safety Net Programmes introduced by the Ethiopian government, those that are able to contribute labour are carefully selected from the poorer section of the community to work for 30–40 days per year in managing community assets and soil and water conservation activities, for which a specified remuneration is often paid in kind or cash. In some villages such as Abraha-We-Atsbaha, people volunteered to work for a higher number of days. Other stakeholders who contributed to the success of the programme included; the regional Bureau of Agriculture and Rural Development,

[38] Reda (2007).

[39] In a particular well-known village in the Easter zone of Tigray know as Abreha-we atsbeha, the head of the village Mr. Gebremichael Gidey (better known as Aba-hawi, literally meaning 'the father of fire', so nicknamed to indicate his stamina to deal with environmental challenges, his skills in rhetorical speech, wisdom and the capacity to influence others) played a key role in mobilising communities and building social capital that was used as an input to change the environment.

Mekelle University, Relief Society of Tigray (ReST), WFP, World Vision Ethiopia, as well as other local, national and international collaborators.

10 Challenges Encountered

The extent of poverty in the region did not allow people to mobilise important financial resources for the common good; labour contribution was largely constrained by the fact that a good part of rural communities have had years of drought and malnourishment. At a country level, soil and water conservation projects in four regions including Tigray has consumed tremendous amount of money amounting to USD 1.2 billion dollars per year during the last 10 years.

In addition, although reforestation programmes were conducted on a massive scale in Tigray, the survival rate of planted seedlings was generally low in the arid areas of the region where evaporation is very high. Additionally, conservation-based programmes for the youth were primarily meant to empower them economically, but growing trees on hillsides had often been tedious and time consuming and revenues obtained from such activities were low. As a result, the youth sought other options for quick economic gain and eventually decided to leave their villages in search of other jobs. Several of them were involved in illicit migration outside the country.[40]

In earlier periods, some coercive measures were taken against farmers who refused to adopt modern inputs, leading to complicated relations between administrators and community members.[41] This was a direct outcome of poor monitoring and evaluation procedures put in place especially during the early phases of the interventions. Occasionally, the inclusion and exclusion criteria for the safety-net programme remained shaky, bureaucratic and often marred with corrupt practices.[42] A number of factors including; agricultural inputs pricing, attitudinal problems and overall risk averse behavior of farmers have contributed to the rejection by farmers of major components of the extension programme. The dearth of well-functioning markets characterized by price fluctuations has particularly resulted in the farmers frustrations.[43]

The absence of a nationally integrated land use policy can also be considered a major gap to promote sustainability in Ethiopia. In some areas, there have been knowledge gaps on various land policies, on one hand, and the age-old customary resource management experiences of the community on the other hand. There were also no clear guidelines on benefit sharing on government protected forest areas, often leading to complete disenfranchisement of local communities. This often leads to a conflict of interest between the state and local people, as the latter want to use

[40]Reda et al. (2018).

[41]Veen and Gebrehiwot (2011).

[42]Dom and Pankhurst (2019).

[43]Wolde-Georgisa et al. (2010).

grass from the protected areas even if it is on a cut-and-carry basis, while the state argues this makes the forest areas vulnerable to soil erosion and moisture losses.[44] Often, there were also some attitudinal problems that relate to people's perception of state land as no man's land, which encouraged free-riding behaviours. In the western part of the Tigray region, settlers often held the aforementioned views, leading to the over-cutting of trees and game hunting. In general, at a higher level, there are challenges related to building resilient stakeholder coalitions to increase multi-sectorial investments in the areas of; land degradation, climate change and biodiversity; thereby rendering opportunities for sustainable land management to deal with issues of climate change adaptation, mitigation and ecosystem services.[45]

At a community level, committees are often established to manage natural resource conservation projects and make sure that by-laws are carefully observed by members. Regular meetings are organized to make and discuss emerging issues as well as, the enforcement of decisions; however, communities often find it difficult to deal with offenders from other areas that trespass such bylaws, for instance; by letting their livestock graze in protected areas.[46]

11 Lessons Learnt from Grassroots NRM Practices

From an anthropological standpoint, the conscious efforts of local rural communities to change the environment to better serve human interests is a major breakthrough in itself- given the fact that traditionally, people had often considered nature as God-given and not liable to interference. The traditionally age-old, held fatalistic beliefs that development is not possible has been significantly challenged. The villagers determination is instrumental in bringing about change; it is the only option for survival in an environment heavily damaged by years of natural and man-made calamities. The creation of a critical mass of productive force and meaningful social capital has been an important milestone in the rural development interventions in the Regional State of Tigray.

Synergy between different stakeholders (i.e., community, policy makers, local and international NGOs and pertinent government bureaucracy) has collectively transformed degraded land into a rehabilitated landscape. Vulnerability to drought, poverty, unemployment and migration gave way to a glimmer of hope and a better future. Depleted natural resources were restored and food security ensured thanks to the wealth of experience in the village that can be expanded to other regions through peer learning and experience sharing platforms.

A largely non-directive, bottom-up approach has been pursued in many of the environmental rehabilitation and food security projects; whereby all activities were

[44]Reda (2007).

[45]Dejene (2009).

[46]Edwards et al. (2010).

managed by the local communities. People participated in the design, implementation as well as monitoring and evaluation of such projects, while other stakeholders facilitated the smooth running of the interventions and provided key resources to ensure sustainability. Often, planning of a new programs or project entails holding a series of consultative meetings with community members in which beneficiaries assess its feasibility and undergo cost-benefit analysis. There are local councils in the grassroots with politico-legal functions operating on the basis of codified by-laws that are enforced by the social courts. Proper coordination between experts, advisors, and development agents facilitate effective planning and implementation of projects at various levels.[47] In addition, the communication and information exchange among various stakeholders was also exemplary.

Coherence in the social protection measures is also reflected in village level political and civil service administrative structures, ensuring that no one is left out of the development agenda. Household food security status and environment-related information is effectively channeled to higher levels of government for evaluation and future interventions. This involves a meticulous relay of information across the formal structures and networks that extend to the lowest administrative unit in the governance system. There are separate entities and associations for the youth, men and women groups that oversee various village level social, economic and political developments as part of the formal government bureaucracy. These on-the-ground interventions have successfully been designed to implement and carry out the mandates in the overarching national strategies and policy measured outlined above. This then resulting in improved land and soil quality as well as enhanced social participation across all levels of community, regions and nationally to address these issues.

12 Conclusion and Recommendations

Years of economic hardship forced the communities in the Tigray region of Ethiopia to look for a way to build a sustainable future by applying an innovative approach involving people at all levels committing to work together and participate in changing their environment. Their efforts to transform degraded areas into productive farmland earned the Tigray regional state in Ethiopia recognition by the UN at the Rio+20. This is basically the result of people's determination to change the environment, functional integration among various stakeholders and a sound policy environment. A non-directive approach to grassroots intervention yielded shared responsibility and accountability that were instrumental in ensuring participatory rural development. A key message remains the integration of sustainable agricultural development with environmental sustainability through effective soil and water conservation programs and revitalisation of the natural resource base in participatory

[47]Ibid, 2010.

ways. Inclusive governance structures in which women, youth and other disenfranchised groups participated in major decision-making processes has been a major driving force behind the achievements. However, the framework which facilitates this to happen depends on the policy makers coming up with appropriate land use policies to support and scale up the good practices at the local level.

Finally, based on the findings, it is recommended that new innovative approaches to reforestation as well as soil and water conservation practices, ought to be attempted to foster agro-forestry development in the Tigray region. In addition, an integrated approach to rural development that combines environmental rehabilitation programs with food security should further be encouraged in order to ensure sustainable development. This is possible in the wake of dialogue between various stakeholders including; community members, research institutions, NGOs and policy makers. Replicable ecological agriculture can, however, be resilient if proper policies and incentives are put in place in ways that are appropriate to changing circumstances at local, national and global levels. Last but not least, although farmland management, hillside management and gully rehabilitation practices, including check dams and cut-off drains have been ventured in the country for the last couple of decades, more studies need to be conducted to determine the impacts of such projects on food security and ecosystems development in the country.

References

Balehegn M, Haile M, Fu C, Liang W (2019) Ecosystem-based adaptation in Tigray, Northern Ethiopia: a systematic review. In: Leal Filho W (ed) Handbook of climate change resilience. Springer Nature Switzerland AG

Bewket W (2007) Soil and water conservation intervention with conventional technologies in Northwestern Highlands of Ethiopia: acceptance and adoption by farmers. Land Use Policy 24:404–416

Dejene A (2009) Combating Desertification and Land Degradation: Sustainable Land Management (SLM). Paper presented at the Intergovernmental Preparatory Meeting for CSD-17 (23–27 February 2009, New York) 27 February 2009, New York, Expert Panel on Desertification

Bewket W (2007) Soil and water conservation intervention with conventional technologies in Northwestern Highlands of Ethiopia: acceptance and adoption by farmers. Land Use Policy 24:404–416

Birhane E, Mengistu T, Seyoum Y, Hagazi N, Putzel L, Mekonen M, Rannest M, Kassa H (2017) Exclosures as forest and landscape restoration tools: lessons from Tigray Region, Ethiopia. Int For Rev 19(S4):37–50

Butzer KW (1981) Rise and Fall of Axum, Ethiopia: a geoarchaeological interpretation. Am Antiq 46:471–495

Corbeels M, Shiferaw A, Haile M (2000) Farmers' knowledge of soil fertility and local management strategies in Tigray, Ethiopia. Managing Africa's Soils Series No. 10. International Institute for Environment and Development, Nottingham

Dom C, Pankhurst A (2019) Economic experiences of rural young people; WIDE Series III Discussion Brief No, 4 of 7. In Pankhurst A, Dome C (eds) Rural Ethiopia in transition, Selected discussion briefs. PDRC, pp 99–116

Edwards S, Gebre Egziabher T, Araya H (2010) Success and challenges in ecological agriculture: expereinces from Tigray, Ethiopia. Food and Agriculture Organization available online at: https://www.ifoam.bio/sites/default/files/page/files/much_greater_resilience_.pdf

Hagos T (2017) Ethiopia: From Barren to Fertile Scaling Up Tigray's Experience, the Ethiopian Herald available online at: file:///C:/Users/DELL3020/Desktop/Gold%20Award/Ethiopia%20From%20Barren%20to%20Fertile%20Scaling%20Up%20Tigray's%20Experience%20-%20allAfrica_com.htm

Meaza H, Nyssen J, Tsegaye D (2016) Allocation of Degraded Hillsides to Landless Farmers and Improved Livelihoods in Tigray, Norsk Geografisk Tidsskrift January 2016, https://www.researchgate.net/publication/298901940. Accessed 20 Feb 2019

Mekuria W, Veldkamp E, Tilahun M, Olscewski R (2011) Economic valuation of land restoration: the case of exclosures established on communal grazing lands in Tigray. Land Degrad Dev 22:334–344

MoFED (2006) Ethiopia: Building on progress: A Plan for Accelerated and Sustained Development to End Poverty (PASDEP) (2005/06-2009/10), Vol. II: Policy matrix. Ministry of Finance and Economic Development (MoFED), Addis Ababa, Ethiopia, September 2006

Munro RN, Deckers J, Haile M, Grove AT, Poesen J, Nyssen J (2008) Soil landscapes, land cover change and erosion features of the central plateau region of Tigrai, Ethiopia: photo-monitoring with an interval of 30 years. Catena 75:55–64

Nyssen J, Haile M, Naudts J, Munro N, Poesen J, Moeyersons J, Frankl A, Deckers J, Pankhurst R (2009) Desertification? Northern Ethiopia re-photographed after 140 years. Sci Total Environ 407:2749–2755

Nyssen J, Poesen J, Veyret-Picot M, Moeyersons J, Haile M, Deckers J, Dewit J, Naudts J, Teka K, Govers G (2006) Assessment of Gully erosion rates through interviews and measurements: a case study from northern Ethiopia. Earth Surf Process Landf 31:167–185

Reda A (2007) Combating Desertification in Tigray, Ethiopia - Field study on the implementation of the UNCCD in the rural region of Tigray, MA thesis. Linköpings University, Sweden

Reda K, Zenawi G, Tsehaye W, Awet H, Kelil D (2018) Causes and consequence of irregular migration in selected Woredas of Tigrai Regional State, Ethiopia, CISP Report, Italian Cooperation, Addis Ababa, Ethiopia

Scoones I (2009) Livelihoods perspectives and rural development. J Peasant Stud 36:171–196

Veen A, Gebrehiwot T (2011) Effects of policy interventions on food security in Tigray, Northern Ethiopia. Ecol Soc 16(1):18

Wolde-Georgisa T, Meles K, Zenebe A, Habtu A (2010) Climate change adaptation and impact mitigation: a case study of Abraha Wa Atsebaha village in Tigray, Northern Ethiopia. J Dry Lands 3(1):150–157

Indigenous Knowledge and Soil Protection: Anthropological Remarks on Experiences in Namibia

Manfred O. Hinz

1 The Challenges Faced by Indigenous Knowledge

The interest in indigenous[1] knowledge entered the scene of action-oriented science in the last years of the past century. While indigenous knowledge has been of interest to anthropologists as they acknowledged the existence of a plurality of societies and their respective world views in their own rights,[2] scholars and actors concerned with development started noticing indigenous knowledge as an issue of relevance when the focus on sustainable development began to influence the conceptualisation of development and development policy. A world-wide noted manifestation of this was the adoption of the Convention on Biological Diversity at the United Nations Conference on Environment and Development in Rio de Janeiro in 1992. The Convention states the obligation to "respect, preserve and maintain" indigenous knowledge "relevant for the conservation and sustainable use of biological diversity" and to "promote" its application.[3] The Convention had far-reaching consequences for the understanding of sustainable development and led to research which enriched

[1] *Traditional* or *local* are other adjectives to qualify this knowledge.

[2] Anthropologists to be named here are e.g. F. Boas, B. Malinowski, B. Whorf, W. Goodenough. Goodenough (1957, p. 167) defined *culture* as "the form of things that people have in mind, their models for perceiving, relating, and otherwise interpreting them".

[3] Article 8(j) of the Convention; see also Article 10(c).

M. O. Hinz (✉)
Public Law, Political Sociology and Sociology of Law, University of Bremen, Bremen, Germany

Law and African Studies, Jacobs University, Bremen, Germany
e-mail: mhinz@uni-bremen.de

H. Ginzky et al. (eds.), *International Yearbook of Soil Law and Policy 2019*, International Yearbook of Soil Law and Policy 2019,
https://doi.org/10.1007/978-3-030-52317-6_6

the discourse on environment and development up to today.[4] Some years after the Convention on Biological Diversity, the World Intellectual Property Organization (WIPO) conducted an international fact-finding mission on indigenous knowledge and the need to protect indigenous knowledge as an intellectual property.[5] The author of an article on indigenous knowledge published in 1998, confirmed the view of the growing interest in indigenous knowledge and the role of anthropology in creating understanding of this knowledge of a "new applied anthropology".[6] This author noted that it is "now increasingly acknowledged beyond anthropology that other people have their own effective 'science' and resource use practices and that to assist them there is a need to understand something about their knowledge and management systems".[7] 1998 was also the year in which the World Bank started its indigenous knowledge programme.[8]

Whether we can now, some 20 years later, observe the working of the then expected new applied anthropology is a matter for/of discussion, which can only be touched upon in this article. It has certainly made possible to establish a platform on which anthropologists and development agents were able to share their views on the practical developmental importance of indigenous knowledge. However, one can also observe that the kind of enthusiasm that prompted e.g. the World Bank, to focus on indigenous knowledge, lost the drive that characterised the 90s of the last century.[9] There are methodological reasons for this. It is easy to call for noting indigenous knowledge; the implementation of the call is not. Indigenous knowledge may offer itself in epistemological constructs that are far from the constructs of the outsider who approaches a given situation with instruments of his or her own culture and may be tempted to ignore the constructs of the visited society because they are seen to be unscientific and therefore irrelevant.[10] Indigenous knowledge can also concern all sorts of societal behaviours and be no less than the expression of local common sense: not attracting attention on such knowledge as what it entailed was known without the attention to it. Methodological problems of this nature have

[4]Cf. here e.g.: Jürgens and Schmiedel (2010); Oldeland et al. (2013): final reports of two international projects focusing on sustainable development and the socio-political conditions for its implementation in Africa in which the author of this article could participate.—The placing of the concept of sustainable development and the concepts related to this, stimulated additional anthropological research in which also non-anthropologists played an important role, f. here e.g. Grim (2001) and other publications of the Religions of the World and Ecology-Project.

[5]WIPO (2001).

[6]Sillitoe (1998); see also Sillitoe (2006).

[7]Sillitoe (1998), p. 223.

[8]Cf. Mohan (2001).

[9]The just quoted web-information of the World Bank on case studies on indigenous knowledge dates to 2009. See also Mohammedbhai (2013).

[10]Cf. Sillitoe (2006), p. 8.

certainly contributed to the state of affairs of the scientific dialogue on the role and function of the use of indigenous knowledge in development.[11]

The following contribution explores aspects of the relationship between indigenous knowledge and soil protection taking Namibia as an example. That a relatively small population of about 2.6 million inhabitants live on an area of 825,000 square kilometres does not mean that soil is not an important factor. At least, half of the population depends on soil-related activities. The very limited rainfall of 370 mm. per year on average—in the southern part of the country it is even less—forces to treat the soil with extreme care. The cases selected from Namibia reflect the mentioned methodological difficulties with respect to assessing the developmental relevance of indigenous knowledge. They are informed by the author's experiences while working with traditional communities in Namibia.[12] The soil and soil protection are not the direct focus of this work but rather natural resources in general, their protection and their sustainable use.

The three submitted cases are related to rural areas of Namibia, which are under the rule of traditional authorities. Although the post-independence state of Namibia is a modern state with a central government and decentralised structures; in the various regions of the country, traditional rules, i.e. the rule of chiefs and traditional councils, are as in many African countries, still a constitutionally sanctioned fact.[13] The land in areas under traditional rule is referred to as so called communal land. It is not owned by individuals. Individuals may have the right to use the land, but not to dispose of it. Land use rights are granted by the respective traditional authorities in accordance with customary law and rules set out by the government of Namibia[14]

The first of the three case studies will give an account of some rules in the customary law of one of Namibia's traditional communities, the community of Uukwambi. These rules have translated indigenous knowledge into environmental rules. Then, a report on the perceptions based on indigenous knowledge relevant to the wildlife management plan of the first Namibian conservancy, the Nyae Nyae Conservancy, will follow. The third case will be discussing the failure of the implementation of the act aimed to regulate customary land rights in the Kavango regions of the country. An interpretation of the three cases will complement the reported cases. Some reflections on indigenous knowledge will reach out to the mentioned methodological problems in dealing with indigenous knowledge.

---

[11]The published article by van Meijl (2019), which illustrates the societal dynamic influencing indigenous knowledge, is a contribution to fertilise the dialogue on indigenous knowledge.

[12]I worked in Namibia from 1990 to 2010. One focus of work were the traditional authorities and their customary law. This work resulted in particular in the publication of three volumes documenting the customary law of the Namibian traditional communities: Hinz (2010, 2013a, 2016a).

[13]Cf. Article 102 (5) of the Constitution of Namibia (providing for a Council of Traditional Leaders and Article 66 (1) (confirming the validity of customary law, the law by the various traditional communities).

[14]See Communal Land Reform Act, No. 5 of 2002. I will come back to concept of communal land and the Communal Land Act, in particular when dealing with the third case.

2 The Three Case Studies

2.1 The Environmental Rules in the Laws of Uukwambi[15]

Uukwambi is one of the Oshiwambo-speaking traditional communities in the North of Namibia.[16] Oshakati, the commercial centre of what used to be called Owamboland, is part of Uukwambi. Owamboland is a highly populated area. The territory of Uukwambi covers the central part of the Cuvelai catchment, an area, which is, during the rainy season, usually flooded by water[17] streaming down from Angola towards the Etosha pan, the Etosha national park. Very often the floods create serious problems to the people whose houses may be destroyed or may be cut off by the water. Almost every year, incidents of death caused by floods are reported. At the same time, water is also precious: while there is an abundance of water in the rainy season, water is scarce during the rest of the year and, therefore, has to be preserved throughout the remaining year until the next rainy season. Water also allows fish to grow, which are added to the diet and the water nourishes the soil.

It is interesting to see how the laws of Uukwambi have taken note of these challenges set by nature. The laws of Uukwambi were one of the first collected of what was termed *self-stated* customary laws. *Self-statement* was used to denote the fact that some of the traditional communities of Namibia had written down parts of their otherwise orally transmitted customary law.[18] What these communities had produced was not a codification of their laws and neither was it a product of a work undergone by experts to *restate* the laws of the communities, as it was done in the customary law restatement project of the School of Oriental and African Studies of the University of London under Antony Allott.[19] Three versions of the laws of the Uukwambi were documented: the first version was published in 1995, the second in 2003 and the third in 2010.[20] The first version of the laws of Uukwambi contains only some basic rules, which set out fines for wrongs like killing a person, rape, adultery or theft. Environmental concerns are only touched upon as part of a list titled *the maintaining of justice*: Section 14 (e) forbids the "use of nets or fish traps to play in water" and in Sub-section (f) the cutting of "green trees without permission".[21]

The second and third versions of the laws of Uukwambi differ substantially from the first version. These later versions respond explicitly to the challenges by means

---

[15]See for the following: Hinz (1995), p. 43ff.; Hinz (2003); Hinz (2010), p. 233ff.

[16]The Oshiwambo-speaking communities live in the central northern part of Namibia and across the border in southern Angola.

[17]In the Oshiwambo language called *efundja*.

[18]See here: Hinz (2016b, 2019).

[19]Cf. here Hinz (2016b), p. 5ff.; Hinz (2019).

[20]The first version can be found in: Hinz (1995), p. 43ff.; the second (however only parts) in: Hinz (2003), p. 62ff.; the third version in: Hinz (2010), p. 233ff.

[21]Cf. Hinz (1995), p. 44f.

of rules on water conservation, the environment with special emphasis on trees, shrubs and animals, the use of certain plants and the allocation of land in general. Section 15 of the third version[22] of the laws of Uukwambi is about water conservation. The section starts in preambular language: *Water is life*.[23] Therefore, the responsibility to protect water falls to the traditional authority and the members of the community. The misuse of water, be it the seasonal water in the *iishana*,[24] or (be it) the water provided by the public water service, both are punishable. Fishing is only allowed after the traditional authority has announced that it is permitted. This restriction does not only give fish the chance to grow but it also secures the quality of water for the consumption by people and animals. The use of fishing nets and, more so, the use of traditional fishing devices, called *iishongo*, cause the water to be muddy.[25] The permission to fish will be issued by the traditional authority when the water level has gone down and the water has become salty.[26] Before building a house, the local representative of the traditional authority has to be consulted for advice about the risk that the area for building is flooded in the rainy season.

Section 16.1 of the laws of Uukwambi calls on the traditional authorities and all members of the society to be responsible for the conservation of nature. The responsibility for conservation is understood to be for the mentioned individuals and "for the nation".[27] Given the fact that the plains of the territory of Uukwambi are lacking trees and shrubs, cutting of trees and shrubs is of particular concern to the laws of Uukwambi. All trees enjoy protection; only dead wood may be used for firewood. Permission to cut trees can be obtained if the wood is needed for building purposes. Otherwise the traditional authority urges to protect "every tree like his or her own wealth, and recommends that for every tree cut, two or three should be planted in order to combat desertification".[28] Protection is also extended to wild animals.[29] The hunting of birds is not allowed beyond the officially announced hunting season. The collection of eggs of birds is prohibited as "this will reduce birdlife".[30]

Apart from plants which are used for healing purposes in Uukwambi and, therefore, specially placed under protection in the laws of Uukwambi, a general rule concerns grass. Grass, which is very much in demand of by the cattle industry in Uukwambi, is at the same time a very principal form of vegetation that covers the

[22]I concentrate my report on the third version, which is mainly an edited version of version two.

[23]Find Section 15 in: Hinz (2010), p. 270f.

[24]Plural of *oshana*, the Oshiwambo word for shallow river beds, which carry water during the rainy season.

[25]*Iishongo*, plural of *oshongo* are fish traps made of reed. The devices are open at the bottom and pressed to ground to catch fish.

[26]See here also: Hinz (2003), p. 65.

[27]Section 16—Hinz (2010), p. 271.

[28]Section 16.1.5—Ibid.

[29]See Section 16.2.3—ibid.

[30]Section 16.2.4—ibid.

soil and protects it from erosion. Section 17.1 states: "Traditional law forbids the cutting of grass with its roots, which degrades grazing land, and the cutting of fresh grass which has not bloomed, which will inhibit new growth".[31]

The rules on land distribution complement the so far reported rules.[32] The competence to allocate land vests in the traditional leader and it will not be allowed to the occupier of the land to extend his or her land into the common land which is set aside for herding. Land that is not used for three years will be free for allocation to another user. The right to have a cattle post on common land will be granted by the traditional leader with the possibility to establish a kraal to protect the herd against the cold and wild animals.

2.2 The Wildlife Management Strategy of the Nyae Nyae Conservancy[33]

The establishment of conservancies in communal areas was provided for by the amendment to the Nature Conservation Ordinance of 1975 in 1996.[34] What is a communal conservancy? A communal conservancy refers to an area of communal land in which wildlife is subject to special management arrangements. Conservancies are to enable the inhabitants of communal land to contribute to the sustainable management and utilization of game. The expectation is to achieve this through the engagement of the local people who will in turn, by accepting responsibility, gain benefits from the income generated by the utilization of the wildlife in their areas.

According to Section 24 of the amended Ordinance, "any group of persons residing on communal land and which desires to have the area they inhabit (...) to be declared a conservancy" have the right to apply for a conservancy. The application for the establishment of a conservancy has to entail a document named *constitution* regulating the work of the body that is to run the conservancy, called the conservancy committee.[35] Sub-section 3 of Section 155B of the regulations relating to nature conservation gives a detailed list of matters this constitution must provide for.[36] Apart from the general objectives of the committee which have to be in line with the just recorded general intention of the Nature Conservation Amendment Act, the procedures for the election and removal of the committee members, the powers

[31] Ibid., 272.

[32] Section 9: see Hinz (2010), p. 266f.

[33] See on the following: Hinz (2003), pp. 35f., 95f.

[34] Act No. 5 of 1996 amending Ordinance 4 of 1975.

[35] Section 24A (2) (a) and (b) of the Ordinance.

[36] Government Notice No. 240 of 1976, as amended. The section referred to in the text was enacted by the Amendment of Regulations relating to Nature Conservation, Government Notice No. 304 of 1996. Cf. here also: Ministry of Environment and Tourism (2013), p. 5ff.

and responsibilities, but also the position of the community members have to be set out in the constitution.

The legislative offer to establish conservancies has been widely accepted by the rural communities of Namibia. 83 conservancies have been approved. The conservancies cover about 19% of the land of Namibia,[37] thus covering one half of the communal land of Namibia.[38] Looking at the way the conservancies were implemented, two matters attract attention: one with respect to the position of traditional authorities and one regarding the scope of activities taking place within conservancies.

Neither the Ordinance nor its Regulations determine who may be a member of the committee.[39] There is in particular nothing said about traditional authorities and the relationship between them and the conservancy committees. Research on some of the established conservancies has shown that the researched conservancies were in one way or the other linked to the relevant traditional authorities, either by way of representations of the traditional authority in the conservancy committee or even by way of conservancies being more or less placed under the authority of the traditional authority.[40]

Although the legislative intention in the amendment to the Nature Conservation Ordinance was the sustainable management of wildlife, the implementation of this in the conservancies' constitutions took a more comprehensive approach. This comprehensive approach can already be found in the constitution of the first conservancy established after the enactment of the Nature Conservation Amendment Act: the Nyae Nyae Conservancy.[41]The Nyae Nyae Conservancy, covers the territory of one of the Namibian San communities.[42] The constitution of the Nyae Nyae Conservancy[43] states in clear terms that the conservancy is "based on the traditional Ju/'hoan land management system": the *n!ore* system.[44] What is the *n!ore* system? The anthropologist Lee has this to answer: *N!ore* is the

---

[37]NACSO (Namibian Association of Community-Based Natural Resources Support Organisations), www.nacso.org.na (accessed 9 February 2018).

[38]Communal land is the land administered by traditional authorities, see for more information on this below: Sect. 2.3.

[39]Cf. Section 24A (1) (a) of the Ordinance and section 155B (3) of the Regulations.

[40]See: Hinz (2003), p. 34ff.

[41]Established under Government Notice 12 of 1998.

[42]The Ju/'hoan community in Tsumkwe East or the eastern part of the area formerly known as Bushmanland.

[43]The conservancy was originally called the Nyae Nyae Farmers Co-operative Conservancy and later renamed Nyae Naye Conservancy. The author has two versions of the constitution of the conservancy on file: The Constitution of the Nyae Nyae Farmers Co-operative Conservancy was adopted in 1996, the amended version of the Nyae Nyae Conservancy in 2009. I(t) will (be) refer (red : "the author" 2 sentences above) to the originally adopted text with respect to an annexure with the title Wildlife Strategy Nyae Nyae Conservancy.

[44]Section 1.4 of the Constitution of the Nyae Nyae Conservancy of 1996. (On file with the author).

> ... !Kung[45] word for locality, land, or territory, *n!ore* is applied to the block of land that surrounds each water hole and provides the resources on which the people of the water hole depend.[46]

In other words, *n!ore* reflects a political and legal concept that takes note of the network of rights and obligations that link different groups of people to everything the *n!ore* entails as important as a resource for life: water, vegetation, wildlife, in short all natural resources. The concept of *n!ore* is holistic in the sense that it does not fragmentize the approach (including statutory rules)[47] to distinguishable natural resources. The concept of *n!ore* reflects an understanding of the interdependence of the different natural resources. Although according to the conservancy's constitution the first objective of the conservancy is the sustainable management of wildlife, the constitution notes "other" objectives. Wild management cannot ignore the fact that wildlife depends on its environment. Therefore, one additional objective of the conservancy is to

> [m]anage and improve natural resources in accordance with game utilization and management plans.[48]

With this, other natural resources than wildlife become the focus of the conservancy. The constitution of the conservancy includes in the objectives of the conservancy to

> [p]romote the economic and social well being of the members of the conservancy through equitable consumptive and non-consumptive sustainable utilization of natural resources and education;[49]

and also in very general terms to

> [d]evelop and maintain basic infrastructure.[50]

The Wildlife Strategy of the Nyae Nyae Conservancy, as annexed to the original constitution of the Nyae Nyae Conservancy,[51] draws consequences from the interdependence of natural resources for the development of a plan for the utilization of (the) land. Section 5.6 reads as follows:[52]

> The community agrees that a land-use plan needs to be developed as part of the Wildlife Management Plan. Areas for livestock, wildlife and joint use will need to be identified. Furthermore, the community agrees that the number of livestock

---

[45]!Kung is the name for San and their language living in Tsumkwe areas and also across the border in Botswana.

[46]Lee (1979), p. 334.

[47]Cf. on this: Hinz (2013b), p. 641ff.

[48]Section 4.3 iii of the Nyae Nyae Conservancy Constitution. See also: 1.2.

[49]Ibid., 4.3 ii.

[50]Ibid., 4.3 iv.

[51]The strategy is subject to annual review (see Section 6.1).

[52]The Wildlife Strategy is attached to the Nyae Nyae Conservancy Constitution.

should be limited to allow for an increase and sustainable management of wildlife. The Ju/'hoan community currently has a holding of 514 cattle. They propose that the number be limited to 2000.

The constitutions of other conservancies reflect the same holistic understanding of the environmental resources of the area in which a conservancy was to be established.[53] While the constitution of the Nyae Nyae Conservancy recognizes that conservancies are primarily about wildlife and places responsibilities towards other natural resources under "other objectives", the constitution of other conservancies employ a language in their objectives, which does not distinguish between the main objective and other objectives. The constitution of the Salambala Conservancy, established in 1998,[54] refers to "Salambala's wildlife and other natural resources", which are to be managed in accordance with an "approved management plan".[55] After this objective, the constitution calls generally

> to protect Salambala's wildlife and plants for future generations of Namibians .....[56]

The constitution of the Uukwaluudhi Conservancy,[57] which was adopted in 2001, uses identical words linking wildlife and other natural resources and placing wildlife and plants under the protection of the conservancy committee.[58]

In addition to what has been quoted above from the constitution of the Nyae Nyae Conservancy where the concept of *n!ore* was said to be the basis of the Ju/'hoan land management, section 12.1.5 of the Constitution deserves reference. Section 12.1.5 deals with a "huntable game" and excludes certain animals and also eggs of ostriches from being respectively hunted and taken. "Huntable" refers to traditional hunting, which lies within the responsibility of the people and is therefore difficult to be administered by management. Traditional hunting is hunting on foot, bow and arrow, spear or bird trap.[59] Only "males of any species may be hunted and only half a nest of ostrich eggs may be taken". The reason for this is added: the restrictions

[53]Search of the author did not discover systematic evaluations of conservancy constitutions.

[54]The Salambala Conservancy covers a part of the traditional territory of the Masubiya traditional community in the Zambesi Region (formerly Caprivi Region). The Salambala Conservancy was established by Government Notice 146 of 1998.

[55]Salambala Conservancy Constitution of 1996: Section 4.2. (On file with the author).

[56]Ibid., 4.3.

[57]The Uukwaluudhi Conservancy occupies a part of the Uukwaluudhi traditional territory in the central North of Namibia. Uukwaluudhi is another Oshiwambo-spekimg traditional community of former Owamboland. Cf. Anyolo (2012). The text of the Uukwaluudhi Conservancy is on file with the author. The Uukwaluudhi Conservancy was established by Government Notice 31 of 2003.

[58]See section 4 of the Uukwaluudhi Conservancy Constitution. (On file with the author).

[59]Which remained permitted: see the Nature Conservation Act, 27 of 1987 and Kabinetsbesluit 1074 of 1987 (Traditionele jag deur die inwoners van Boesmanland—Traditional hunting by the inhabitants of Bushmanland—reprinted in Hinz (2003), p. 25f.

Table 1 Rating of societal importance of animals

Criteria for ranking importance	Roan	Elephant	Buffalo	Eland	Giraffe	Gemsbok	Leopard & wild dog
Healing	0	3	1	8	7	6	0
Meat	5	0	3	6	7	4	0
Household items: shoes arrows/water containers	4	0	0	8	8	5	0
Photographic safaris	4	5	4	2	4	1	5
Professional hunting	6	5	5	2	0	1	6
National biodiversity	5	4	3	3	4	2	4
Total[a]	24	17	16	29	30	19	15

[a]This line was added by the author

Table 2 Rating of benefits from photographic safaris and professional hunting

	Roan	Elephant	Buffalo	Eland	Giraffe	Gemsbok	Leopard & wild dog
Total direct benefits	9	3	4	22	22	15	0
Total indirect benefits	15	14	12	7	8	4	15

are made "to ensure ongoing reproduction and success of all species in the conservancy." Further animals are excluded from traditional hunting, namely roans, giraffes and elands. This follows from an investigation conducted in preparation of the constitution for the conservancy and reported in the above-mentioned Wildlife Strategy of the Nyae Nyae Conservancy. Members of the Ju/'hoan community were asked to value roans, elephants, buffalos, elands, giraffes, gemsboks, as well asleopards and wild dogs. The criteria to allocate marks for the importance of the animals were healing, meat, household use, attraction to take photos, professional hunting and national biodiversity. Each interviewee could allocate 25 points.

The result is shown in Table 1.[60]

The high values given to roans, giraffes and elands explain the exemption of these animals from being allowed to be hunted traditionally. The separation of marks given to the first three criteria on the one hand, which can be seen to note the direct benefits to the members of the Ju/'hoan community; and to the last three criteria on the other hand, which relate to indirect benefits, gives the picture shown in Table 2.

[60]See Section 3 of the Wildlife Strategy.

2.3 *The Failed Reform of Customary Land Rights in the Kavango Regions*[61]

Namibia inherited from its colonial past two land tenure systems: The traditional system according to which land was a matter of the community and its government, the traditional authority, and in which individual ownership of land did not exist. This system applied to the communal areas, i.e. this part of the country that was left for the non-white population after the settlement of the white colonizers.[62] Their land tenure followed the land tenure system in which the concept of individual ownership had prominence. After the independence of the country, the government was faced with the difficult task to enact legislation, which would respond to the many problems the two land tenure systems posed to the new democratic order. With respect to the traditional land tenure system, the post-independence government enacted the Communal Land Reform Act in 2002.[63] This act harmonized the statutory law on communal land. It accepted that land in communal areas are (to be) principally administered by the respective traditional leaders.[64] Land boards were established to secure that this administration was subjected to the rule of law.[65] In order to contribute to the security of land use rights under customary law, the act introduced the requirement of registering customary land rights.[66]

While the registration of customary land rights was basically accepted in accordance with the legal expectation in all parts of the country, it was not in the two Kavango regions.[67] Here, the traditional authorities and their members refused to register their customary land rights. It is worthwhile to take note of the arguments in the debate about the registration of customary land rights as it prevails in the Kavango regions up to today.[68] Namwoonde and Mendelsohn report about the

[61]The reference is to Kavango regions after the Kavango Region, one of the original 13 regions of Namibia, was divided into two regions, Kavango East and Kavango West in 2013.

[62]On the pre-independence situation of land tenure in communal areas, see Hinz (1998).

[63]Act 5 of 2002.

[64]See Sections 20 and 17 of the Act.

[65]See Section 24 of the Act according to which the land boards have to ratify land allocations by the chief or a traditional authority. Only in limited cases, as listed in Section 24 (4) (c) of the Act, a land board can object to an allocation.

[66]Section 25 of the Act.

[67]See the reports on the situation in the different areas of Namibia in: Mendelsohn (2008), p. 33ff.

[68]See on this: Namwoonde (2012)—summary of a master thesis under the supervision of the author containing empirical findings collected in 2008. Namwoonde did not consider the research done by Mendelsohn (2008) who published his work on the management of communal land in December of 2008. A report in the Namibian newspaper *New Era* referred to a workshop held for the Kavango regions in preparation of the national land conference of October 2018 during which the call for registration of customary land rights was part of the governmental statement (See: New Era of 20. July 2017—electronically accessed on 10 April 2029). Cf. further: Mandimika and Mulofwa (2018) and also the governmental reports for the national land conference of 2018: Republic of Namibia (2018a, b).

many technical problems the processing of applications for registration caused in many traditional offices, owing to the lack of capacity.[69] However, the move against the registration of customary land rights in the Kavango regions was founded on substantial reasons. In a petition submitted to the Minister of Lands and Resettlement the five traditional authorities operating in the Kavango regions[70] expressed their dissatisfaction with Section 25 of the Communal Land Reform Act.[71]

A number of reasons were noted to explain the resistance to register customary land rights.[72] The traditional authorities stated that the registration was not in the interest of the Kavango people: the registration ignored the lifestyles of the Kavango people and, with this, the traditional practice(s) of land management practice (s) successfully experienced in the Kavango regions. Land management in the Kavango regions employed a special form of management aimed at the protection of the soil: shifting cultivation. People have fields close to their settlements, which are mainly close to the Kavango river. These fields are primarily used as garden fields. In addition, many people have fields inland, i.e. field away from the river. There, they also keep their cattle. When the inland fields lose their fertility, the farmers move to another field while the field that has so far been occupied is cleared by burning.[73] The burned field remains uncultivated for some time so that the soil is able to recover for new cultivation. This shifting cultivation is seen to be affected by the registration as expected by the Communal Land Reform Act. Namwoonde summarized the opinions collected in interviews with people in the Kavango regions:[74]

> On the question whether registration is in conflict with shifting cultivation, the respondents had mixed feelings. While some respondents held that "registration is not in conflict [with traditional cultivation practices] because a person can still move if s/he so wishes", the majority of the interviewees were of the opinion that registration was an impediment of the practice of shifting cultivation.

Reason for the majority to believe that the registration is an impediment to the practice of shifting cultivation is certainly also that the registration in a land management system which practices shifting cultivation will prompt additional bureaucratic hurdles. To comply with these will add more difficulties to the process of registration than where the shifting of cultivation is not practiced. Why then invite such unnecessary hurdles?

---

[69]Namwoonde (2012), p. 122f.; Mendelsohn (2008), p. 59.

[70]Namely: Ukwangali, Mbunza, Shambyu, Gciriku and Mbukushu.

[71]Petition of 4 June 2007 on file with the author.

[72]Cf. Namwoonde (2012), p. 115ff.; Mendelsohn (2008), p. 57ff.; Mandimika and Mulofwa (2018), p. 7ff.

[73]Land management using this kind of shifting cultivation is not special to Kavango. It has widely been used in many parts of the world, see e.g. Yale School (2019).

[74]Namwoonde (2012), p. 117.

Apart from this, the registration of customary land rights was also seen to individualize the customary land management:[75]

> If all the land is registered, then the chief will not have control over such land and this will also interfere with the community's sense of sharing. This is because, right now, if you have a *maguni*[76] tree on your field, the whole community can eat its fruits; but if the land is registered in my name, then it will be my *maguni* tree and I will not be willing to share it.

The fact that the registration will only concern customary rights allocated to use the communal land for housing and cultivation but will not concern the collective rights on non-allocated land, commonage land, which is "traditionally used for the common grazing of stock",[77] was also forwarded as an argument against the registration. Will this commonage be open for alternative use after the registration, open for interventions by the state?[78] Interviewees had further difficulties with the fee to be paid when registering their rights. What about those who did not have the money to pay? Furthermore would payments not contribute to the understanding that one bought the land which, under customary law, is not possible?[79]

People interviewed about the registration of customary land rights were also reluctant to accept the view that the registration would contribute to more security. It was stated that the customary procedures in place had proven that these procedures were able to solve arising disputes about customary land rights.[80] The mere registration of customary land rights would also not prompt banks to accept these registered rights as mortgage for loans.[81]

In concluding the report on why registration was refused, Namwoonde quotes the following from one of her interviewees who said:[82]

> ... the input given by the Traditional Authority during the consultations that took place in the process of preparation of the Bill [MOH: on Communal Land] .... was not taken on board. Consequently the final Act did not provide for certain prevailing circumstances in our communal areas.

In conclusion, registration of customary land rights will lead to a change in the customary concept of communal land. The registration stands for a form of land management lacking the flexibility of the traditional land management, the governing principle of which is the collective use of land.[83]

---

[75]Ibid.

[76]Monkey orange tree or *strychnoscocculoides*.

[77]Section 1 of the Communal Land Reform Act.

[78]See on this: Namwoonde (2012), p. 117f.

[79]Ibid., 120.

[80]Ibid., 118f.

[81]Ibid., 120.

[82]Ibid., 121.

[83]See: Namwoonde (2012), p. 123f. with further references to the discussion in other parts of Africa.

3 Indigenous Knowledge: Local Common Sense

The three cases submitted have in common that the people involved in these cases perceived the respective local conditions and translated them into legal or at least social norms, which were meant to guide the behaviour of the members of the respective societies. All the norms took note of the relevance of natural resources for the survival of the members of the communities. While soil and soil protection were a direct concern in the Kavango regions, the same concern surfaced in Uukwambi and the conservancy of the Ju/'hoan with norms on the management of natural resources through which the soil indirectly received protection. In the laws of Uukwambi, this was done by/through rules resulting from the perception of the *efundja* of the Cuvelai catchment. The members of the Ju/'hoan community drew consequences from the Ivaluation of the animals in their region for the Wildlife Strategy to be implemented in the conservancy. The members of the traditional communities in the Kavango regions decided to resist the governmental policy of the registration of customary land rights on grounds of their traditional land management practice(s). Thus, in all three cases, it was locally perceived knowledge that motivated the behaviour of the people.

The perceptions in the three cases follow local observations and express local common knowledge. The discovering and the understanding of such local common knowledge does not require very much special expertise. In other words, it may rise attention without being pointed out by specialists of whatever "new applied anthropology", at least as (long as?) the person coming from the "outside" is open for what the local people will say. This applies to the case of the floods in the Cuvelai catchment, which expects, indeed, a balanced treatment of water. The focus on the sustainable use of certain wildlife in the Ju/'hoan conservancy is only an exception to this where the interest in certain animals for healing purposes comes in. Apart from this, the basis for the Wildlife Strategy was founded on immediately convincing interests in the use of wildlife. The same is true for the Kavango case. Even if one has objections to the system of shifting cultivation operated in the Kavango regions or one may not favour the arguments against the registration of customary land rights, both local positions are reasoned in the sense that the reasons can be noted without the explanation of an anthropological expert.

It was explained above that the self-stated laws of Uukwambi changed over the years by being extended from a set of rules dealing with traditional wrongs and the legal consequences applying to cases of violations to matters of environmental protection. Although no research exists about this process of change, it can be assumed that the change was supported by the increasing public debate about the environment. Demands from this debate found obviously the way into the amendment of the laws of Uukwambi.[84] In other words, the indigenous knowledge

[84]Personal communication with the supreme leader of the Uukwambi Traditional Authority, *Elenga Enene* Herman Iipumbu (*Elenga Enene* is the traditional title of the leader) supports this view. See also Section 3 (2) (c) of the Traditional Authorities Act (Act 25 of 2000), which lists among the

reflected in the laws of Uukwambi was amended by new knowledge that came from outside and, with this, reshaped the traditional indigenous knowledge as it existed so far. That indigenous knowledge became subjected to reshaping to meet the changed circumstances in which a society operates, leads to more general observations on what indigenous knowledge can mean in the implementation projects: Indigenous knowledge is to be understood as a set of statements that are open for change and, with this, for dialogue. At least, as long as the dialogue takes note of the setting of the knowledge and the framework it is part of![85]

The fact that the customary law of a traditional community, such as the community of Uukwambi, contains rules that are of environmental relevance, that rules of environmental relevance were integrated into the self-stated rules in a process of amendment offers an excellent entry point for any development effort; strengthening the scope of environmental protection is only one field to benefit from. The indigenous knowledge offers a foundation on which further development offers can easily be built. The existing set of indigenous knowledge forms a societally accepted basis for such offers. Policies and legislative interventions that make use of what is in place will find positive societal responses.[86]

The enquiry undertaken in preparation of the establishment of the Nyae Nyae Conservancy had two developmental effects. The first was that the enquiry made collective memory generally visible. The people who had the relevant indigenous knowledge in their memories expressed it in their responses to those who conducted the enquiry and thus contributed to preserving the memory. At the same time, the enquiry prompted the people to link what they had in mind to the expectation behind the conservancy: the sustainable management of wildlife which included making benefits by using wildlife for economic purposes. This resulted again in a process of reshaping the indigenous knowledge of the people by opening up their focus to the interest for photo safaris and professional hunters beyond their traditional interest in healing procedures, household items and meat, leading them to an impressive the reported mixture of direct and indirect benefits. What could be said about the potential of indigenous knowledge reflected in the laws of Uukwambi can be repeated here. The exploration of indigenous knowledge in the case of the Ju/'hoan

duties of the traditional authority the duty "to ensure that the members of his or her traditional community use the natural resources at their disposal on a sustainable basis and in a manner that conserves the environment and maintains the ecosystems for the benefit of all persons in Namibia".

[85]Cf. here again the above-mentioned article by van Meijl (2019).

[86]It would even be possible to avoid formal legislative interventions, i.e. by an act of parliament by supporting less formal amendments to customary law by adding something to the self-stated laws of the communities. Suh an approach could save money and time, but would expect confidence by governments in the policy- and law-making capacity of traditional rulers! Whether such a confidence is or could be in place will differ from situation to situation. In a recent publication on the law of environment in Cameroon, an article on customary law and the environment quotes information according to which "sustainable development" is not within the vision of the traditional leaders of Cameroon (Kahler 2017, p. 687). Having stated this, the author asks: "But does customary law imply a soil-sustainable land use?" The question is left open although what was said about the positions of the chiefs suggests a no as answer.

Conservancy contributed to the acceptance of the conservancy—acceptance, from which the people of the community have, indeed, profited up to today.[87]

The Kavango case is an example for what can result from by-passing or better: ignoring indigenous knowledge. The resistance to the land policy of the government and its translation into an act of parliament has been maintained since the Communal Land Reform Act had come into force, now some 17 years ago. Instead of acknowledging the objectives of the people in the Kavango regions and, with this, respecting the indigenous knowledge as the platform to argue, the governmental policy restricted itself to calls for accepting the policy of the government and the Communal Land Reform Act. The alternative could have been to engage the traditional leaders and members of the Kavango communities in a debate, which would have given enough space to the arguments of the Kavango people as informed by their indigenous knowledge. This would have included the position of the competence of traditional leaders to manage communal land and also the state of customary land law to set the framework for the fair handling of the management and achieving legal certainty. Questions about maintaining shifting cultivation in times of an increase in population and thus growing scarcity of land would be part of the dialogue, and in this respect a challenge to the traditional leaders who will in all probability not close their eyes faced with these developments.[88]

To conclude: There are—and this is partly a repetition of what has been incidentally noted in the previous parts of this article—one can learn two general lessons one can be learnt from the three cases in which indigenous knowledge was highlighted. The first is that indigenous knowledge is in many instances normal in the sense that it expresses something which any outsider would be able to discover if he/she would just go the extra-step to unpack it and accepts that locals usually know their environment better than newcomers. Locals will know when winds blow, when rain comes or does not come, they will know what the soil can or cannot produce. Most of the information reported in the three cases belongs to this "normal" type of indigenous knowledge, knowledge informed by local common sense, therefore, not being something exotic and, with this also attracting economic interests as it is the case with special medical plants on which a great deal of research concentrated.[89]

The second is that, although indigenous knowledge is mainly knowledge informed by local common sense, it nevertheless may contain elements which are due to challenges the knowledge outsiders may arrive with. The most obvious in the three cases reported for this article appears in the Kavango case. The core part of the objections to the registration of customary land rights is less the feared inroad into the traditional management of land with shifting cultivation but rather the

[87]See here e.g. Hitchcock (2017).

[88]The author's experience in consulting traditional authorities when working on projects with traditional authorities supports this statement.

[89]This statement does not mean that the mentioned research—I refer for the Namibia to Chinsembu (2015)—is not of importance: it is! However, it should not lead away from the fact that indigenous knowledge is more than knowledge about medical plants.

traditionally informed view that registration would be the first step to change the collective replacing the collective approach to land management with an approach leading to the concept of individual land ownership. It is in this respect that indigenous knowledge informs the continued resistance against registration. The second obvious challenge by indigenous knowledge is the reservation of certain animals for healing in the Ju/'hoan community. That healing will play a role in exempting animals from hunting was not to be expected form a mere economy-oriented development approach. A third challenge is contained in the holistic approach as expressed in the constitutions of conservancies. Although the purpose of conservancies is to provide for the sustainable use of wildlife, the constitutions of conservancies went beyond this purpose and included other natural resources, thus accepting an approach which has difficulties (in) fragmentizing natural resources: wildlife, plants, water, soil etc. and their protection, as it is basically practiced in the so-called modern approaches of policy and law.[90]

These concluding remarks give the above-quoted statement about the "new applied anthropology" at least partly a new orientation: There is still the expectation that anthropology, better: empirical anthropological research, assist the discovery of indigenous knowledge in the broadest possible way. However, in addition to this, political and legal anthropology is in demand for research on indigenous knowledge that reflects societal concerns, meaning indigenous knowledge which is "normal" common sense but also knowledge based on views that are in conflict with the views prone to import.[91]

References

Anyolo PN (2012) The Uukwaluudhi conservancy: does it contribute to poverty alleviation? In: Hinz MO, Ruppel OC, Mapaure C (eds) Knowledge lives in the lake. Case studies in environmental law from southern Africa. Namibia Scientific Society, Windhoek, pp 29–39

Chinsembu KC et al (2015) Indigenous knowledge of Namibia. University of Namibia Press, Windhoek

[90]Cf. the above-quoted reference to Hinz (2013b), p. 641ff.

[91]From a legal perspective, it could even be held that this conclusion is not only a practically expected, as there cognition of and respect for indigenous knowledge will support local acceptance, but is legal a requirement. Without going into the details of interpretation, there are good reason for the position according to which the right to culture entails the right to customary law (See: Articles 19 and 66 (1) of the Constitution of Namibia). Following from this, one could argue that such a right to customary law would entail the right to recognition of indigenous knowledge. The fact that Namibia supported the Swakopmund Protocol on the Protection of Traditional Knowledge and Expressions for Folklore of 2010 (Electronically retrieved from the African Regional Intellectual Property Organization on 24 April 2019—see on the Protocol: Hinz 2011) and enacted the Access to Biological and Genetic Resources and Associated Traditional Knowledge Act (Act 2 of 2017) supports the suggested constitutional extension of the protection to traditional knowledge in the two quoted legal documents to the said right.

Goodenough W (1957) Cultural anthropology and linguistics. In: Garvin PL (ed) Report of the seventh round table meeting on linguistics and language study. Georgetown University, Washington D.C., pp 167–173

Grim JA (ed) (2001) Indigenous traditions and ecology. The interbeing of cosmology and community. Harvard University Press, Cambridge

Hinz MO (1995) Assisted by S. Joas, Developing customary law: self-stated laws of Namibian communities and customary law consultative meetings with traditional leaders. Centre for Applied Social Sciences, Windhoek

Hinz MO (1998) Communal land, natural resources and traditional authority. In: d'Engelbronner FM, Hinz MO, Sindano JL (eds) Traditional authority and democracy in Southern Africa. New Namibia Books, Windhoek, pp 183–227

Hinz MO (2003) Without chiefs there would be no game. Customary law and nature conservation. Out of Africa, Windhoek

Hinz MO (ed) (2010) Assisted by N. E. Namwoonde, Customary law ascertained. Vol. 1: the customary law of the Owambo, Kavango and Caprivi communities of Namibia. Namibia Scientific Society, Windhoek

Hinz MO (2011) The Swakopmund protocol on the protection of traditional knowledge and expressions of Folklore. Namibia Law J 3(1):101–112

Hinz MO (ed) (2013a) Assisted by A. Gairiseb, Customary law ascertained. Vol. 2: the customary law of the Bakgalagari, Batswana and Damara communities of Namibia. University of Namibia Press, Windhoek

Hinz MO (2013b) Agenda 21 and climate protection: the development of global and local governance for environment and development – observations from research in Namibia. In: Ruppel OC, Roschmann C, Ruppel-Schlichting K (eds) Climate change: international law and global governance, Vol. II: policy, diplomacy and governance in a changing environment. Nomos Verlagsgesellschaft, Baden-Baden, pp 621–659

Hinz MO (ed) (2016a) Assisted by A. Gairiseb, Customary law ascertained. Vol. 3: the customary law of the Nama, Ovaherereo, Ovambanderu, and San communities of Namibia. Namibia Scientific Society, Windhoek

Hinz MO (2016b) The ascertainment of customary law: what is *ascertainment of customary law*, what is it for and what is the work ahead? In: Hinz (2016a), pp 3–20

Hinz MO (2019) The recording of customary law: what to learn from customary law ascertainment in Namibia. Potchefstroom Electron Law J 528–529 (in print)

Hitchcock RK (2017) Sustainability, socioecology, and resilience: communal conservancies and the impact of development north eastern Namibia. Paper presented at the 2017 INSAKA International Symposium, University of Namibia. Windhoek (on file with the author)

Jürgens N, Schmiedel U (eds) (2010) Biodiversity in southern Africa, vols. 1–3. Klaus Hess Publishers, Göttingen, Windhoerk

Kahler A (2017) The significance of customary law for environmental conservation in Cameroon. In: Ruppel OC, Yogo EDK (eds) Environmental law and policy in Cameroon. Towards making Africa the tree of life. Konrad Adenauer Foundation, Yaoundé, pp 675–691

Lee RB (1979) The !Kung San. Men, women, and work in a foraging society. Cambridge University Press, London

Mandimika P, Mulofwa J (2018) Securing customary land rights for development in Namibia: learning from new approaches, opportunities and social settings. Paper for Land Governance in an Interconnected World. Annual World Bank Conference on Land and Poverty. The World Bank, Washington (Electronically available, retrieved on 15 April 2019)

Mendelsohn J (2008) Customary and legislative aspects of land registration and management on communal land in Namibia. Report for the Ministry of Land & Resettlement and the Rural Poverty Reduction Programme of the European Union. Research & Information Services of Namibia, Windhoek (Electronically available, retrieved on 5 April 2019)

Mohan, P. C. (2001): Indigenous knowledge for development. Africa Region Findings of Good Practice Infobriefs. No. 70. Washington, DC. https://openknowledge.worldbank.org/handle/10989/979License.CCBY3.0IGO (Retrieved: 15 November 2020).

Ministry of Environment and Tourism (2013) Guidelines of conservancies and standard operating procedures. Ministry of Environment and Tourism, Windhoek

Mohammedbhai G (2013) Indigenous knowledge must be harvested for development. In: University World News of 9 March 2013. http://www.facebook.com/UniversityWorldNews (Retrieved: 6 April 2019)

Namwoonde NE (2012) Registration of customary land rights in the Kavango Region: a rejected import. In: Hinz MO, Ruppel OC, Mapaure C (eds) Knowledge lives in the lake. Case studies in environmental law from southern Africa. Namibia Scientific Society, Windhoek, pp 105–129

Oldeland J, Erb C, Finckh M, Jürgens N (eds) (2013) Environmental assessments in the Okavango Region, vol 5. Biodiversity & Ecology, Hamburg

Republic of Namibia (2018a) Kavango West Regional Council. Regional consultations in preparation for the second national land conference, 26–27 July 2018. Republic of Namibia, Windhoek

Republic of Namibia (2018b) Kavango East Regional Council. Regional consultations in preparation for the second national land conference, 26–27 July 2018. Republic of Namibia, Windhoek

Sillitoe P (1998) The development of indigenous knowledge. A new applied anthropology. Curr Anthropol 39(2):223–235

Sillitoe P (2006) Introduction: indigenous knowledge in development. Anthropol Action 13 (3):1–12

Van Meijl T (2019) Doing indigenous epistemology. Internal debates about inside knowledge in Mãori society. Curr Anthropol 60(2):155–173

WIPO (2001) World Intellectual Property Organisation. Intellectual property needs and expectations of traditional knowledge holders. WIPO report on fact-finding missions on intellectual property and traditional knowledge (1998–1999). WIPO, Geneva

Yale School (2019) Yale School of Forestry & Environmental Studies. Global forest atlas. Traditional land use and shifting cultivation. Yale: http://globalforestatlas.yale.edu (Retrieved: 27 April 2019)

Land Rights, Land Use Patterns and Soil Fertility Significantly Contribute to the Two-Decade Long Regional Conflagration in Eastern Congo

Bila-Isia Inogwabini

1 Introduction

The Democratic Republic of Congo (DRC) is the second largest country in Africa and has massive land area. From an ecological point of view, DRC is among the most varied country in terms of its ecosystems, which includes nearly half the African rainforests.[1,2] The country comprises forest-savannah ecotones, savannahs, afro-mountainous forests, large and small lakes, rivers and swampy forests.[3] Yet, from a socio-economic perspective, these varied ecosystems offer assets, among which the most important is the land, which constitutes the foundations of all social and economic development. Not only does land provide pavement for the forest types and space for agriculture, but it lies at the heart of mining activities that have made DRC so well-known since the historical times. This is so true that the curse of natural resources that plagues the country since being opened to international exploitation in the late nineteenth century can rightly be attributable to question of land ownership and land use patterns.

Unfortunately, however, the vexing question of management of natural resources and the bleak consequences it has on the Congolese populations have always been

[1]IUCN (1992).

[2]Ernst et al. (2010).

[3]Inogwabini et al. (2005).

B.-I. Inogwabini (✉)
The Jesuit Loyola University of Congo, Faculty of Agricultural and Veterinary Sciences, Center for Research and Communication in Sustainable Development (CERED), Kinshasa, Democratic Republic of the Congo

Department of Aquatic Sciences and Assessment, Swedish University of Agricultural Sciences (SLU), Uppsala, Sweden, Uppsala, Sweden
e-mail: bila_inogwabini@biari.brown.edu

H. Ginzky et al. (eds.), *International Yearbook of Soil Law and Policy 2019*,
International Yearbook of Soil Law and Policy 2019,
https://doi.org/10.1007/978-3-030-52317-6_7

presented as an issue of its own value without factoring the most fundamental questions of ‘who owns which piece of land’ under what conditions and ‘for the benefits of whom’ in DRC. Hence, this chapter tries to shed some lights on this niggling question; it does focus on the eastern part of DRC (which is just one part of the country). To make sense of the entire context, however, the chapter begins by looking at the overall land ownership details and its evolution over the recent past. In the final paragraph, the chapter accords attention to the critical issue of galloping demographics of the entire country and how that would affect the overall land ownership and the consequences that are likely to be expected.

2 Materials and Methods

Materials used in this chapter mostly come from published records. Some of these published materials have gone through peer-review processes, but others are grey literature. One reason to use this later type of sources is that acquiring recent statistics from DRC is always a tedious exercise. Indeed, most of the estimates on human populations are projections done using the last census, which was carried out back in 1984. Despite this caveat, however, some estimates do exist and have been widely used because they are felt to fit the reality on ground. The populations being used in this chapter has been reported by Population Data.Net in 2019,[4] where a serious attempt was made to come up with consolidated human statistics. Territorial land statistics came from the same source, but were checked against the ones that the United Nations Development Program (UNDP) produced for the same provinces.[5,6,7] Overlaying population estimates from Population Data.Net, led to crude population densities per province while effectively available land was derived from factoring that the Kivu area (covering the territories of the current three provinces in the DRC, namely North and South Kivu and Maniema) exhibits altitudes that are higher than those of other provinces of DRC. Areas like the Itombwe Mountains, which comprise the territories of Fizi and Mwenga, are located at 2000 m above the sea level. Furthermore, this geographic layout implies that slopes are steep and predispose land to soil erosions. For example, 23%, 30% and 47% of soil around Bukavu had respective slopes ranging between 0–25%, 25–50% and above 50%.[8] Evidently, where the land slope is greater than 50%, it is not, obviously, usable, under normal conditions. So, using this data and setting the limits of cultivatable land to be at the 14% slope, Wils et al. Mathematically modeled that only 30% of the

[4]Population data.Net (2019). https://www.populationdata.net/pays/republique-democratique-du-congo/.

[5]PNUD (2009a).

[6]PNUD (2009b).

[7]PNUD (2009c).

[8]Wils et al. (1986).

land in the Kivu was readily functional for communities. An additional 35% was added when the model was expanded to include slopes ranging between 15% and 35%. Hence, the crude densities calculated for each of the three provinces are aberrant while discussing issues of land. This means that normally, the readily functional land area in the Kivu is about 65% of the current flat-mapped zones and therefore, should be adjusted accordingly. Effective human population densities per province and per other territorial units were reached by dividing population estimates by 65% of the land accessible to people, as suggested by Wils et al. Types of land use patterns were collected from provincial monographs produced by the UNDP on North and South Kivu and Maniema.[9,10,11] Apart from smallholder's agriculture, major land uses included those that were brought up by Kejuni et al., Wabubyula et al. and Khete et al. and were the presence of protected areas, large industrial agriculture's fields, logging concessions and cattle raising ranches. Mining sites, which are known to constitute a major part of the land use patterns in the region, were not included in this analysis because of lack of coherent data. Even with the other land use patterns, it has not been possible to conduct a thorough assessment as to how these forms of land use patterns are statistically linked to the political unrest in the Eastern DRC region. This was due to lack of a source providing consistent data on each of these categories. However, indirect inferences were made as to how the presence of protected areas, large industrial agriculture's fields, logging concessions and cattle raising ranches impacted the availability of cultivatable lands to different communities.

Crude human densities were used as surrogate for land availability; they were checked against the civil unrest intensities felt by communities. Civil unrest was expressed in intensities and measured by the levels of insecurity that appropriate security bodies used to express how safe they felt a place was; they range from '*safe*' through '*open crisis*' and were converted into alert scales. These security alert scales (felt intensities) were the 2017 annual averages published by the UN mission in DRC and which ranged from 0.5 (totally safe) to 5 (open crisis or totally incontrollable insecurity). The data used was averages cumulated over a period of 2-year preiod. Exponential regressions were run on densities plotted against felt unrest intensities. Logarithmic regressions were chosen because they fitted best with the shape of data. The Island of Idjwi, which is an administrative territory of South Kivu, in its own right, has been excluded from the analysis because, despite exhibiting the highest human density of all territories (889.70 individuals/km^2), its annual average felt unrest intensity was too low and consequently brought an unpredictable effect in the analysis.

The chapter pursues the objective to sort out the role played by land availability in Eastern DRC against the issues of land ownership and land use patterns. Beyond sorting out the specific role of land availability, the chapter goes as further as to look

[9]Kejuni et al. (1998).

[10]Wabubyula et al. (1998).

[11]Khete et al. (1998).

at which role would be played by the fertility of available soil in the mixed multiple components that explain the conflict in the country. Looking at the soil fertility helps to see how soil health and its fertility can conduce to heightened conflicts. Acting as a positive back-loop mechanism, higher human densities lead to conflicts over land and impoverish land; impoverished land further escalates in more human conflicts. To link the chapter with the overall objective of this Yearbook Volume, which is about protecting soil fertility and health, equitable land use planning process and ecological agricultural practices would contribute to preserving good soil (in health and fertility) and easing social tensions across the region.

3 Land Ownership in Democratic Republic of Congo: Evolution Over the Recent Past

There is no written record on land ownership in DRC that can be tracked beyond the colonial era. The first land tenure ordinance was passed on 1 July 1885, about lands acquired by Stanley on behalf of King Leopold II'.[12] In 1886, King Leopold II signed another decree that set clarified conditions of accessing and possessing lands throughout the current space of DRC.[13] However, despite these new conditions, stating that written records only go back to the aftermath of the Berlin Conference' does not equate to the absence of legal system in charge of managing land rights, land occupations and land use before the invasion of the space that is known today as DRC by the colonialists. To state the obvious, this massive space was occupied by small, yet contiguous kingdoms such as; the Kongo, Kuba, Tshokwe, etc. These are the most well-known political entities and were structured in such a way that external claims over their territories were causes of wars. Apart from these well-known polities, the remaining space of DRC was occupied by ethnic groups that were centered on smaller chiefdoms ranging from kingdom-like entities to more or less headless, but still ordered social structures of early village societies of 700–1000.[14] From this perspective, it can be genuinely argued that, despite varieties in social and political organizations, land has never been a vacant entity in the space that has been called Congo since the end of Berlin Conference in February 1885.[15] This claim has been attested by wars over space that were documented throughout the western Bantu expansion,[16] which were observed in the forested areas, some centuries back before the current boundary-demarcated shape of DRC.

[12]Jeal (2008).

[13]Etat Indépendant du Congo (1886).

[14]Vansina (2004).

[15]Inogwabini (2014).

[16]Vansina (1990).

Table 1 Forest and game reserves of North and South Kivu

Administrative territory	Number of protected areas	Total Area (km^2)
Beni	22	227.4
Lubero	10	61
Masisi	12	69.4
Nyiragongo	1	4
Rutshuru	23	1,150.3
Total North Kivu[a]	*68*	*1512.1*
Shabunda	4	298
Uvira	2	44[b]
Walungu	4	107[b]
Total South Kivu[a]	10	449

[a]Totals are respectively 68 and 10 because the 3 National Parks are not included in the case of North Kivu while for South Kivu protected areas with no data were omitted
[b]One protected area has no data in the compilation done by Wabubyula et al. (1998), which served as source for this chapter

4 Land Use Patterns in Kivu

The intent of this section is far from providing a detailed map of the land use in Kivu; it rather offers a global outlook of how land is functionally shared across the different major stakeholders in the region. To start with, it must be recognized that the presence of protected areas in DRC dates back to the creation of the Virunga National Park, back in 1925; Virunga is the first African protected area. National parks and other types of reserves have, since then, become part of the physical and psychical landscapes throughout the entirety of the Kivu. Overall, North Kivu has 71 protected areas including parks, forest and game reserves of different categories. There are three national parks that bisect the territory of North Kivu. The 7600-km^2 (95% of the 8000-km^2) of the Virunga National Park is spread across the administrative territories of Beni, Lubero and Rutshuru. Two other national parks are Kahuzi-Biega and Maiko that respectively take 1480-km^2 and 1083-km^2 in the respective administrative territories of Walikale (entirely) and Lubero (entirely). Rutshuru should, in principle, have a 1000-km^2 area to buffer the Virunga National Park, which is why its relative figure on other types of protected areas is higher compared to other administrative territories (see Table 1).

In South Kivu, apart from the 75% of Kahuzi-Biega National Park (4500-km^2) and the relatively recently created Itombwe (7000-km^2) natural reserve, there are 14 other protected areas, of which 9 have cartographic data totaling 449-km^2 (see Table 1). There are no data on protected areas in Maniema, but UNDP (2009a) indicated that there were 24 forest natural reserves in Maniema without giving any details about where they are located and what their respective sizes were. In addition to a small part of the Maiko National Park that is located in Maniema, a 8879-km^2 Lomami-Lualaba national park has been recently created, which brings Maniema in

Table 2 Land areas in the three provinces of Kivu and human populations

Province	Area (km^2)[a]	Populations[a]	Density (Humans/km^2)
North Kivu	59,483 (38,664)	6,655,000	112.0 (172.0)
South Kivu	65,070 (42,296)	5,772,000	89.0 (136.0)
Maniema	132,250 (85,963)	2,333,000	18.0 (27.0)

[a]Data in parentheses are effectively available land and effective human population densities modeled after Wils et al. and populations are from Population Data.Net

the same footing as other three provinces of the Kivu complex, in terms of protected areas as being a major landmark in terms of functional land uses.

The Kivu complex covers a territory of 256,803-km^2, which is the total area of North and South Kivu and Maniema (see Table 2). Adding areas of national parks and other categories of reserves (forest and game reserves, principally), it gets to a total of 32,523.1-km^2, which, as a quotient over the total area of the Kivu, leads to 12.66% land falling under the status of protected areas. This figure is not spread evenly across the three provinces. North Kivu is the region most dominated by protected areas that cover 21% of its entire territory. Factoring the caveat that 24 forest natural reserves of Maniema (UNDP 2009a), are not included in this analysis would lead to the conclusion that the figure of 12.66% is a minimal chunk of the areas occupied by protected areas in the Kivu complex.

The second most important land use pattern in the Kivu is the presence of industrial agriculture's fields. As it is the case with protected areas above, the availability of statistics suffers from its own paucity and it is variable across the region. Officially, classified industrial agricultural fields were those of size >50-ha. In North Kivu, they occupied a 28.39-km^2. Of this total, with 60.5%, coffee plantations take the largest share (17.2-km^2) located in Beni, Lubero, Rutshuru and Masisi. The remainder of the 39.5% of the land falling within this functional land use category is taken up by palm oil plantations in Beni (5.69-km^2) and Walikale (5.50-km^2). In South Kivu, a 95.95-km^2 surface belongs to the industrial agricultural fields serving to grow coffee, quinine trees, tea and sugar cane. This surface comes from plantations of Kabare (2 fields), Kalehe (2 fields), Uvira (1 field) and Walungu (7 fields). With 60-km^{2}, over 95.95-km^2 of the total (62.5%), the single field in Uvira is, by far, the largest and the most used plantation. The field in Uvira is for the cultivation of the sugar canes that produce sugar at Kiliba. Figures produced by Walemba and Mungu,[17] are that a 211.36-km^2 in Maniema is occupied by plantations of palm trees whereas coffee plantations are found in a total surface of 8.46-km^2. Given the political situation that prevails in the Kivu since more than 20 years now, some of these plantations have been abandoned, but they remain properties of their owners.

The third functional usage of land in the Kivu is logging concessions on which data is difficult to obtain. However, Kejuni et al. were able to glean a figure of 45.7-km^2 of logging concessions in North Kivu, with the largest (16.3-km^2) being located

[17]Walemba and Mungu (2016).

in Walikale, followed by Lubero (15.79-km^2). Other territories that have logging concessions, in a decreasing order, are Beni (9.35-km^2), Rutshuru (2.8-km^2) and Masisi (1.46-km^2). A total of 5.19-km^2 was found to be in logging concessions in South Kivu, which was divided, in a decreasing order, as follows: Kalehe (2.49-km^2), Fizi (2.0-km^2), Uvira (0.5-km^2) and Kabare and Shabunda (0.1-km^2 each). Again, comparable data from Maniema lack and even though Khete et al. suggested that industrial logging was nearly absent of Maniema since the end of colonization back in 1960, it is plausible to argue that logging must be on-going even at smaller scale. Stating that smaller scale logging must be on-going in Maniema conforms to the finding by Walemba and Mungu that 226 logging concessions were illegally held by local elites that traded wood with international timber merchants. Unfortunately, Walemba and Mungu, gave no magnitude of the size of what the 226 logging concessions would represent in terms of surface.

Ranches for are the fourth functional land use for cattle growing in Kivu. While indicating how difficult it was to get fine-tuned figures on space occupied by ranches in North Kivu, Kejuni et al., came up with 260-km^2 for professional cattle growing enterprises in Lubero, Masisi and Rutshuru. Apart from these professional enterprises, Kejuni et al. also quoted the figure of the 130,000 other people who invested in the small scale cattle growing activities and revealed that each of these people had six cows on average, with each reaching up to 900-kg. Kejuni et al. gave the ratio of 350-kg for 1-ha. If one uses these details, he would come up with about 2005.71-km^2. Therefore in total, ranches to grow cattle occupy about 2,265.71-km^2 in North Kivu. To get an idea of what this figure represents, it is larger than the town of Goma and the territory of Nyiragongo combined; it represents 48% of the total area of Masisi. The same calculation methods applied to the 120,416 heads of cows owned principally in the administrative territories of Uvira, Mwenga, Kalehe and Fizi, conduce to the total of 3,096.41-km^2 in South Kivu being used to grow cattle. Unfortunately, data on this land use category is not available for Maniema.

Smallholders' agricultural lands are also present; they are made of those pieces of land owned by households throughout the region of Kivu (North and South Kivu and Maniema) for diverse livelihood activities, including farming for food and commerce. In principles, this should be the remaining lands after subtracting all other exclusive functional land uses: protected areas, industrial agriculture's fields, logging concessions, ranches for cattle and mining logging concessions. With this in minds, one can already see that densities discussed in the next section of this chapter are far from being reflective of the reality on ground.

5 The Land Availability in the Mountains of Kivu and Human Densities

Adjusted for steep slopes, of the total of 256,803-km^2 (see Table 2) the usable land accessible to human populations is 166,922-km.^{2} The total population of the Kivu as estimated by the National Institute of Statistics (2015) is 14,760,000, which lead to a mean crude density of about 57.5 individuals/km^2, within the range of 18.0 individuals/km^2 (Maniema) and 112.0 individuals/km^2 (North Kivu). To fit the readily available land, meaning excluding slopes higher 35% (total territory = 166,922-km^2), human densities accrue to 88.4 individuals/km^2 within a range of 27.0 individuals/km^2 (Maniema) and 172.0 individuals/km^2 (North Kivu). The above relatively lower figures veil a much natural yet complex reality, which is that some areas are more abundantly populated than others. Hence, Bukavu, Goma and Kindu, are respectively homes to 17.5% (1,012,053 individuals), 16.5% (1,101,306 individuals) and 19.4% (453,941 individuals) of the total populations of the provinces they headquarter. Converted in terms of human densities, using respective surfaces of 60-km^2, 75.72-km^2 and 150-km^2, Bukavu, Goma and Kindu have the staggering densities of 16,868 individuals/km^2, 14,544 individuals/km^2 and 3026 individuals/km^2.

The variability is rendered even much more rugged when administrative territories are considered. In North Kivu the highest densities occur in Nyiragongo (437.7 individuals/km^2), Masisi (152.8 individuals/km^2) and Beni (122.4 individuals/km^2). In South Kivu, the highest densities happen in Idjwi (889.7 individuals/km^2), Walungu (400.4 individuals/km^2), Kabare (398.3 individuals/km^2), Kalehe (159.1 individuals/km^2) and Uvira (120.3 individuals/km^2). Human densities remain relatively lower in Maniema but variability also predominates; with the highest densities being found in Pangi (37.7 individuals/km^2), Lubutu (35.8 individuals/km^2), Kasongo (32.2 individuals/km^2) and Kabambare (28.7 individuals/km^2).

Staggering densities of 16,868 individuals/km^2, 14,544 individuals/km^2 and 3,026 individuals/km^2, respectively for Bukavu, Goma and Kindu should not be surprising, particularly if they are viewed from historical perspectives. Indeed, Wils et al., found that already 1970 Bukavu and Goma, with respectively 2,255 and 249 individuals/km^2, had already attained unsustainable densities. Furthermore, the high human densities generated in the above calculations do not factor respective areas of these provinces that are already inside protected areas (see Table 1 and subsequent discussion) that are, by law, prohibited of being accessed by communities. Factoring that single fact would still decrease land readily available to communities and, ipso facto, increase human densities.

6 Human Densities and Felt Levels of Insecurity

Figure 1 conveys several messages about using human density to predict the insecurity in the provinces of Kivu. The first of these messages shows that, none of the annually felt insecurity levels can be solely and significantly explained solely by accessibility to land. However, nearly 80% (R^2 = 79.19%) of all the cases in South Kivu, Maniema and North Kivu felt insecurity levels are correlated with human densities in these provinces. A second interesting feature of the Fig. 1 is that Beni (North Kivu), where the felt unrest intensity had been the highest for the entire 2017 and 2018, has lower human density than other territories that offered a

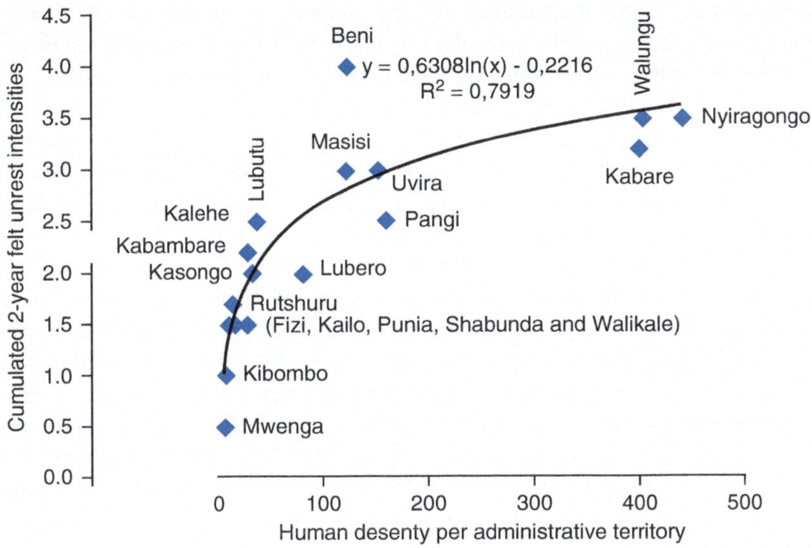

Fig. 1 Logarithmic regressions of densities plotted against felt unrest intensities

relatively safer environment. This can also be said of Uvira (South Kivu) and Kabambare (Maniema). This finding, obviously, suggests that the felt unrest, despite being in greater percentages explainable by human densities, has other reasons, which are briefly discussed in the sections to come beneath.

However, the situation of human densities continued to worsen over the years with steadily increasing birth rates in the Kivu, which are higher even when compared to the overall means of the country. The upward population spirals in these provinces was also positively fed the loop of rural-urban migrations caused by multiple factors. These factors included, not only traditional reasons such the need and search for better public services (schools, health facilities, clean water, etc.), but also newly emerging issues such as security and personal safety and lack of opportunities to sustain a decent livelihood. Decent livelihood in rural areas across Africa, it must be admitted, is primarily a function of access to fertile land. Fertility of land in the region is discussed in the section below.

7 Soil Fertility and Current Levels of Felt Insecurity

It is common to hear that the soils of Kivu are fertile because of their proximity to volcanoes. This generalized statement does not factor the diversity of environmental conditions in a 256,803-km^2 area. Stating the obvious, generalizations are yet understandable given the sketchiness of data available on the soil of the region.

Even Baert et al.,[18] who indicated that a large amount of soil data was available for the country recognized that the same data set was essentially scattered throughout the entire DRC. That this is the case is understandable because of the geographic extent of the country, which is a subcontinent of its own right.

Despite being sketchy, a general overview of the soils of North and South Kivu and Maniema can be found in the work of Sys.[19] According to Sys, the most important soil types of the region of Kivu (North and South Kivu and Maniema) consist of crude mineral soils on lava, Kivu volcanic ashes, Kivu basalts, allophane-type clays (non-hydromorphous soils), brown allophane soils on volcanic ashes, humic ferrisols on basic rocks, xeroferrisols, tropical black clays associated with recent soils and solonetzic on alluvial deposits. Crudely divided and summarizing Sys, it can be argued that the current three provinces of the former Kivu mirror three rough soil divisions even though provincial boundaries does not overlap with soil divisions. The first zone, the area of North Kivu, is constitutes of crude mineral soils on lava, Kivu volcanic ashes, Kivu basalts, allophane-type clays (non-hydromorphous soils), brown allophane soils on volcanic ashes and tropical black clays associated with recent soils and solonetzic on alluvial deposits. The allophane brown soils on volcanic ashes occupy vast expanses in the region of Masisi and Rutshuru, they are characterized by the absence of kaolinite in the clay. Tropical black clays associated with recent soils and solonetzic on alluvial deposits stretch from the northern part of Rutshuru throughout the entire territory of Beni. The allophane present in the soils of Kivu has a high exchange capacity and it is found in the forest at 1600-m above the sea level. The second soil division is that of a large part of South Kivu, which is generally built on humic ferrisols on basic rocks present on basalt flows; this is particularly the cases of regions located west and south-west of Bukavu situated at 2500-m above the sea level. This area consists mostly of the mountain bamboo forests. The third rough soil division comprises any portion of Kivu (North and South) which is then located within the ranges of 1500-m and 850 m of altitude and Maniema. This is an area where ferrisols on undefined rocks and ferrisols on Karroo rocks are the most prevalent. Ferrisols on undefined rocks are dominant in Walikale while ferrisols on Karroo rocks are found in extensive quantities in the valley of the middle Lomami in Maniema. This is the zone of the humid dense forest.

One important detail to notice is that all the three soil divisions receive an annual rainfall of about 2000 mm. Therefore, differences in rainfall regimes should not cause differences in fertility across the region. According to Sherman,[20] soils developed on recent volcanic ash depositions and the alluvial soils represent some of the most productive areas of the world. An illustration of this type of soil is the allophane brown soils on volcanic ashes found in Masisi and Rutshuru, and tropical black clays associated with alluvial deposits that happen in parts of Rutshuru and the

[18]Baert et al. (2012).

[19]Sys (1960).

[20]Sherman (1971).

entirety of Beni and Nyiragongo. Sherman attributed to the high potential fertility of the soil types to the weathering conditions, which are in the early stage and under the full expression of the influence of the factors of chemical weathering, thus facilitating the rapid decomposition of primary minerals and the release of essential elements. This high soil fertility should not be extrapolated for the entire Kivu region, but it can plausibly count as one explanation as to why areas such as Beni, Masisi, and Uvira (see Fig. 1) are more prone to violence. Indeed, despite holding lower human densities, Beni, Masisi and Uvira where this type of soil is prevalent harbor most of the fertile land where pastoralists have erected fences for herds of cows they raise. Large expanses of Masisi have now been converted into ranches; the ubiquitous presence of ranches excludes local communities from accessing fertile lands where they can grow staple species to support their livelihoods.[21] In fact, if these ranches were included in the density calculations, areas being shared by all would shrink down, which would have quadrupled human densities. Arguing this way lends partial support to the point raised by Bucyalimwe,[22] when he suggested that cattle raising and struggle for power caused the land tenure issue in North Kivu.

Humic ferrisols on basic rocks and ferrisols on undefined rocks prevail in administrative territories of Kabare (partly), Kalehe (in most part), Walungu (totally), Walikale, Fizi (partly), and Pangi. According to studies (e.g. Sys), humic ferrisols on basic rocks are good agricultural soils while ferrisols on undefined rocks offer excellent terrain for agriculture, specifically for growing cocoa and coffee. The presence of these soil types may explain why administrative territories such as Kalehe, Fizi Pangi and Walikale are also ranked high in the feeling of the intensities of the civil unrest (see Fig. 1). That humic ferrisols on basic rocks are very sensitive to erosion, increases the fragility of the situation as eroded soil is lost to cultivation and contributes to increased human density. Obviously, Kabare and Walungu being humanly packed and harboring fertile soil have both ingredients of the felt social and political unrest, which occur in significant magnitude (see Fig. 1).

8 Beyond Figures and Beyond North and South Kivu and Maniema

The last section is not the academically conventional discussion of a study. The intentions is to provide what the figures given above in Tables 1 and 2, Fig. 1 and throughout the sections may be telling if brought together. As one, obviously, would have to admit, the 2-decade conflict in Eastern DRC, particularly in the North, South Kivu and Maniema is a multi-facet one and has been caused by a cocktail of different

[21] Mathieu and Tsongo (1998).

[22] Bucyalimwe (2001).

reasons. This mixture includes, among others, political reasons such as identities and nationality,[23] and power and regional hegemonic struggle,[24] etc. There are also other reasons such as structural organizational reasons such as simple failure of the state,[25,26] as well as ethnic reasons and the scramble over mining and other resources, etc. It is also obvious that ethnic causes of the crisis include or rather, are exacerbated by conflicts over land, as it has been indicated by other authors[27,28,29,30]. These studies focuse on land tenure policies in the country, human densities and hint on other causes considered in this study. The contribution, if anything, of this study is in the fact that it demonstrates that densities alone, are not the only cause in relationship to land rights; human densities need to be placed in the context of other land usages, which compress communities in very limited marginal lands. Indeed, even though nearly 80% of felt insecurity is correlated with higher human densities, the cases of Beni, Masisi and Uvira (see Fig. 1) show that the zones with the highest felt unrest are not necessarily those with the highest human densities. This single fact calls for other explanations. A possible plausible explanation is soil fertility in these areas. Soil fertility acts as a magnet for large industrial agricultural concessions and cattle growing ranches that are present in those zones and further decrease the physical space available to small household stakeholders. That the soil fertility is proposed as one of the most important causes of the conflicts in Kivu is not that new; Kujirakwinja et al.[31] had also suggested that one of the reasons why local communities interfered with the limits of the Virunga National Park was because they were in search of more fertile lands to cultivate on, the land outside of the park having been impoverished after years of exploitation.

Some studies (e.g. Pèlerin), have emphasized the prevalent presence of protected areas in the Kivu as one other possible explanation of conflict over land. Indeed, a number of protected areas given above may provide and do provide support this claim. Also, it is important to note the fact that, most of these protected areas, particularly forest and game reserves were found to be infringed by local communities in search of cultivatable lands. According to Kejuni et al. and Wabubyula et al., of the long list of game reserves identified in both North and South Kivu, 100% were permanently squatted to the point that they would no longer deserve the categorization of protected areas. That said, however, the conflicts between protected areas and local communities should not be over-emphasized to the point of eclipsing other forms of conflicts over lands. In fact, the recent evolution in North Kivu has

[23]Malengana (2001).

[24]Young (2003).

[25]Lukusa (1999).

[26]Renton et al. (2007).

[27]Cox (2012).

[28]Mathieu et al. (1997).

[29]Pèlerin (2010).

[30]Vlassenroot (2004).

[31]Kujirakwinja et al. (2006).

demonstrated that communities have come to think of parks as part of their psychological landscape. At least some communities have been demonstrating against the idea of de-classifying some parts of the Virunga national park to drill for oil. This is not negating the existence of conflict between local communities and protected areas in Kivu, but rather contextualizing that conflict with the perspectives of other aspects of the issues around lands.

Finally, one thing merits to be noted: once more, densities discussed in the preceding section, are not reflective of the reality on ground; they should be used as an indication of how land is rare in Kivu. These densities are not inclusive of populations that moved into the region between 1994 and 1996 when the civil war broke out in Rwanda and the continuous in-migrations from Burundi and, in a lesser extent, South Sudan. There have been also internal movements of people due to frequent crises' that have moved some significant numbers of Congolese from Northern Katanga (now known as the province of Tanganyika) into South Kivu. Obviously, if these influxes of people were included in the analysis, available space and fertile soil would have decreased even more. Placed in the context of the region, higher human densities, imply, as it has been modeled by Wils et al. decreased soil fertility, erosion and human conflicts.

Stating that if refugees and internally displaced people were included in the analysis, human numbers over available lands would significantly increase, brings us to what should have been done to release the pressure over the lands in Kivu. Discussions, whether serious or not, about 'encouraged internal west-ward movements of populations' as a palliative solution to the cycles of crises in Eastern DRC, are not the solution for long-term sustainable peace. This is so because, as Inogwabini (2014) wrote some years back, seemingly 'empty' lands are not vacant lands. Even with the state proclaimed ownership over all the lands, the *de facto* powers over lands are held by traditional authority, *i.e.* by communities. With this in minds, to consider a peace framework whereby people should be encouraged to occupy the land laying west of mountains of Kivu would be simply displacing conflict from one location to another. Of course, as all citizens of DRC, each person in Kivu has the right to establish where it pleases everyone within the national territory. However, this should not be an encouraged mass movement instead of individual decisions. This caution is worth when one considers the fact that beyond the provinces of North and South Kivu and Maniema, land ownership has led to multiple lethal conflicts in other provinces of the country as well. Recent cases to sustain this affirmation include cases in the current provinces of Maindombe, North Ubangi and Tanganyika. To state the evident, the mid-term solution in conflicts over land in DRC, whether in Eastern DRC or anywhere else in the country, is to build a stronger state that strives for an inclusive, open and democratic society. It is this type of society that will ensure transparent, simplified and secured land tenure that avoids overlaps between conflicting functions over one piece of land. It is also that type of society that will ensure that communities participate in the land use planning processes whereby they agree on different functional land unite; which in the end will pre-empt potential future conflagrations. Of course, a stronger state in DRC today also means being able to work with the country's neighbors toward a collective regional peace that should be cherished more than any benefit stemming out of the orchestrated chaos. Neighboring countries that cherish peace more than any other booty of the chaos will, together, stop floods of cross-border refugees that increase pressure over land in eastern DRC. Over the long term, it will be a stronger inclusive, open and democratic DRC that will have to curve down its demography, which is increasing in an alarming pace. Increased demography and impoverished soil will fuel still more conflicts in the long run future, even if proper land use planning is achieved and even if the neighborhood is pacified.

Finally, to ensure that the fertility and that the health of the soils of Kivu are protected over the future, a proper land use planning process needs to engage all different stakeholders, to agree on what function to attribute to which land piece. Equally, agricultural practices that ensure higher productivity without jeopardizing soil attributes need to be brought in and applied by communities. Both will stabilize people in the region and will ultimately be likely to contribute to the easement of decades-long social tensions across the region. Of course, these would be of positive impacts only, if the current demography of the Kivu is kept under control, which is a task on which the national government should plan and lead.

References

Baert G, Van Ranst E, Ngongo M, Verdoodt A (2012) Soil survey in DR Congo – from 1935 until today. Paper presented at the meeting of the Section of Natural and Medical Sciences held on 27 March 2012 by the University of Ghent, Belgium

Bucyalimwe MS (2001) Pouvoirs, élevage de bovin et la question foncière au Nord-Kivu. In: Marysse S, Reyntjens F (eds) Annuaire de l'Afrique des Grands 2000-2001. L'Harmattan, Paris, pp 219–250

Cox TP (2012) Farming the battlefield: the meanings of war, cattle and soil in South Kivu, Democratic Republic of the Congo. Disasters 36(2):233–248

EIC [Etat Indépendant du Congo] (1886) Département des finances: Régime Foncier. Bulletin Officiel de L'État Indépendant du Congo. 2e Année, (4): 53–55

Ernst C, Verhegghen A, Mayaux P, Hansen M, Defourny P (2010) Central African forest cover and cover change mapping. In: De Wasseige C (ed) The Forest of the Congo Basin: state of the forest 2010. Publications Office of the European Union, Brussels, pp 23–41

Inogwabini BI (2014) Conserving biodiversity in the Democratic Republic of Congo: a brief history, current trends and insights for the future. Parks 20(2):95–104

Inogwabini BI, Omari I, Mbayma AG (2005) Protected areas of the Democratic Republic of Congo. Conserv Biol 19(1):15–22

IUCN (International Union for the Conservation of Nature) (1992) Protected areas of the World, vol 3. IUCN Publishing Unit, Gland

Jeal T (2008) Stanley: the impossible life of Africa's greatest explorer. Yale University Press, New Haven

Kejuni K, Katoyishi K, Zibuka KA, Vavi D, Makwala L, Musapa (1998) Monographie de la Province du Nord Kivu. Programme National de Relance du Secteur Agricole et Rural (PNSAR) : 1997 – 2001. PNUD/UNOPS. Ministères de l'Agriculture et de l'Elevage, du Plan, de l'Education Nationale et de l'Environnement, Conservation de la Nature, Forets et Pêche de la République Démocratique du Congo

Khete MM, Katoyishi K, Lukusa M, Mieme U (1998) Monographie de la Province du Maniema. Programme National de Relance du Secteur Agricole et Rural (PNSAR) : 1997 – 2001. PNUD/UNOPS. Ministères de l'Agriculture et de l'Elevage, du Plan, de l'Education Nationale et de l'Environnement, Conservation de la Nature, Forets et Pêche de la République Démocratique du Congo

Kujirakwinja D, Bashonga G, Plumptre A (2006) Etude Socio-économique des populations environnant le secteur Nord du Parc National des Virunga. Rapport soumis à l'Institut Congolais pour la Conservation de la Nature, Wildlife Conservation Society, World Wide Fund et l'Union Européenne

Lukusa MG (1999) Congo-Zaïre: la faillite d'un pays. Déséquilibre macro-économique et ajustements. Institut Africain (CEDAF)/L'Harmattan, Bruxelles/Paris

Malengana NNC (2001) Nationalité et citoyenneté au Congo Kinshasa: cas du Kivu. Collection Etudes Africaines. L'Harmattan, Paris

Mathieu P, Tsongo AM (1998) Guerres paysannes au Nord-Kivu (République démocratique du Congo), 1937–1994. Cahiers d'études africaines (38): 150

Mathieu P, Laurent PJ, Mafikiri T, Mugangu SM (1997) Compétition foncière, confusion politique et violences au Kivu: des dérives irréversibles? Institut d'études du développement. Université catholique de Louvain, Louvain-la-Neuve

Pèlerin E (2010) Etude sur la problématique foncière au Nord-Kivu, RDC. Comité catholique contre la Faim et pour le Développement (CCFD), Forum des Amis de la Terre (FAT), Groupe de recherche et d'échanges technologiques (GRET) et Pôle Alimentation et agriculture durable

Population data.Net (2019) République démocratique du Congo. Available at: https://www.populationdata.net/pays/republique-democratique-du-congo/. Accessed on 19 Mar 2019

Programme des nations unies pour le développement (UNDP) (2009a) Province du Maniema, profil résume: Pauvreté et conditions de vie des ménages. Unité de lutte contre la pauvreté, PNUD RDC

Programme des nations unies pour le développement (UNDP) (2009b) Province du Nord Kivu, profil résume: Pauvreté et conditions de vie des ménages. Unité de lutte contre la pauvreté, PNUD RDC

Programme des nations unies pour le développement (UNDP) (2009c) Province du Sud Kivu, profil résume: Pauvreté et conditions de vie des ménages. Unité de lutte contre la pauvreté, PNUD RDC

Renton D, Seddon D, Zeilig L (2007) The Congo plunder and resistance. Zed Book

Sherman GD (1971) Mineral weathering in relation to utilization of soils. In: UNESCO (ed) Soils and tropical weathering. Proceedings of the Bandung Symposium 16 to 23 November 1969, pp 51–57

Sys C (1960) Carte des sols et de la végétation du Congo Belge et du Ruanda-Urundi: Notice explicative de la carte des sols du Congo Belge et du Ruanda-Urundi. Publications de l'Institut National pour l'Etude Agronomique du Congo Belge (I.N.E.A.C.)

Vansina J (1990) Paths in the rainforest: toward a history of political tradition in Equatorial Africa. Wisconsin University Press, Madison

Vansina J (2004) How societies are born: governance in west Central Africa before 1600. University of Virginia Press, Charlottesville

Vlassenroot K (2004) Households land use strategies in a protracted crisis context: land tenure, conflict and food security in eastern DRC. Conflict Research Group, Ghent, Academia Press

Wabubyula M, Abiza NI, Kethe MM, Mingedi M, Chiromod A, Lofonga I, Wabukangama MG (1998) Monographie de la Province du Sud Kivu. Programme National de Relance du Secteur Agricole et Rural (PNSAR) : 1997 – 2001. PNUD/UNOPS. Ministères de l'Agriculture et de l'Elevage, du Plan, de l'Education Nationale et de l'Environnement, Conservation de la Nature, Forets et Pêche de la République Démocratique du Congo

Walemba NV, Mungu SP (2016) Rapport d'étude: Accès à la terre par les communautés pauvres et marginalisées de la Province du Maniema, RDC. Produit pour l'Institut Supérieur de Développement Rural de Kindu, UWAKI Maniema et Christian Aid (RDC)

Wils WM, Carael MM, Tondeur MG (1986) Le Kivu Montagneux. Surpopulation–sous nutrition–érosion du sol: Etude Prospective par Simulations Mathématiques. Université Libre de Bruxelles et Bukavu, CEMUBAC/IRS, Bruxelles

Young C (2003) Contextualizing Congo conflicts: order and disorder in postcolonial Africa. In: Clark JF (ed) The African stakes of the Congo war. Fountain Edition, Kampala, pp 13–32

Workshop "Implementing Land Degradation Neutrality in Africa: Means, Legal Instruments and Institutional Challenges?": Outcome Document

Harald Ginzky, Oliver Ruppel, Robert Kibugi, and Walter Engelberg

The international workshop "Implementing Land Degradation Neutrality (LDN) in Africa: Means, legal instruments and institutional challenges?"—Organized in cooperation between the German Environment Agency (UBA), the Konrad-Adenauer Stiftung—Climate Policy and Energy Security Program for Sub-Saharan Africa—, the GIZ and University of Nairobi—was held on 14/15 August 2018 at the Hotel Windsor in Nairobi, Kenya. The workshop was attended by about 40 experts from African countries and abroad. Representatives of the African Soil Partnership, of the New Partnership for Africá's Development (NEPAD) and the African Union as well as of International Union for the Conservation of Nature (IUCN), the United Nations Environment Programme (UNEP) and the United Nations Convention to Combat Desertification (UNCCD) attended the workshop.

The workshop was officially opened by Prof. Kiaie Mwarua, Dean of the Law Faculty of the University of Nairobi. He underlined the timeliness of the workshop due to increased illegitimate land acquisition practices and the interfaces of land degradation with climate change and food security.

H. Ginzky (✉)
German Environment Agency, Dessau, Germany
e-mail: harald.ginzky@uba.de

O. Ruppel
Development and Rule of Law Programme [DROP], University of Stellenbosch, Stellenbosch, South Africa
e-mail: ruppel@sun.ac.za

R. Kibugi
School of Law, University of Nairobi, Nairobi, Kenya
e-mail: rmkibugi@uonbi.ac.ke

W. Engelberg
Gesellschaft für internationale Zusammenarbeit, Bonn, Germany
e-mail: walter.engelberg@giz.de

H. Ginzky et al. (eds.), *International Yearbook of Soil Law and Policy 2019*,
International Yearbook of Soil Law and Policy 2019,
https://doi.org/10.1007/978-3-030-52317-6_8

During the opening podium discussion keynotes were provided by:

- Hon. Jude Njomo, Pan African Parliament on behalf of Hon. Senator Janet Ongera, Pan African Parliament
- Prof. Patricia Kameri-Mbote, Professor of Law, University of Nairobi
- Dr. Francis Tetteh, African Soil Partnership
- Alexander Juras, UNEP
- Mamadou Diakhite, NEPAD African Union
- Hon. Justice Antony Ombwayo, Environment & Land Court, Kenya

In the keynotes the following aspects were addressed and emphasised:

- Sustainable soil management is a cross-cutting issue and has interfaces with i.e. environment conservation in general, climate change, property ownership, agriculture, chlorophyll influence, public health-pests, wastes and extractives.
- Market demands and trends could greatly impact the healthy status of soils. There is a particular need to promote technological innovation based on the assessment of those demands and trends.
- Access to land and land property rights need to be fully understood for the establishment of an effective regime on sustainable soil management.
- A stronger involvement of women is required in the planning, decision making and monitoring of land use activities such as agriculture.

The overall objective of the workshop was to discuss the current state of affairs and how existing deficiencies and shortcomings in terms of institutional and organisational aspects, knowledge and science as well as capacity for an effective soil protection policy and legislation in Africa could be overcome. The workshop addressed several dimensions in light of this question. The major outcomes are briefly summarised below:

1 Sustainable Development Goals: General Aspects

1. Sustainable soil management is a precondition for sustainable development in Africa and worldwide as the degradation of soils is a cause for hunger, migration and even wars. Twenty-three percent of the terrestrial surface is already affected by degradation which actually concerns about 1.5 billion people.
2. The world community has committed itself to the 2030 sustainable development (SDG) agenda. Target 15.3 obliges states to strive to achieve a land degradation neutral world by 2030. As soil is an elementary part of land, this objective currently is the most important anchor in international law for sustainable management of soils.
3. UNCCD has established itself as the international lead organisation for the implementation of the LDN objective. Projects on LDN voluntary target setting have been launched in more than 110 countries.

4. UNCCD has furthermore worked out the "Scientific Conceptual Framework for Land Degradation Neutrality", which should provide a conceptualised model approach on how to best implement the LDN objective at national or regional level.
   https://www.unccd.int/sites/default/files/documents/18102016_Spi_pb_multipage_ENG_1.pdf
5. The following principles should, inter alia, govern application of the framework, and thus prevent unintended outcomes during implementation of LDN:
   - Maintain or enhance land-based natural capital. Neutrality is thus only the minimum objective.
   - Protect the rights of land users.
   - Integrate planning and implementation of LDN into existing land use planning processes and apply the response hierarchy: Avoid degradation > Reduce degradation >Reverse degraded land.
   - Establish an effective participatory process for all relevant stakeholders.

2 Factual Challenges on Soils in Africa

6. About 65% of arable lands, 30% of grazing lands and 20% of forests are already degraded in Africa.
7. Land, or soil, is the main resource base for many people in Sub-Saharan Africa, especially the rural population.
8. With an estimated population growth for SSA from the current 900 million to 1.4 billion by 2030, the region's soils are expected to experience increasing pressure as a natural resource to provide for the needs of its people.
9. As the main challenges for the African continent and the soil protection were identified: demographic growth, deforestation, urbanization, weak land/soil governance and governmental coordination and interaction between the different sectors resulting in non-effective soil management, lack of awareness of importance of soil, lack of institutional capacities and inefficient enforcement.
10. The aforementioned challenges have in particular negative impacts on migration of population, mainly women and youth, on orientation on exports crops and monocrops.
11. It has to be taken into account that the level of land degradation significantly varies between the African countries and regions.
12. For small scale farmer reduction in soil loss or long-term environmental degradation are not tangible inducements to adopt sustainable soil management practices when their immediate concern is simply feeding their families.
13. Adoption of sustainable soil management technologies by smallholder farmers is conditioned by risk, thus their commitment must ensure both short-term and long-term economic benefits.

14. Sub-Saharan Africa nevertheless has the potential to position itself as champion in terms of increasing food production and security, achieving land restoration, and increasing agricultural resilience to climate change.

3 Soil Protection Regulatory Concepts and Challenges in Africa

15. The outcomes of the four presentations held during the workshop are subsequently documented. The presentations concerned:
    - Kenyan water and soil law;
    - Land cover policies in Cameroon;
    - Regulation on persistent organic pollutants in Kenya;
    - German concepts on LDN implementation.
16. The Kenyan Constitution of 2010 promotes sustainable and productive management of natural resources.
17. In addition, Kenya has established a specific environment and land court (ELC).
18. There is no specific constitutional, policy or legal tool that deals with soil protection in Kenya. It is embedded within the environmental, agricultural and water statues including policies and legislation. There is however a constitutional obligation on the government to take action to eliminate harmful environmental practices.
19. Although land in Kenya belongs to people collectively as a nation, communities or individuals, the Land Registration Act confers absolute proprietorship with all rights and privileges on the land owner. This has created the perception that the land owner is not obligated to abide by certain legal requirement concerning the use of his land. However, the police power of the Kenyan state to regulate land use is founded in the constitution, and this could, with modifications in legislation, be applied directly to govern soil health. The 2009 National Land Policy has clarified that the dual existence of the freehold and the absolute proprietorship causes confusion.
20. Soil as natural resource is thus not explicitly and sufficiently integrated in Kenyan legislation.
21. Land cover policy in Cameroon could set incentives for sustainable soil management. It needs to be supported by economic instruments and tools in order to raise awareness for the importance of the maintenance of fertile soils.
22. The use of persistent organic pollutants (POPs) could have detrimental effects on soils. Sources of emissions of POPs are very diverse requiring to regulate the whole use chain from the production, the use of POPs in industrial facilities and in agricultural activities and finally the waste management.
23. The Rotterdam and the Stockholm Convention set appropriate international standards for the production, use, management and trade on POPs. Some open

questions remain. Kenya has ratified the Stockholm Convention in 2004 and sent its first National Implementation Plan 2007.

24. Kenyan law is not prepared to thoroughly deal with these challenges in order to avoid particularly negative effects on soils by POPs.
25. A better coordination in Kenya is needed in order to effectively deal with the potentially negative effects of POPs.
26. The analysis of German environmental law shows that there are not sufficient legal instruments in place to implement thoroughly the LDN objective. However, there are some positive regulatory concepts which may potentially be used by African legislation too.
27. First, the so-called intervention clause of the German Nature conservation act requires that negative effects on nature by human activities needs to be offset. This has to be implemented on project level. This would fit with the "compensation idea" of the LDN objective.
28. Second, according to the EU industrial emissions directive, an operator of an industrial facility is obliged to re-establish the initial status of soil and groundwater under and close to his slot in case of significant detrimental effects during the operation of the facility. By this obligation the law establishes a strong economic incentive to avoid such negative effects (=land degradation).
29. German environmental law contains several planning instruments including spatial planning, urban planning as well as planning instruments for nature conservation, water management and what is more. A specific planning instrument to implement the neutrality concept on land degradation is yet to be established.
30. In addition, a banking concept of already finalised restoration measures in order to compensate future degrading activities such as soil sealing due to habitation and infrastructure has been proposed.

4 Deficiencies with Regard to an Enabling Environment for More Effective Soil Protection Regimes in Africa

31. The workshop focussed on the question which deficiencies and shortcomings exist in Africa with regard to establish an enabling environment for more effective soil protection regimes in Africa and how these could be overcome.
32. Several aspects were differentiated and intensively discussed:

    - Legislation;
    - Institutional and organizational arrangements, capacity;
    - Awareness raising;
    - Funds/resources;
    - Science-Policy Interface;
    - Traditional knowledge;
    - Data availability.

4.1 Legislation

33. Generally, there is no dedicated, comprehensive and coherent soil policy in place, which could inform law. No comprehensive soil protection law has been issued. Soil relevant provisions are very much fragmented.
34. The land ownership concept in most African states poses a significant barrier to sustainable soil management as land ownership is perceived as an absolute use (and misuse) right on the respective land slot. Soil related provisions are thus not effectively implemented and enforced.
35. Gender inequality in land right issue needs to further be addressed to prepare for an enabling environment.
36. Overarching principles for land and soil protection legislation such as prevention, precaution, obligation to rehabilitee and planning approaches need to be worked out and agreed upon.

4.2 Institutional and Organisational Arrangements, Capacity

37. Competent institutions are in a relatively weak position, in particular with regard to implementation and enforcement.
38. The competences are often shared by several ministerial entities. However, the communication and cooperation with regard to legislation, implementation and enforcement between these entities needs to be developed further.
39. Usually these entities face a substantial under-staffing which hinders the necessary implementation and enforcement.
40. The division of labour between central and local entities needs to be clarified in order to avoid duplication of work and the challenge of "no one being in charge".
41. Public participation and access to environmental information is not sufficiently developed.

4.3 Funds/Resources

42. Capacity building is required on all levels: government level including legislators, policy makers, executive enforcement and judiciary as well as civil society organizations and right holders.
43. Drafting, implementation and enforcement of future soil protection legislation require personal resources. These resources must be in place to be successful.
44. It could be particularly effective to train the trainers in order to increase the knowledge basis. Continuous measures on capacity building are required. It is not a one-off exercise.

45. Moreover, platforms which bring together all levels, all perspectives and stakes could be an appropriate format, in particular if run in a participatory and interdisciplinary manner.
46. Materials to be disseminated should be adequate for the audience to be addressed taking into account i.e. language, culture and context. Information on ecological background and "economics of land degradation" are of crucial importance.
47. Judges and other court officials require training and tailored workshops or initiatives to help develop the requisite capacity because of the wide jurisdiction of the ELC.

4.4 Awareness Raising

48. The effect and economic cost of land degradation is underestimated at all level of government mainly because it is a slow steady process and its effects acknowledged only at the point of where rehabilitation costs are high.
49. Adequate information on sustainable soil management are of eminent importance with regard to education in schools and universities, preferably integrated in the respective curricula.
50. Information on status of soils and appropriate measures to achieve sustainable soil management are key.
51. Policy makers need to be involved from the beginning. The overall benefit of sustainable soil management for the society as a whole must be discussed and determined from the beginning, involving the relevant policy makers.

4.5 Science-Policy Interface

52. The cooperation of science and competent authorities needs to be developed further. An institutional form of uptake of new scientific knowledge needs to be established.
53. Scientific information on sustainable soil management, in particular with regard to agricultural measures, is usually not made available for farmers, in particular small-scale farmers.

4.6 Traditional Knowledge

54. Traditional knowledge is often neglected and not appropriately valued. Traditional knowledge opportunities are not factored into soil protection legislation.
55. A conceptual approach how to collect, evaluate, synthesize and make publicly available traditional knowledge is yet to be established. Research institutions should be involved in such an endeavour.

4.7 Data Availability

56. Soil information at national level is often inadequate, outdated, not in digital format and not georeferenced.
57. Data availability is further restricted by intellectual property often held by private institutions that are not willing to share data for national use, or data needed to be paid for prior to use.

5 Assistance by Regional: African—Institutions, Including UNEP

58. In the following section concepts are discussed how regional bodies such as the African Union or UNEP could assist African states to promote an "enabling environment".
59. It was recommended to develop an assessment toolkit which would enable African states to evaluate whether and how international obligations and guidelines are transposed into national law and whether and to which extent the national law provides for an effective regulation on soil protection, sustainable management of soils and the implementation of the LDN objective.
60. UNEP has worked the report "Towards a pollution-free planet". UNEA 3 also dealt with this theme in 2017 and approved a resolution on "mapping of soil pollution". Insufficient dumpsites, misuse of pesticides and fertilizers as well as inadequate industrial installation may be sources of soil/land pollution.
61. According to the UNEP resolution of 2017, Member states are invited to develop and implement appropriate policies, stakeholder should be in particular involved in gathering and delivering the required information. UNEP should draft a report on soil pollution, work out a technical guidance how to deal with soil pollution and should assist research, cooperation and awareness raising.
62. The New Partnership for Africa's Development (NEPAD) is running several projects whose overarching objective is to maintain and increase the status of natural resources in Africa in quantitative and qualitative terms. The following programmes and initiatives could be mentioned in this regard: The comprehensive Africa Agricultural Development Programme (CAADP), TerrAfrica, African Forest Landscape Restoration Initiative (AFR100) and Great Green Wall for the Sahara and Sahel Initiative (GGWSSI).

6 Land Degradation and NDC by African States

63. Climate change and soils are strongly interlinked. Climate change will negatively affect the quality and the availability of soils. At the same time, soils are a major sequester of carbon.
64. A significant turn to—permanent—sustainable management of soils could potentially allow for an important uptake of carbon—both to mitigate climate change and to improve soil quality.
65. Improved soil quality and sustainable management of soils could be an important vehicle for ensuring food security on the African continent. It might additionally offer the chance to increase significantly the export of food products thus raising additional income.
66. UNFCCC only tackles land use issues marginally. Although the Paris Agreement does not expressively mention "land" or agriculture" there seems to be a stronger focus on land use as a potential source and sink of CO2 emissions.
67. The Nationally Determined Contributions (NDC)—requested by the Paris Agreement—are the tool to document the current national activities on climate change. It is recommended to involve the land use sector in order to achieve two or even three objectives by one means: to mitigate climate change, to maintain and protect fertile soils and to ensure food security in African states.
68. When drafting respective national climate change legislation several aspects concerning existing deficiencies (as mentioned above) might be addressed in a "soil friendly" manner.

7 Outlook and Next Steps

69. Participants felt that the workshop was crucially important, and work should be continued.
70. A need was seen to reach out to policy makers and gain their support for better legislation and for an enabling environment for sustainable soil management.
71. A coherent soil policy and law was seen as crucial. Soil as natural resource should be put at the heart of such a policy and legislation.
72. In addition, a change of the overall narrative is required in order to explain the costs of land degradation and to highlight the economic benefits of sustainable soil management.
73. Soil policy and law must be sector specific. The interests of small-scale farmers must be particularly taken into account.
74. An integrated approach on such legislation involving all expertise, all perspective and stakes is of paramount importance.
75. LDN as an internationally agreed objective should be integrated in the legal networks using the approved terminologies, nomenclature and methodologies.

76. In order to support African societies to strengthen sustainable soil management, the development of a model legislation on sustainable soil management could be an important step forward.
77. Awareness raising is key. Appropriate curricula in universities are of great importance in this respect.
78. Scientific information must be the basis of sustainable soil management. An appropriate organizational structure of research and science as well as an elaborated networking amongst African research institutions are key.

Part II
Recent International Developments

The Future of International Soil Governance

Ralph Bodle, Heidi Stockhaus, Franziska Wolff, and Sebastian Oberthür

1 Introduction

This chapter explores whether and how international governance for the purpose of sustainable soil management can be improved in the short, medium and long term.[1] First, we take stock of and assess existing international instruments,[2] institutions and steering mechanisms that are relevant for soil protection and its governance at the international level, including an evaluation of their actual and potential steering effect. From this assessment we develop options for improving international soil governance.

Our analysis is based on a broad understanding of international governance which includes binding as well as non-binding instruments and steering mechanisms. Binding international law comprises international treaties, international customary law that states[3] consider to be binding and follow in practice, and general principles

---

[1]This chapter is based on the results of research funded by German Environment Agency—"Improving International soil governance - Analysis and recommendations"—, which developed options for the German government (FKZ 3716 71 210 0).

[2]We use the term "instrument" in a broad sense that includes binding treaties as well as non-binding documents, declarations etc.

[3]For ease of reference, references to "states" in this study also include the EU unless otherwise stated.

R. Bodle (✉) · H. Stockhaus
Ecologic Institute, Berlin, Germany
e-mail: ralph.bodle@ecologic.eu

F. Wolff
Öko Institut, Berlin, Germany

S. Oberthür
Institute for European Studies, Brussels, Belgium

H. Ginzky et al. (eds.), *International Yearbook of Soil Law and Policy 2019*, International Yearbook of Soil Law and Policy 2019,
https://doi.org/10.1007/978-3-030-52317-6_9

of law. The latter are not relevant for soil governance. Other elements of the international legal framework are not legally binding in the strict sense, but as so-called "soft law" nonetheless influence or provide guidance to states' conduct.

In examining international soil governance, it is important to keep in mind that the term "soil" is not generally synonymous with "land". Generally, the term "land" is used to comprise not only soil-related issues, but more dimensions (above ground) and interactions with vegetation.[4] This interpretation is shared, at the international level, by the CCD.[5] Soil protection is often closely linked to and partly overlapping with the use and management of land. This is particularly relevant with regard to the United Nations' Sustainable Development Goal (SDG) 15.3: achieving land degradation neutrality.

2 Bird's Eye View of Existing International Soil Governance

This section provides a bird's eye view on key aspects of the existing international governance of soil protection.[6] As we show below, international soil governance is fragmented. Different instruments and institutions at the international and regional level cover different aspects of soil protection. The following main existing elements also include instruments, guidance, initiatives etc. adopted under their mandate:

Binding treaties

- Convention to Combat Desertification (CCD)
- Biodiversity Convention (CBD)
- Ramsar Convention on wetlands
- Paris Agreement on climate change
- Soil Conservation Protocol to the Alpine Convention
- Maputo Convention.

Institutions and legally non-binding instruments

- Agenda 2030 and the Sustainable Development Goals (SDGs)
- UN Food and Agricultural Organisation (FAO)
- UN Environment Programme (UNEP) and the United Nations Environmental Assembly (UNEA)
- Intergovernmental Panel on Climate Change (IPCC)
- Intergovernmental Science-Policy Platform on Biodiversity and Ecosystem Services (IPBES).

[4] Stavi and Lal (2015).

[5] Cf. the definition in Art. 1(e) CCD.

[6] The detailed research and arguments are set out in the research report for the German Environment Agency in footnote 1, which will be available as Bodle et al. (2020) at https://www.umweltbundesamt.de/en/publications.

None of these instruments and institutions provides an overarching or comprehensive framework for soil governance. Attempts to promote an international treaty on soil or a protocol on soil under one of the existing treaties have not gained political momentum so far. However, the Sustainable Development Goals, adopted in 2015, provide a new reference point in international policy which is also relevant for soil protection. We begin the stocktake with this instrument.

2.1 The SDG Process As a Global Point of Reference for International Soil Governance

Agenda 2030, adopted by the UN General Assembly in September 2015,[7] provides a framework for sustainable development and sets 17 Sustainable Development Goals that also cover several issues relevant for international soil governance. Notably, with target 15.3 under SDG 15 all UN member states agreed to strive to achieve a **land degradation neutral (LDN) world** by 2030. It builds on groundwork by the CCD and the Rio+20 outcome document of 2012, which had set the LDN objective "in the context of sustainable development" and without a target year.[8]

In scientific and political contexts, there is no universally accepted definition or a uniform or consistent use of the term "land degradation".[9] Yet regardless of how exactly it is defined and measured, degradation of soil is always also degradation of land in respect of LDN. The LDN target **includes all soil threats** and drivers of land degradation and is therefore highly relevant for the international protection of soil.

While the SDGs are not a binding treaty, their adoption by the General Assembly establishes a strong political commitment. It is now up to the states to **implement** the SDGs according to their national circumstances. Guidance for understanding and implementing LDN is provided by the Scientific Conceptual Framework developed by the Science-Policy Interface under the auspices of the CCD.[10] To track progress towards LDN, the UN Statistical Division (UNSD) has cooperated with the CCD and in March 2016 presented indicator 15.3.1: "Proportion of land that is degraded over total land area".[11] There is also a minimum consensus on three sub-indicators: land cover and land cover change, land productivity and carbon stocks in and above the soil. While these indicators may not be sufficient to reflect soil protection, they are an important first step.

---

[7]UNGA Res. A/RES/70/1 "Transforming our world: the 2030 Agenda for Sustainable Development" of 25.09.2015.

[8]UNGA Res. A/RES/66/288 "The future we want" of 11.09.2012, para 206. See also below on the CCD.

[9]See analysis and examples in Bodle and Stockhaus (2019), pp. 20–21; Wunder et al. (2018), section 3.2; Ehlers (2017), p. 73.

[10]Orr et al. (2017), p. 59 et seq.

[11]See https://sustainabledevelopment.un.org/sdg15 (last accessed 15.05.2019).

The UN assists implementation of the SDGs and provides a forum for exchange—in respect of data, implementation and progress review. Several institutions are involved in the implementation of the SDGs. The main political forum for **follow-up and review** at the global level is the UN High-level Political Forum on Sustainable Development (HLPF). It meets every year under the auspices of the Economic and Social Council and every 4 years at the level of heads of state under the auspices of the General Assembly.[12] Follow-up and review are based on the UN Secretariat's annual reports to the HLPF on overall trends. The follow-up and review process for SDG implementation has two pillars: Reporting of data by custodian agencies and voluntary reporting from all countries. While reporting by the custodian agencies—the CCD for the LDN target—is supposed to enable the international community to track the progress made, it does not hold states accountable. The voluntary reporting from countries fills this gap only to a limited extent: While various countries have submitted and presented their Voluntary National Reviews since 2016, they are neither subject to peer review of performance nor to monitoring. Currently, countries are using the country-led and country driven reporting mainly to showcase best practices. However, VNRs are the only global mechanism for follow-up and review on SDGs and their implementation.

Apart from the LDN target, the SDGs include other targets relevant for the protections of soil ecosystems and soil biodiversity.[13] On their own, the SDGs and the LDN target in SDG 15.3 are not a comprehensive soil policy and have shortcomings in terms of normativity and institutional anchoring as well as follow-up and review. But at least the SDGs provide a global consensus on soil in general and a political basis for further work in existing regimes and institutions. In this sense, the SDGs and in particular the target 15.3 on LDN have established the only global political point of reference specifically on land and soil. This political weight is arguably its main value, as it opens and widens other avenues for action.

[12]For format and organisational aspects see General Assembly Resolution 67/290. For guidance on the follow-up and review see General Assembly resolution 70/299.

[13]These include, notably, SDG 15.1 ("By 2020, ensure the conservation, restoration and sustainable use of terrestrial ... ecosystems and their services (...)",SDG 15.5 ("Take urgent and significant action to reduce the degradation of natural habitats, halt the loss of biodiversity (...)", also SDG 2.4 ("By 2030, ensure sustainable food production systems and implement resilient agricultural practices that ... progressively improve land and soil quality"), SDG 3.9 ("By 2030, substantially reduce the number of deaths and illnesses from ... soil pollution and contamination") and SDG 12.4 ("By 2020, achieve the environmentally sound management of chemicals and all wastes t... and significantly reduce their release to air, water and soil ...)".

2.2 *The UN Convention to Combat Desertification: CCD*

The CCD of 1994 is a binding instrument, a multilateral treaty whose currently 197 parties[14] make it basically universal. Its objective to combat desertification and mitigate the effects of drought makes it the only global treaty specifically addressing land-related issues.[15] However, there are two important constraints on its role in international governance: The content of the CCD's obligations and its scope of application.

The content of the CCD's obligations are structured along different categories of parties: all parties, affected parties,[16] developing country parties and developed country parties. Some obligations are further specified by regional implementation annexes. Despite their value for guiding affected countries towards general good governance, the obligations are rather general and toothless, lacking in precision and prescriptiveness.[17] The core obligation under the CCD requiring specific action is to prepare, publish and implement National Action Programmes (NAPs). The obligation is on affected developing countries and affected parties covered by an implementation annex; other parties may voluntarily prepare NAPs. Affected parties are not obliged, but "urged" to align their NAPs with the CCD's strategic framework which guides implementation. However, NAPs appeared to have little effect in terms of implementation at national level and integration into planning and decision making due to practical constraints and structural problems.[18] In addition, by 2017 only 20% of affected parties had finalized the alignment with the strategic framework for 2007–2018.[19] The current strategic framework for 2018–2030 has also been made the reference point for reporting under the CCD. But with the NAP alignment process only making slow progress, reporting and review currently provides little information on NAP implementation.

With regard to its scope of application, the CCD is in principle limited to drylands[20] through an intricate mixture of geographical and subject-related parameters in the CCD's objectives, definitions and in the action to be taken. Since there is no list of drylands, it can be a matter of scientific as well as legal and political debate how the CCD applies to individual countries. The issue became relevant when the

[14]See https://treaties.un.org. After withdrawing with effect of 28.03.2014, Canada re-joined the CCD with effect of 21.03.2017.

[15]Cf. Boer et al. (2017), p. 53 and the CCD's website http://www2.unccd.int/convention/about-convention.

[16]I.e. parties whose lands include, in whole or in part, areas affected or threatened by desertification, Art. 1 (i) CCD. "Affected countries" means countries whose lands include, in whole or in part, affected areas.

[17]Cf. Art. 4(2) (a), (c) and Art. 5(a)-(e) CCD. For a brief analysis of the Convention's shortcomings see Montanarella and Alva (2015), p. 44.

[18]Fritsche et al. (2015), p. 43; Smith (2015), p. 3 et seq.

[19]CCD decision 2/COP.12, preamble.

[20]We use the term "drylands" as a shorthand for what the CCD refers to as arid, semi-arid and dry sub-humid areas.

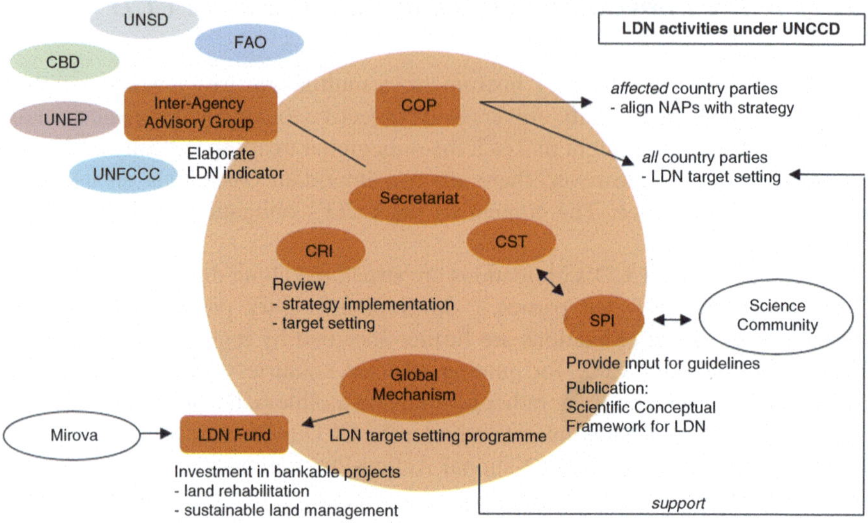

Fig. 1 LDN activities under the CCD. Source: Ecologic Institute

CCD started to focus on integrating and implementing the LDN target in SDG 15.3.[21] This raised concerns that the CCD in practice appeared to abandon its objective to protect drylands and headed towards addressing land degradation *everywhere*.[22] There is also a political subtext that the limited resources available to combat land degradation should not be diverted from the most vulnerable areas.[23] The 2017 COP decision on this matter is non-prescriptive and pragmatic and leaves the issue open.[24]

Despite these limitations, the CCD can be credited for laying the groundwork for developing and establishing the concept of the LDN target in SDG 15.3. The CCD Secretariat to some extent claims leadership for its implementation. The CCD decided to integrate LDN in its strategies, work programmes, financial support and other processes, and has engaged in various activities such as a target setting programme and elaborating guidance material (for an overview see Fig. 1).[25] In

[21] IISD (2015).

[22] This was the reason for some countries to oppose setting LDN targets for the CCD as a whole, IISD (2017), p. 17.

[23] IISD (2015), p. 10.

[24] While preamble notes that a significant proportion of land degradation occurs beyond drylands, the operative paragraphs recognise that parties (without limitation to "affected" parties) may use the CCD in pursuing their policies towards LDN, CCD COP decision 8/COP.12, preamble and operative paras 1 and 2.

[25] See for instance CCD decision 1/COP.13, Annex, strategic objective 1.

particular, the CCD published a Scientific Conceptual Framework[26] that is intended to apply to all land and guide all parties in implementing LDN.[27]

Although the legal and political constraints make the CCD's potential difficult to assess, it could continue to pursue a leading role in implementing the LDN target and serve as forum for discussing soil-related issues between developing and developed countries. Even if the CCD's motivation behind claiming leadership for LDN was the opportunity to regain political relevance, that is not necessarily bad, as long as the CCD is up to the task. Specifically with regard to LDN, the CCD has potential because it can integrate it into its existing treaty rules. It is a basically universal regime with an institutional setup[28] and mandate to address LDN and has in practice done so, notably with the decisions at COP12 in 2015 and at COP13 in 2017, integration of the LDN target in the CCD 2018–2030 Strategic Framework and the alignment of reporting with the LDN indicators. Notably, the Science Policy Interface (SPI) was established in 2013 with a mandate including, inter alia, developing proposals to the COP based on findings and recommendations from the science community and providing thematic guidance on implementation.[29] The CCD also contains several provisions and general obligations as well as institutional arrangements with regard to financial and other support. While the CCD's "Global Mechanism" is essentially a broker and not an institution providing financial resources, the Global Environmental Facility (GEF) has been funding activities regarding land degradation since in 1994 and is now formally available as a financial mechanism to the CCD. In the GEF-7 replenishment period 2019–2022 funds allocated for the land degradation focal area amount to USD 475 million. In addition, the LDN Fund was established in 2017, which is managed by a private sector investment management firm and leverages public money to raise private capital.[30] Launched with a capital commitment of around USD 100 million, it aims at raising USD 300 million and claims to have invested in its first project in 2019.[31]

Despite its limited mandate, the CCD could be slowly moving towards addressing LDN generally, i.e. not only in "affected" countries. The CCD's practice appears to walk a thin line between reassuring parties about staying within its mandate and

[26]Orr et al. (2017). See also Cowie et al. (2018), p. 25. See generally, https://knowledge.unccd.int/knowledge-products-and-pillars/guide-scientific-conceptual-framework-ldn/about-scientific (last accessed 15.05.2019).

[27]CCD decision 18/COP.13 paras 1–2.

[28]See http://www2.unccd.int/convention/about-convention (last accessed on 15.05.2019).

[29]CCD decision 23/COP.11, para. 3 and CCD decision 19/COP.12, para. 2.

[30]https://www.unccd.int/actions/impact-investment-fund-land-degradation-neutrality.

[31]https://www.im.natixis.com/en-institutional/news/news/first-investment-and-first-strategic-board-meeting-for-the-ldn-fund. However, apart from press releases, information such as founding documents, Board documents and operational policies do not seem to be publicly available. Some documents are available on the website of IDH, the manager of the LDN Fund's Technical Assistance Facility.

addressing LDN in general.[32] Legally speaking, most recent action by the COP regarding LDN is weak in the sense of non-prescriptive, non-specific or both. But this does not necessarily mean that implementation will be weak if there is strong political buy-in by parties. The commitment of 120 parties to participate in the LDN target setting programme could be a positive sign.[33] However, it is unclear to what extent parties could be willing to accept such a shift.

In any event, the CCD does not claim, and is unlikely to be suitable for, *sole* leadership on soil governance. It acknowledges the necessary role of other actors.[34] These include, for instance, the FAO and the CBD, which both have overlapping mandates and are in potential competition and conflict with the CCD.

2.3 UN Convention on Biological Diversity: CBD

Adopted in 1992, the Convention on Biological Diversity (CBD) is the major international treaty dealing with biological diversity.[35] While soil biodiversity undoubtedly forms a crucial part of biodiversity and its conservation is a sub-aspect of sustainable soil management, the CBD itself does not deal with soil biodiversity or the role of soils for non-soil biodiversity in a comprehensive or ambitious way.

Historically, soil biodiversity was identified as an area requiring particular attention soon after the adoption of the Convention. Both the CBD's overarching Ecosystem Approach, developed since 2000, and its Addis Ababa Principles of 2004 promote a holistic approach towards the conservation and sustainable use of biodiversity, interfacing nature and ecosystems with the human sphere. The Aichi Targets of 2010 furnish both the Ecosystem Approach and the principles regarding sustainable use with a framework of targets, timelines and indicators.

In 2002, the CBD parties also agreed to establish the International Soil Biodiversity Initiative, managed by FAO, as part of the Programme of Work on Agricultural Biodiversity. This is the CBD activity that most explicitly calls upon parties to integrate the conservation and sustainable use of soil biodiversity into their national biodiversity strategies. However, parties have, for the most part, failed to follow through and implementation at national level with regard to soil biodiversity remains weak.[36] Soil or related issues, such as erosion or pollution from excess nutrients, also

[32] Ginzky (2015), p. 18. Cf. the frequent repetition in CCD COP12 decisions referring to the scope of the CCD COP decision 3/COP.12 paras 3, 4, 10.

[33] See https://www.unccd.int/actions/ldn-target-setting-programme (last accessed on 15.05.2019).

[34] CCD decision 3/COP.12, preamble.

[35] Almost universal membership with 196 parties, not including the USA.

[36] Balakrishna and Prip (2015) and Orgiazzi et al. (2016).

do not feature prominently in the regular reports that parties have to submit to the CBD.[37]

The CBD is linked to the more recently established Intergovernmental Platform on Biodiversity and Ecosystem Services (IPBES), an independent scientific body mandated with the preparation of assessments on biodiversity and ecosystem services at both regional and global level, as well as policy support, capacity building, and outreach.[38] The Thematic Assessment on Land Degradation and Restoration and the Global Assessment Report on Biodiversity and Ecosystem Services provides valuable and comprehensive scientific data on the state of soil in different regions around the world. However, the link between the treaty and the scientific body is less normative than in the climate regime.

While the CBD and its Secretariat's activities are generally conducive to sustainable soil management, the CBD does not deal with soil biodiversity or the role of soils for biodiversity in a comprehensive manner and implementation by parties remains relatively weak. With the exception of the International Soil Biodiversity Initiative, soil is mostly addressed indirectly. While there would be potential for an increased contribution of the CBD to sustainable soils, there is currently no evidence that its parties are willing to push for international soil (biodiversity) governance.

2.4 The Paris Agreement and the Climate Regime

The international climate regime is also relevant for soil protection and governance since soil as well as land use, land degradation and sustainable land management are closely linked to climate change. Land-based ecosystems absorb and store CO2 and are amongst the most significant sinks of greenhouse gases, while land use and land use change accounts for one of the most important source of anthropogenic greenhouse gas emissions.[39] The Paris Agreement of 2015 under the United Nations Framework Convention on Climate Change (UNFCCC) is also almost universal with its currently 186 parties. It does not replace but supplements the existing 1992 Climate Change Convention and the 1997 Kyoto Protocol and incorporates existing elements of the climate regime.[40]

The Paris Agreement sets out the goal of holding global warming well below 2° and pursuing efforts to limit it to 1.5°. The core obligations to achieve these

---

[37] See in detail at https://www.cbd.int/reports/search/ (last accessed on 15.05.2019) or the syntheses of the forth National Reports: UNEP/CBD/COP/10/INF/2 (2010), UNEP/CBD/COP/10/8 (2010), UNEP/CBD/WG-RI/3/INF/1 (2010); on the third National Reports: UNEP/CBD/WG-RI/2/INF/1 (2007) and UNEP/CBD/WG-RI/2/INF/1/Add.3 (2007).

[38] See www.ipbes.net and the Memorandum of Cooperation with Convention on Biological Diversity (CBD), https://www.ipbes.net/memorandum-cooperation-convention-biological-diversity-cbd.

[39] La Vina and de Leon (2017), p. 166; see also Streck and Gay (2017), pp. 106–108; https://unfccc.int/topics/land-use/the-big-picture/introduction-to-land-use.

[40] For a general analysis of legal form and nature of the obligations see Bodle and Oberthuer (2017).

overarching objectives are mainly procedural: Parties are required to prepare and present individual climate plans (nationally-determined contributions, NDCs) every 5 years that set out how the party intends to contribute to the collective objectives. Parties have to take measures with the aim of achieving these NDCs but the climate regime does not require specific measures from individual parties. There are also provisions on adaptation planning. Parties have to report regularly GHG emission inventories as well as on progress in implementing their NDCs and on climate finance. Every 5 years, a "Global Stocktake" is to assess collective progress towards the Paris Agreement's objectives. The climate regime including the Paris Agreement also contains several provisions and general obligations as well as institutional arrangements with regard to financial and other support. Developed countries have committed to jointly mobilize USD 100 billion per year by 2020 and through to 2025, from a variety of sources.

The UNFCCC and Kyoto Protocol adopted rules on reporting and accounting for emissions from land use, land use change and forestry (LULUCF). These rules determine how parties have to report LULUCF in their regular emission inventories, which under the KP is also relevant for accounting of whether parties meet their emission reduction targets. In addition, verified actions on "Reducing emissions from deforestation in developing countries and approaches to stimulate action" (REDD+) are supposed to lead to "results-based payments".[41]

However, the Paris Agreement does not contain specific soil or land-related obligations,[42] apart from a reference to deforestation. But it has relevant provisions on sinks and reservoirs and adaptation, as well as reporting and accounting. In addition, in 2017 the first COP decision on agriculture brought this issue into the negotiation process.[43] Scientific input could also increase the need to address soil more specifically. Science indicates that reaching the Paris Agreement's temperature goals of keeping global warming under 1.5 °C or at least 2 °C will most likely require addressing soil and land use, given their important role as sinks as well as emissions sources.[44] This is part of the discussion of the "net-zero" objective in Article 4.1 and so-called "negative emissions" that offset remaining emissions.

The climate regime is linked to the Intergovernmental Panel on Climate Change's (IPCC), the United Nations' independent scientific body for assessing the science related to climate change, which has 195 Member countries.[45] The IPCC does not conduct its own research, but thousands of scientists review the scientific literature in an elaborate transparent process[46] and prepare regular and comprehensive assessment and other reports. For instance, the "good practice methodologies" accepted by

[41] See overviews at https://unfccc.int/topics/land-use/the-big-picture/introduction-to-land-use and http://redd.unfccc.int/.

[42] See also Boer et al. (2017), p. 59.

[43] "Koronivia joint work on agriculture", Decision 4/CP.23.

[44] See the overview at https://unfccc.int/topics/land-use/the-big-picture/introduction-to-land-use.

[45] UN GA Resolution 43/53 of 6.12.1988; see generally www.ipcc.ch.

[46] See https://www.ipcc.ch/about/preparingreports/ (last accessed on 15.05.2019).

IPCC are mandatory for accounting for the national inventories and its reports are a source of input for the regular global stocktake.[47]

The IPCC's 2018 scientific special report on 1.5 °C accordingly includes afforestation and reforestation, land restoration and soil carbon sequestration in the options for carbon dioxide removal measures that it considers necessary to achieve this goal.[48] An IPCC special report on climate change, desertification, land degradation, sustainable land management, food security, and greenhouse gas fluxes was published in 2019.[49] As the role and importance of soil in the climate regime is likely to increase, the role for the climate regime in international soil governance should be clearer.

2.5 *Food and Agricultural Organisation: FAO*

FAO was established in 1945 as the United Nations' specialized agency on nutrition and food security. It has been engaged in technical work on soils since the 1950s, both globally and bilaterally, and today describes itself as "the lead United Nation's agency concerned with soils".[50]

FAO initiated the development of the non-binding World Soil Charter (1981), implemented numerous soil-related projects world-wide over the last 30 years, funded both by FAO regular programme and by extra budgetary (voluntary contributions) resources.[51] Its technical assistance included implementing soil survey projects, preparing soil survey reports, carrying out soil assessments, setting up land resources inventories as well as promoting integrated land-use planning and the sustainable management of soils, e.g. through training and institution building.[52]

Although FAO's current strategic objectives[53] do not explicitly mention "soils", all its objectives to a greater or lesser extent depend on healthy soils. In practice, it can be considered a central hub for soil-related activities, albeit with a traditional focus on agricultural soils. In recent years, FAO made efforts to shift focus and adopt a more holistic approach on soils as a provider of ecosystem services. This shift in perception has led to the revision of legally non-binding instruments such as the

[47]Art. 13.7(a) Paris Agreement; decision 1/CP.21, para. 31.

[48]IPCC (2018), pp. 14, 19.

[49]https://www.ipcc.ch/srccl/.

[50]http://www.fao.org/soils-portal/resources/en/ (last accessed on 15.05.2019).

[51]FAO (2014).

[52]FAO (1983).

[53]The objectives include to: (1) Contribute to the eradication of hunger, food insecurity and malnutrition; (2) Increase and improve provision of goods and services from agriculture, forestry and fisheries in a sustainable manner; (3) Reduce rural poverty; (4) Enable more inclusive and efficient agricultural and food systems at local, national and international levels; (5) Increase the resilience of livelihoods to threats and crises. Cross-cutting themes are gender and governance. Cf. FAO (2013).

World Soil Charter and the development of new ones such as the Voluntary Guidelines for Sustainable Soil Management (VGSSM). FAO also established and hosts the Global Soil Partnership (GSP), a multi-stakeholder partnership which promotes sustainable soil management through various activities. These include information dissemination and awareness rising, e.g. preparing the "Status of the World's Soil Resources" Report,[54] establishing the Intergovernmental Technical Panel on Soils and the preparation of the development and revision of the above mentioned normative instruments.

One shortcoming is that membership in the FAO does not entail legally binding obligations. Compared to treaties such as the CCD, this weakens FAO's role in international soil governance. In addition, given the apparent perception among its members that soil is a national resource and their concerns, for instance, of being exposed to intensified soil monitoring, it is also uncertain whether soil protection will play a more explicit role in the future. However, the FAO and in particular the GSP could become crucial in disseminating knowledge and developing non-binding technical guidance on soil.

2.6 UN Environment: UNEP

Established in 1972 to serve "as an authoritative advocate for the global environment" and to promote "the coherent implementation of the environmental dimension of sustainable development within the United Nations system",[55] UNEP focuses on environmental conditions and instruments on a global, regional and national level. Its work on soils covers industrial land and soil pollution as well as land degradation and restoration. While it does not explicitly cover "soils" as an individual major topic different from "air" and "water", soils form an important part within various topics such as "ecosystems". Issues pertaining to soil conservation are thus, inter alia, covered in UNEP's environmental information and assessment activities such as the Global Environment Outlook, technical assistance to member states, its recent campaign "Towards a pollution-free planet" as well as support to implementing the SDGs and the CCD.[56] Although it does not create legally binding obligations on UN members, UNEP was also involved in norm development, publishing the World Soils Policy in 1981 which complement the World Soil Charter.[57] In 2017, the United Nations Environmental Assembly, UNEP's highest-level decision-making

[54]FAO and ITPS (2015).

[55]https://www.unenvironment.org/about-un-environment (last accessed on 15.05.2019). The mandate was first established in 1972 in the UN General Assembly Resolution 2997 (XXVII) which established UNEP; other resolutions reinforced this mandate, including the Nairobi Declaration on the Role and Mandate of the United Nations Environment Programme of 7 February 1997 and the Malmö Ministerial Declaration of 31 May 2000.

[56]https://www.unenvironment.org/explore-topics (last accessed on 15.05.2019).

[57]Hannam and Boer (2002), p. 61.

body, adopted a resolution specifically on soil pollution—the first international instrument which covers soil pollution more broadly and not in relation to specific pollutants. UNEP also co-funds the Global Soil Partnership (GSP) and hosts several secretariats of relevance to soil conservation such as the Secretariat of the Basel, Rotterdam and Stockholm Conventions. While the visibility of UNEP with regard to soil matters has not been very high in the past years, this could change with its recent soil-related activities and policies and the policy guidance by UNEA. In relation to the CCD and FAO, the future role of UNEP in international soil governance could lie in a continued focus on soil pollution.

2.7 Ramsar Convention on Wetlands

Adopted in 1971 to protect wetlands as waterfowl habitats, the Ramsar Convention with its 171 parties is now the foundation for a comprehensive and sophisticated policy framework for the management of wetland areas in general. It does not tackle specific drivers of land degradation, but aims at the conservation and wise use of wetlands through local and national action as well as international cooperation. Parties are required to formulate and implement plans to promote the wise use of wetlands. In addition, parties have to promote the conservation of those wetlands that they included in the Ramsar List and monitor them. In case adverse ecological changes occur or are likely to occur, these wetlands will be added to the Montreux Record, an act that triggers assistance and increases the level of protection. Parties have to prevent or mediate damage to the wetland.

Although the Ramsar Convention has been criticised for its weak and vague obligations, it has gained influence via the recommendations and resolutions adopted by the COP and triggered the development of wetland policies and legislation worldwide.[58] The drivers for land degradation are addressed through strategic plans, the last of which has been adopted in 2015 and guides implementation between 2016 and 2024.[59] It has been developed within the context of the SDGs and links target 15.3 to the targets of the strategic plan, such as the restoration of degraded wetlands.

Despite the importance of wetlands in achieving LDN and the experience with prevention of land degradation or restoration of degraded wetlands, the role of the Ramsar Convention in SDG implementation is weak. The CCD with its mandate for drylands and the Ramsar Convention with its mandate for wetlands are complementary, but cooperation between the two is rudimentary and could be improved, e.g. on the LDN Target Setting Programme.

---

[58]Ferrajolo (2011), p. 243 et seq.

[59]Resolution XII.2: The 4th Strategic Plan 2016–2024, Punta del Este, 2015; see for an overview of the resolutions adopted at COP12 Laina and Tsioumani (2015), p. 190 et seq.

2.8 *Soil Conservation Protocol to the Alpine Convention*

The 1998 Soil Conservation Protocol has few parties,[60] but it is the only legally binding international treaty regulating exclusively soil protection at a regional level.[61] It was adopted to implement the Alpine Convention and has been ratified by all of its parties. Its provisions address specific drivers of land degradation (e.g. tourism, mineral extraction, agriculture and industries) and certain soil threats (e.g. erosion and soil contamination). Besides, it provides for measures of more cross-cutting nature.

It aims to safeguard and preserve the ecological functions of soil both qualitatively and quantitatively as well as to promote the restoration of impaired soils.[62] Despite these ambitious objectives, the Soil Conservation Protocol is largely preventive.[63] Although taking the measures is strictly required, parties still have some discretion in most areas. However, while some provisions are rather vague and general, others are sufficiently defined and unconditional and therefore even considered directly enforceable. These self-executing norms have to be applied by authorities and courts and can therefore have a strong impact.[64]

The Soil Conservation Protocol responds to the specific needs of the parties and the Alpine region as an important and fragile ecosystem. The measures reflect a compromise between the parties—all developed countries—that balances interests in the use and conservation of the Alpine region. It is tailored for the Alpine region and cannot—at least not in content—be applied to other regions.

2.9 *Maputo Convention*

Intended to modernise the Algiers Convention[65] of 1968, the Maputo Convention was adopted in 2003 as an African regional treaty. It took until 2016 to enter into force, without attracting much attention,[66] and currently has 17 parties. Its text is based on a proposal developed and submitted by the International Union for

[60]The Alpine countries Austria, France, Germany, Italy, Liechtenstein, Monaco, Slovenia and Switzerland, as well as the European Union, See http://www.alpconv.org/en/convention/ratifications/default.html.

[61]Markus (2015), p. 214; Fromherz (2012), p. 104.

[62]Article 1 (2) Soil Conservation Protocol.

[63]Brandon (2013), p. 49.

[64]Wissenschaftliche Dienste des Bundestages (2016), pp. 5, 11; BMU (2004), p. 14.

[65]African Convention on the Conservation of Nature and Natural Resources, Algiers, 15 September 1968, in force 16 June 1969, 101 *United Nations Treaty Series* (1976), p. 3.

[66]The Maputo Convention, according to its article XXXVIII.1, requires 15 ratifications to enter into force. Even the African Union, as depositary, lists 17 ratifications, but does not list the Convention as being in force. https://au.int/treaties.

Conservation of Nature.[67] Notably, it has one article dedicated to land degradation and soil conservation. Its instruments overlap with those contained in the CCD, but increase the level of commitment. Key instruments are long-term integrated strategies for land resources and land-use plans. As agricultural activities are an important driver for land degradation in Africa, the Convention formulates more detailed requirements for the implementation of agricultural practices and agrarian reform. Other forms of land use are only addressed on a more general level—they shall not result in erosion, pollution, or any other form of land degradation. Besides these requirements focusing on prevention, the Convention also establishes obligations related to rehabilitation. For areas affected by land degradation, parties have to plan and implement mitigation and rehabilitation measures. The land and soil provision also picks up on the conflicts around land tenure in many African countries and requires parties to develop and implement land tenure policies that are able to facilitate the measures to prevent land degradation and to conserve and improve the soil.

The Maputo Convention could become an important regional instrument that takes up land and soil protection.[68] However, it remains to be seen to what extent the parties will identify with and implement the Convention. Up to now, the institutions have not picked up work and parties have not started with implementation.[69]

3 Assessment, Conclusions and Options for Improving International Soil Governance

3.1 Assessment and Conclusions: The Big Picture

Our assessment of the big picture of current international soil governance uses different functions of international governance as a starting point. These governance functions include (a) agenda-setting; (b) visions and goals that provide guidance and policy signals; (c) setting of rules and governance architecture; (d) means of implementation; (e) international transparency and accountability; and (f) knowledge and learning:[70]

(a) While the last few years have seen an increase in activities attempting to set the agenda for international soil governance, there is **no general consensus that soil is an issue that calls for or requires *international* policy** and governance

[67]IUCN (2006), p. 5.

[68]Beyerlin and Marauhn (2011), p. 208.

[69]The African Ministerial Conference on the Environment that took place in September 2018, called in its final declaration for the convening of the first COP as a next step for implementation, https://www.iucn.org/news/world-commission-environmental-law/201810/maputo-convention-protection-nature-gets-a-boost-african-ministerial-conference-environment.

[70]Oberthür et al. (2017).

efforts. However, the SDGs might have the potential to change this to some extent.

(b) Although they are not binding, the SDGs and in particular the "land degradation neutrality" (LDN) target in SDG 15.3 have established **a central, global political point of reference**[71] that should be supported. Although the LDN target on its own is not a comprehensive soil policy and has shortcomings, it is a useful starting point for guiding national policies and further work. While it might appear vague in terms of specific individual actions, government actions endorsing the SDGs and the follow-up process are politically relevant and help maintain the SDG's political weight.

(c) **There are almost no international binding obligations specifically regarding soil** that apply more or less universally to all states. The relevant existing global treaties have a scope and provisions that would implicitly cover many soil types, components and threats. The extent to which these treaties actually address them depends, inter alia, on how closely the soil threat is linked to the general mandate and objective. Some soil threats such as compaction or salinisation are basically not addressed.

At the same time, **there does not appear to be political appetite for a new treaty** specifically on soil at least in the short term. There have been various initiatives over the past decades that promoted the idea,[72] but they did not make it to the political agenda. One option is to work towards creating the right moment for a new treaty in the medium term.

(d) With regard to **means of implementation**, the existing channels for finance and other support seem to work well by and large, and are complemented by new channels. Substantial finance for implementing LDN and soil governance is channelled through bilateral aid and the GEF as well as relatively new channels such as the Green Climate Fund and the LDN Fund. However, better capacity is needed e.g. regarding knowledge about existing support channels, project and programme design and application procedures. Moreover, opportunities should be explored for reducing misaligned subsidies.

(e) While there is an abundance of international knowledge and guidance and growing work on indicators and monitoring, the **main substantive governance gap** is the absence, or inadequacy of **transparency and review** mechanisms at the *international* level. SDG monitoring, and CCD and CBD reporting are not sufficient in their present form. The main opportunity for improving international governance *within* exiting regimes relates to requirements and guidance regarding communication of national strategies and plans, reporting on implementation and review. Each of the relevant regimes and institutions provides further specific options.

---

[71] See the ENB's comment that focusing on LDN was the CCD's "saving grace" at COP 13 in 2017, IISD (2017), p. 17.

[72] See Boer and Hannam (2015), Fritsche et al. (2015), Altvater et al. (2015) and overview in Bodle et al. (2020).

(f) Regarding **guidance for national land and soil policies**, the existing framework and international guidance can already be used to assess and strengthen national strategies and policies. The LDN target could be regarded as the generic core of what is required to operationalise a soil policy at national level. In particular, achieving LDN requires a forward-looking planning element. The existing international guiding documents could be checked for duplications, coherence and gaps, and be consolidated. Gaps to be addressed could include a fresh look at which soil policies and measures are feasible even where land (tenure) rights are an issue.

There a number of governance elements that are not binding but do address soil specifically and have political clout, mainly including the work of institutions such FAO, UNEP, IPBES and IPCC. But apart from the SDGs, there is no general mandate or central point of reference for soil governance at the international level, either in political or normative terms. Current governance of soil at the international level is piecemeal and spread over parts of different mandates. There is significant overlap of mandates and activities of relevant institutions, while at the same time each of them has limitations. Improving soil governance at the international level therefore includes options for enhancing coordination and coherence. **A clearer division of labour** between the institutions addressing soil holds significant potential for improving international soil governance. While a certain degree of a rudimentary division of labour is emerging, there is scope and a need for advancing this further. A more robust coordinating forum could be developed in the medium- to long-term.

3.2 Options for Improving International Soil Governance

Based on this assessment of the big picture, we suggest the following **specific options** for improving international soil governance.

Since the options do not always directly and unequivocally match just one of the governance functions set out above, we cluster the options slightly differently as follows:

- Overarching issues: Improving international framework conditions for soil policy
- New treaty or institutions
- Improving existing governance
- Means of implementation
- Enhancing co-ordination and coherence.

The options were developed mainly with a view to consideration by government and vary in terms of time frame, level of political buy-in etc.

3.2.1 Overarching Issues: Improving International Framework Conditions for Soil Policy

- Deepen the recognition at the international level that soil is an international issue and not a purely domestic matter. It should be noted that such recognition and international co-operation do not necessarily mean regulation, prescriptive legal obligations or having to take specific courses of action. On the other hand, consistent action over time in this respect can also support *legal* developments.
- Support the SDG process politically, notwithstanding its shortcomings, as a political reference point for other fora and processes. While it might appear vague in terms of specific individual actions, government actions endorsing the SDGs and the follow-up process are politically relevant and help maintain the SDG's political weight.
- Consider how to better recognise the significance of the global land footprint for soil policy in order to increase the incentive for countries in the Global South to participate in the strengthening of international soil governance.
- Explore and identify, e.g. through studies, soil policies and measures that are feasible even where land (tenure) rights are an issue. Although land tenure rights are a serious and politically hot topic in many countries, they do not automatically have to be a reason for postponing any discussion on effective soil governance.

3.2.2 Create New Treaty or Institutions?

- In the medium to long term, work towards creating the political conditions for a new binding instrument such as a treaty on soil protection. While binding rules are not an end in itself, a new treaty on soil protection could be useful to address gaps and shortcomings in current governance and focus international commitments. Besides amending an existing instrument, options for a new binding instrument include a stand-alone treaty or a new instrument under an existing treaty. A standalone instrument could require more political effort and its added value vis-a-vis existing instruments would have to be justified. On the other hand, a protocol to for instance the CCD is difficult for legal reasons because the CCD, in contrast to many other treaties, does not provide for protocols.

3.2.3 Improve Existing Soil Governance Within Existing Fora

- Gradually shape the CCD towards strengthening transparency and review: Specific requirements, decisions and guidance on transparency for all parties regarding national strategies and plans, reporting on implementation, and for reviewing these plans and the reports at the international level. Despite its shortcomings, the CCD is the most obvious point for improvement, because it is a binding treaty with basic transparency and reporting obligations and an institutional framework for adopting detailed guidance.

- Increase participation in the CCD's voluntary LDN Target Setting, which would actively endorse a more comprehensive understanding of the CCD's mandate and approach.
- FAO to allocate greater strategic importance to soils, carry out an internal consistency check on potential conflicts and synergies between its policies and programmes and sustainable soil management, and scale up efforts (including donor funding) for implementing the Voluntary Guidelines on Sustainable Soil Management.
- UNEP to step up its work and capacities on soil protection, using the UNEA Resolution on soil pollution as entry point, in particular on contamination.
- In the climate regime, use the nexus between land use and soil management to further address sustainable *soil* protection *within* the climate regime. Following on from the findings of the UNFCCC special report and the Koronivia joint work, this might be part of the discussion of the "net-zero" objective in Article 4.1 and so-called "negative emissions" that offset remaining emissions. In addition, sustainable soil management and LDN can be a crucial part of adaptation efforts that fit in with developing countries' priorities. In the other direction, the increasing recognition of importance of land use and soil management can also spawn related climate initiatives by other international institutions such as the FAO, sub-national authorities and non-state actors which can also add political weight to international soil policy.[73]

3.2.4 Strengthen Means of Implementation

- Improve capacity building for accessing international support, in particular through information and technical assistance. While means of implementation, including financial support, capacity building and technology development and transfer for developing countries, is a recurring theme in international environmental law and governance, so are difficulties by developing countries in accessing available funding and other international support.
- Promote political commitments that governments as well as international organisations should rationalise, reduce and eliminate subsidies that are incompatible with sustainable soil management. The approach could build on the SGDs which address specific types of unsustainable subsidies. As a first step, the Global Soil Partnership could analyse data regarding the volumes of subsidies potentially detrimental to soils.

[73] Cf. https://www.thegef.org/news/global-business-government-and-agricultural-leaders-announce-land-focused-commitments-mitigate.

3.2.5 Enhance Coordination and Coherence

- Increase coordination between the relevant institutions addressing soil, with a view to expanding soil governance to cover all biomes, soil types and drivers of soil degradation. Overlapping mandates between international institutions are neither rare nor *necessarily* problematic. In practical terms, the potential rivalry between FAO and CCD might be resolved by each focusing on different issues. The existing linkages could pragmatically develop into de facto division of labour, e.g. between the climate regime and FAO on agriculture or between the FAO and the CCD regarding indicators.
- Develop a more robust coordinating forum in the medium-term, with a mandate designed to coordinate and promote international soil governance. While some division of labour appears to be evolving in the wake of the SDGs, this does not seem *satisfactory*. One argument against waiting for 'organic development' is that existing fora so far have not shown a clear potential to fulfil this function, since their focus to date is not on the governance of soils. We suggest an option in between the organic development of existing fora and a high-level formalised approach.
- While there is no lack of general substantive guidance for national soil policies, check the existing guiding documents for duplications and coherence and, if necessary, consolidate them, for instance in a non-binding instrument. Additional guidance could be added on certain issue areas that are currently not addressed, such as land degradation by industry or urbanization. This could be discussed by the CCD's Inter-Agency Advisory Group, which might require an adjustment of its mandate.
- Prepare an international assessment of options for international soil policies on what would be needs and options for international soil governance. This could be carried out e.g. by FAO and the Global Soil Partnership.

References

Altvater S et al (2015) Legal instruments to implement the objective 'land degradation neutral world' in international law. German Federal Environment Agency (Umweltbundesamt), Dessau-Roßlau. Available at: http://www.umweltbundesamt.de/publikationen/legal-instruments-to-implement-the-objective-land

Balakrishna P, Prip C (2015) Interim assessment of revised National Biodiversity Strategies and Action Plans (NBSAPs). UNEP WCMC/Fridtjof Nansen institute, Cambridge/Lysaker. https://www.cbd.int/doc/nbsap/Interim-Assessment-of-NBSAPs.pdf

Beyerlin U, Marauhn T (2011) International environmental law. Hart Publishing, Oxford

BMU (2004) Alpenkonvention konkret. Ziele und Umsetzung. Alpensignale 2. Ständiges Sekretariat der Alpenkonvention, Innsbruck

Bodle R, Oberthuer S (2017) Legal form of the Paris Agreement and nature of obligations. The Paris agreement on climate change: analysis and commentary

Bodle R, Stockhaus H (2019) Geeignete Rechtsinstrumente für die nationale Umsetzung der bodenbezogenen sustainable development goals, insbesondere des Ziels einer 'land degradation neutral world.'. Umweltbundesamt, Dessau-Roßlau

Bodle R, Stockhaus H, Wolff F, Scherf C-S, Oberthür S (2020) Improving international soil governance – analysis and recommendations. Umweltbundesamt, Dessau-Roßlau

Boer B, Hannam I (2015) Developing a global soil regime. Int J Rural Law Policy 2015:1–16

Boer B et al (2017) International soil protection law: history, concepts and latest developments. In: Ginzky H et al (eds) International yearbook of soil law and policy 2016. International yearbook of soil law and policy. Springer International Publishing, Cham, pp 49–72. https://doi.org/10.1007/978-3-319-42508-5

Brandon E (2013) Global approaches to site contamination law. Springer, Dordrecht

Cowie AL, Orr BJ, Castillo Sanchez VM et al (2018) Land in balance: the scientific conceptual framework for land degradation neutrality. Environ Sci Policy 79:25–35

Ehlers K (2017) Chances and challenges in using the sustainable development goals as a new instrument for global action against soil degradation. In: Ginzky H, Heuser IL, Qin T, Ruppel OC, Wegerdt P (eds) International yearbook of soil law and policy 2016. Springer International Publishing, Cham, pp 73–84

FAO (1983) Soils bulletin 50: keeping the land alive. Rome

FAO (2013) Reviewed strategic framework: Doc. C 2013/7, Available at: http://www.fao.org/3/mg015e/mg015e.pdf

FAO (2014) Healthy soils facility of the global soil partnership. Programme Document PGM/MUL/2014-2018-GSP. Rome, Available at: http://www.fao.org/3/a-az925e.pdf

FAO and ITPS (2015) Status of World's Soil Resources (SWSR) – main report. Food and Agriculture Organization of the United Nations, Rome

Ferrajolo O (2011) State obligations and non-compliance in the Ramsar system. J Int Wildlife Law Policy 14:243–260

Fritsche U, Eppler U, Iriarte L, Laaks S, Wunder S, Kaphengst T, Wolff F, Heyen D, Lutzenberger A (2015) Resource-efficient land use – towards a Global Sustainable Land Use Standard (GLOBALANDS). Umweltbundesamt, Dessau-Roßlau

Fromherz NA (2012) The case for a global treaty on soil conservation, sustainable farming, and the preservation of Agrarian culture. Ecol Law Q 39:57–121

Ginzky H (2015) Bodenschutz weltweit – Konzeptionelle Überlegungen für ein internationales Regime. ZUR 26:199–208

Hannam I, Boer B (2002) Legal and institutional frameworks for sustainable soils: a preliminary report. IUCN, Gland. Available at https://portals.iucn.org/library/sites/library/files/documents/EPLP-045.pdf

IISD (2015) Earth negotiations bulletin: summary of the twelfth session of the Conference of the Parties to the UN Convention to Combat Desertification, Available at http://www.iisd.ca/desert/cop12/

IISD (2017) Earth negotiations bulletin: summary of thirteenth session of the Conference of the Parties to the UN Convention to Combat Desertification, Available at Online at http://www.iisd.ca/desert/cop13/

IPCC (2018) Summary for policymakers. In: Masson-Delmotte V, Zhai P, Pörtner HO, Roberts D, Skea J, Shukla PR, Pirani A, Moufouma-Okia W, Péan C, Pidcock R, Connors S, Mat-thews JBR, Chen Y, Zhou X, Gomis MI, Lonnoy E, Maycock T, Tignor M, Waterfield T (eds) Global warming of 1.5 °C. An IPCC special report on the impacts of global warming of 1.5 °C above pre-industrial levels and related global greenhouse gas emission pathways, in the context of strengthening the global response to the threat of climate change, sustainable development, and efforts to eradicate poverty. World Meteorological Organiza-tion, Geneva

IUCN Environmental Law Programme (2006) An introduction to the African convention on the conservation of nature and natu-ral resources, IUCN Environmental policy and law paper No. 56, 2nd edn. IUCN, Gland

La Vina A, de Leon A (2017) Conserving and enhancing sinks and reservoirs of greenhouse gases, including forests (article 5). In: Klein D, Carazo MP, Doelle M, Bulmer J, Higham A (eds) The Paris agreement on climate change: analysis and commentary. Oxford University Press, Oxford, pp 166–177

Laina E, Tsioumani E (2015) Wetlands conservation in progress. Environ Policy Law 45:190–193

Markus T (2015) Verbindlicher internationaler Bodenschutz im Rahmen der Alpenkonvention in Zeitschrift für Umweltrecht, pp 214–221

Montanarella L, Alva IL (2015) Putting soils on the agenda: the three Rio Conventions and the post-2015 development agenda. Curr Opinion Environ Sustain 15:41–48

Oberthür S, Hermwille L, Khandekar G, Obergassel W (2017) Strengthening international climate governance: the case for a sectoral approach, IES Policy Brief Issue 2017/04, Available at https://www.ies.be/policy-brief/strengthening-international-climate-governance-case-sectoral-approach

Orgiazzi A, Bardgett RD, Barrios E et al (2016) Global soil biodiversity atlas. European Commission, Publications Office of the European Union, Luxembourg

Orr BJ, Cowie AL, Castillo Sanchez VM et al (2017) Scientific conceptual framework for land degradation neutrality. a report of the science-policy interface. United Nations Convention to Combat Desertification (UNCCD), Bonn

Smith J (2015) Evaluation of the effectiveness of national action programmes to implement the United Nations Convention to Combat Desertification, commissioned by the CCD Evaluation Office. Bonn. https://www.unccd.int/sites/default/files/relevant-links/2017-01/NAP%20evaluation_0.pdf

Stavi I, Lal R (2015) Achieving zero net land degradation: challenges and opportunities. J Arid Environ 112:44–51

Streck C, Gay A (2017) The role of soils in international climate change policy. In: Ginzky H, Heuser IL, Qin T, Ruppel OC, Wegerdt P (eds) International yearbook of soil law and policy 2016. Springer International Publishing, Cham, pp 49–72

Wissenschaftliche Dienste des Bundestages (2016) Zur unmittelbaren Anwendbarkeit der Durchführungsprotokolle zur Alpenkonvention, Sachstand: WD 2 – 3000 – 080/16

Wunder S, Kaphengst T, Frelih-Larsen A (2018) Implementing land degradation neutrality (SDG 15.3) at national level: general approach, indicator selection and experiences from Germany. In: Ginzky H, Dooley E, Heuser IL, Kasimbazi E, Markus T, Qin T (eds) International yearbook of soil law and policy 2017. Springer, Cham, pp 191–219

Critique of the Report "Improving International Soil Governance: Analysis and Recommendations"

Robert John Fowler and Ian Hannam

1 Introduction[1]

This chapter provides a critique of the report, "Improving International Soil Governance: Analysis and Recommendations" prepared for the German Environment Agency by the Ecologic Institute in cooperation with the Oko Institute and Professor Dr. Sebastian Oberthur (hereafter "the Report").[2] The purpose of the report is "to examine whether and how international cooperation between states for the purpose of sustainable soil management can be strengthened and improved in the short, medium and long term."[3] A distinction is made at the outset between the terms "soil"

[1]The Report notes (at p. 35) that its "cut-off date" was 31.12.2018, but that it has included some information that is more recent. In particular, the final version of the Report references both the IPBES Global Assessment approved in May 2019 and the outcomes of discussions concerning a proposed Global Pact on the Environment held in Nairobi from 20 to 22 May 2019. For the purposes of this critique, we have assumed that the authors of the Report had access to all relevant material available up to and including May 2019.

[2]Bodle, R et al. (2018) Improving International Soil Governance, Report for the German Federal Environment Agency, 29 May 2019.

[3]Id at 35. While the concept of sustainability has increasingly been used as a desired goal for soil conservation worldwide, there has been a general failure of the soil literature to adequately explain the precise context and limitations of sustainability and the appropriate type of soil legislation needed for its successful implementation; see Hannam and Boer (2002). However, the 2017 FAO

R. J. Fowler
Law School, University of South Australia, Adelaide, SA, Australia
e-mail: rob.fowler@unisa.edu.au

I. Hannam (✉)
Australian Centre for Agriculture and Law, School of Law, University of New England, Armidale, NSW, Australia
e-mail: ian.hannam@ozemail.com.au

H. Ginzky et al. (eds.), *International Yearbook of Soil Law and Policy 2019*,
International Yearbook of Soil Law and Policy 2019,
https://doi.org/10.1007/978-3-030-52317-6_10

and land", with a corresponding observation that "soil protection is often closely linked to and partly overlapping with the use and management of land".[4] This provides a justification for examining the concept of land degradation neutrality as a starting point in the study.

A substantial proportion of the Report is devoted to a detailed examination of existing international instruments and institutions that have some connection to soils. This provides an insightful and refreshingly frank assessment of the limitations and deficiencies of the existing international framework for sustainable soil management. The authors are to be commended for both the detail and perceptiveness of their assessment, and for their conclusion that:

> ...there are almost no binding obligations for all states specifically regarding soil. Current governance of soil at the international level is piecemeal and spread over parts of different mandates.

Given our broad agreement with the Report's assessment of the current situation, this critique focuses on its final section, which presents "options and recommendations to improve international soil governance that the German government could pursue".[5] The options identified in the Report are clustered under five headings: overarching issues; new treaty or institutions; improving existing governance; means of implementation; and enhancing co-ordination and coherence. This critique is structured to reflect these five headings.

At the outset, we note that the evaluation of potential international soil governance reforms appears to have been influenced substantially by a "realpolitik" judgement by the authors of what initiatives may be palatable for the German government. As a result, while the options and recommendations presented in the Report are generally sound and reasonable, they are also relatively conservative and postpone some more far-reaching initiatives to the medium or long term. We will suggest that, in adopting this approach, the Report has missed a valuable opportunity to promote a more immediate and urgent option that might substantially improve international soil governance within a shorter time-frame. Specifically, we will present a proposal to develop a new international instrument on soil based on the innovative model provided by the Paris Agreement (2015).[6] We also set out the

---

Voluntary Guidelines on Sustainable Soil Management (VGSSM) provide some recent, useful technical and policy recommendations. The basis of the VSGSSM is Sustainable Soil Management, which it defines as follows: "*Soil management is sustainable if the supporting, provisioning, regulating, and cultural services provided by soil are maintained or enhanced without significantly impairing either the soil functions that enable those services or biodiversity. The balance between the supporting and provisioning services for plant production and the regulating services the soil provides for water quality and availability and for atmospheric greenhouse gas composition is a particular concern*" (at 3).The concept of SSM could form the basis of a new instrument for soil. The Report does not attempt to address this matter and instead focuses on the subject of governance arrangements through instruments and institutions at the international level.

[4]The Report, at 13.

[5]The Report at 126–140.

[6]Paris Agreement, available at: http://unfccc.int/paris_agreement/items/9485.php; adopted by consensus on 12 December 2015.

reasons why we believe this could provide a more immediate and effective reform of international soil governance than is likely to result from the options presented to the German government in the Report.

2 Overarching Issues: Improving International Framework Conditions for Soil Policy

A recurring theme in the Report is encapsulated in its conclusion on overarching issues. This is that "there is no general consensus that soil is an issue that calls for, or requires, international policy and governance efforts".[7] Accordingly, this section of the Report identifies as a medium to long term option the need to "maintain and actively support recognition at the international level that soil is an international issue and not purely a domestic matter".[8] It also calls for greater recognition of the global land footprint for soil policy in order to incentivise action by countries in the Global South.[9]

The Report bases this significant conclusion on the view that an appropriate international dimension is lacking in relation to global soils management, as evidenced by a perception that soils are not a 'global commons' and that there is not any direct transboundary impact that requires an international response.[10] Whilst it acknowledges at one point that: "Soil degradation is no longer perceived as a purely local phenomenon and has slowly been moving onto the international political agenda,"[11] it concludes that raising the profile of soils as a serious international issue is a medium to long-term challenge that will require considerable further effort.

This cautious assessment is questionable, particularly given some recent technical reports that have generated a fresh understanding and awareness with respect to the global dimensions of land degradation and the need for collective action to address this problem. The Report references the UN Food and Agriculture Organizations (FAO) "*Status of the World's Soil Resources*" [SWSR] Report (2016), in particular its finding that 33% of land is moderately to highly degraded,[12] and rightly notes that "the lack of policy recommendations, including overarching, global-level ones, can be counted as a weakness in the SWSR."[13] The Report also provides a short summary of the more recent Intergovernmental Science-Policy Platform on Biodiversity and Ecosystem Services (IPBES) *Assessment Report on Land Degradation*

[7]Id The Report at 126.

[8]Ibid.

[9]The Report at 128.

[10]The Report at 126–127.

[11]The Report at 34.

[12]The Report at 95 (and seemingly also at 34).

[13]Ibid.

and Restoration (2018),[14] compiled over a period of 3 years by 100 experts from 45 countries. This Assessment provides a much stronger policy perspective and presents some disturbing findings about the likely global impacts of land degradation between now and 2050, including that:

- less than 25% of the Earth's land system has escaped substantial impact from human activity, and the estimate that that this will be less than 10% by 2050;
- by 2050, land degradation will have forced 50–700 million people to migrate;
- by 2050, a combination of land degradation and climate change is predicted to reduce crop yields by 10% on average, and up to 50% in some regions (especially central and south America, sub-Saharan Africa and Asia); and
- land degradation is affecting the well-being of at least 3.2 billion people and is pushing the planet towards a sixth massive species extinction.[15]

Surprisingly, the Report does not acknowledge the IPBES Assessment in its introductory section and the short overview of the IPBES activities was added only in its final revision. We suggest the Report has failed to appreciate the significance of the IPBES Assessment Report, which emphatically rebuts the view that land degradation is essentially a national rather than an international issue, and also makes a compelling case for immediate and urgent international action. The international dimensions are succinctly summarised in the following statement in the IPBES Assessment:

> Worsening land degradation caused by human activities is undermining the well-being of two fifths of humanity, driving species extinctions and intensifying climate change. It is also a major contributor to mass human migration and increased conflict....[16]

The Chair of the IPBES, Sir Robert Watson, also has identified the need for international action to be taken in the near future to address these challenges:

> Delaying the implementation of proven actions to combat land degradation will result in the necessary steps becoming progressively more difficult and costly. Existing multilateral environmental agreements, coupled with coordinated policy agendas that encourage

[14]IPBES (2018) available at https://www.ipbes.net/assessment-reports/ldr.

[15]See IPBES Media Release: Worsening Worldwide Land Degradation Now Critical, Undermining Well-Being of 3.2 Billion People, available at https://www.ipbes.net/news/media-release-worsening-worldwide-land-degradation-now-%E2%80%98critical%E2%80%99-undermining-well-being-32.

[16]Ibid. For a similar assessment, see UNEP (2019), p. 94: "The degradation of soil and land continues due to heighten[ed] competition for land use, *undermining the long-term security and development of all countries.* From 1999 to 2013, approximately one-fifth of the Earth's land surface covered by vegetation showed persistent and declining trends in productivity, primarily due to poor land and water management. . .Reversing these worrying trends through sustainable land management is key to improving the livelihoods and resilience of over 1.3 billion people living off degraded lands." (emphasis added).

> sustainable production and consumption, provide a platform for action to avoid and reduce land degradation and promote restoration.[17]

The evidence with respect to both the global extent and urgency of land degradation is mounting rapidly at present. To take one further example, in January 2019, the Committee for the Review of the Implementation (CRIC) of the UN Convention to Combat Desertification (UNCCD)[18] reviewed a global assessment of land degradation which found that 169 countries across the globe are affected by land degradation, desertification or drought.[19] It also found that almost one-fifth of all land within countries who are Parties to the UNCCD is subject to land degradation.[20]

We suggest that there is an emerging "common concern" about land degradation that might well provide a platform in the immediate future for the development of a new international instrument specifically related to sustainable soil management and/or the goal of Land Degradation Neutrality (LDN)[21] that has been identified in Sustainable Development Goal (SDG) target 15.3.[22] As has proven to be the case with respect to biodiversity protection, the absence of direct transboundary environmental impacts from land degradation should not be regarded as an immoveable obstacle to such an initiative. However, generating the necessary political appetite for a new instrument is a genuine challenge, which the Report acknowledges, and it is to this aspect of the Report that we turn next.

---

[17]Ibid. For a similar call for action, see Institute for Public Policy Research (2019), available at https://www.ippr.org/research/publications/age-of-environmental-breakdown (arguing that land degradation is a top-level global issue that is being under-played).

[18]UNEP (1994).

[19]See IISD, SDG Knowledge Hub (2019), available at http://sdg.iisd.org/news/unccd-cric-reviews-global-assessment-of-land-degradation/, for a report on the CRIC review.

[20]Ibid.

[21]See UNCCD (2016), available at http://www2.unccd.int/sites/default/files/relevant-links/2017-01/18102016_Spi_pb_multipage_ENG_1.pdf, defining LDN as 'a state whereby the amount and quality of land resources necessary to support ecosystem functions and services and enhance food security remain stable or increase within specified temporal and spatial scales and ecosystems'.

[22]Transforming our World: the 2030 Agenda for Sustainable Development, UN A/RES/70/1, https://sustainabledevelopment.un.org/post2015/transformingourworld. The general aim of SDG 15 is to 'conserve and restore the use of terrestrial ecosystems such as forests, wetlands, drylands and mountains by 2020'. Note that fifteen soil scientists, in Keesstra et al. (2016), outline in a landmark paper how to reach the recently adopted UN Sustainable Development Goals (SDGs) in the most effective manner. The authors point out (at 111) that "soil science, as a land-related discipline, has important links to several of the SDGs, which are demonstrated through the functions of soils and the ecosystem services that are linked to those functions".

3 New Treaty or Institutions

The Report includes, alongside its review of existing instruments and institutions related to soil, a survey of initiatives for a new international soil instrument, and concludes that:

> Considering the number of initiatives over the years that never made it on[to] the political agenda, there does not appear to be an appetite for a treaty specifically on soil – whether stand-alone or under one of the existing conventions, at least in the medium term.[23]

Its recommendation therefore is to "work towards creating the political conditions for [a] new binding instrument such as a treaty on soil protection" in the medium to long term.[24] As an immediate, short-term option, it recommends exploring how soil might be recognised as part of a possible political declaration on the environment, should the UN General Assembly mandate the 5th UN Environment Assembly in 2021 to prepare such a declaration.

The relegation by the Report of the option of a new binding instrument to a medium to long-term option is based on two observations by the authors. First, there is a general assessment that there is no "political appetite" for such a measure at present; second, it notes that "states might consider land and soil to be special because it defines their main territory and is relevant for food production."[25] This diplomatic language in the Report reframes the view held by some nations that soil management is a national issue and that "external" requirements for land-use are a violation of their sovereignty and authority. The Report presents the following conclusion:

> No group of states regard themselves as potential 'winners' from the international regulation of sustainable soil management, and thus there is little incentive for governments to become leaders and push the process forward.[26]

This is a significant conclusion that reflects a reluctance by the authors of the Report to urge the German government to take up a leadership role with respect to an international soils instrument, on the basis that there is no incentive for it to do so. However, it is submitted that the Report has overlooked the possibility of a potential, new approach to the development of an international soil instrument that has been demonstrated by the Paris Agreement on climate change. This approach may be more attractive to states than the conventional, regulatory approach that has been canvassed previously. To further explain this idea, it is necessary to advance two preliminary propositions.

First, as emphasized strongly in the IPBES Report mentioned above, the most extensive direct global driver of land degradation is "the rapid expansion and

---

[23]The Report at 121.

[24]The Report at 129.

[25]The Report at 126-7.

[26]The Report at 133.

unsustainable management of croplands and grazing lands".[27] However, as extensively detailed in the Report, all initiatives to date to address this problem through the development of an international instrument for soil, either as a stand-alone agreement or as a protocol to an existing convention such as the UNCCD or the UN Convention on Biological Diversity (CBD),[28] have proven fruitless. The reality (as recognised by the Report) is that the political climate for an instrument that would directly regulate soil management at the national level does not exist. However, we submit that this is not necessarily the end of the conversation about a new global instrument, as has been assumed in the Report. Based on the model provided by the Paris Agreement, it may be possible to contemplate a more immediate initiative to develop a similar international agreement on soils.

The second proposition, which is also advanced strongly in the Report, is that:

> In [the] absence of specific obligations on soil protection, the main substantive governance gap is the absence or inadequacy of transparency and review mechanisms at the *international* level.[29]

We agree with this observation but question the option recommended in the Report to address this gap, the Report recommends action to "gradually shape the UNCCD more towards the model of the Paris Agreement through strengthening transparency and review"—for example, by way of specific requirements, decisions and guidance on transparency and review in relation to national strategies and plans.[30] It also recommends voluntary LDN target setting by governments who have not yet done so.[31]

This recommendation is questionable, given the extensive criticism levelled at the UNCCD by the Report and its general, damning conclusion that there is "almost universal agreement that desertification governance has been a failure".[32] It constitutes either an expression of unjustifiable optimism with respect to the future implementation of the UNCCD or a clutching at straws to find a suitable strategy to recommend to the German government for future diplomacy on international soils governance. In advancing this option, an opportunity has been missed to promote the alternative strategy of developing a new international instrument that focuses on comprehensive reporting and review obligations, rather than those of a direct,

---

[27]Supra fn 18.

[28]UNEP (1992a, b).

[29]The Report at 129.

[30]The Report at 130.

[31]Ibid.

[32]The Report at 46. These criticisms include (see at 26–39) that the UNCCD applies to only 40% of the terrestrial surface of the Earth; that its definition of land degradation could include virtually any change of land-use; that although National Action Plans (NAPs) are core legal obligations, they appear to have had little effect in practice; that eight years after adoption of a 10-year strategic plan, only 20% of parties to the UNCCD have aligned their NAPs with it; and that the "CCD's obligations are rather general and toothless, lacking in precison and prescriptiveness, and its potential for a global approach on LDN is limited by its geographical scope" (at 37).

regulatory nature that appear to be unpalatable to a number of states. In this regard, a clear model has been provided by the Paris Agreement in the context of climate change that could readily be adapted to the protection of soil under the concept of sustainable soil management.

The Paris Agreement represents an historically innovative use of international law that involves a less regulatory and more information-focused approach. It establishes several global goals and also a collective review mechanism that requires the level of ambition of states to increase over time and to align with the achievement of the global goals. Most importantly, states are required to make voluntary commitments in the form of Nationally Determined Contributions (NDCs) that are complemented by obligations with respect to regular evaluation and review.[33] In terms of enforcement, its principal mechanism is the "naming and shaming" of parties who are found to be lagging behind in their level of ambition or in the implementation of their commitments.[34]

The architecture of the Paris Agreement involves a new, hybrid form of global instrument that embodies both top down global goals and bottom up national commitments. Whilst the climate goals are of a political rather than a legal nature, the reporting and review requirements are mandatory under international law".[35] This architecture could be readily adopted and adapted in the context of soil protection to provide a similar combination of political goal identification backed up by firm obligations to report and review in relation to state commitments that would be implemented through national soil strategies and plans.

With respect to the establishment of a political goal, the LDN target in SDG 15.3 provides a ready-made and widely-accepted point of reference that could be incorporated within a Paris Agreement style of instrument. This would be a far more forceful action than that proposed in the Report of supporting the SDG process "as a political reference point for other fora and processes".[36] Likewise, the need for globally-based transparency and review mechanisms could be addressed through the prescription of "bottom-up" obligations based on a similar approach to the NDC mechanism established by the Paris Agreement. States would need to identify LDN targets and then to report on their efforts to accomplish these targets. As has proven necessary with respect to the implementation of the Paris Agreement, a "rulebook" would need to be developed to flesh out exactly how these commitments would be measured and evaluated. As the report notes, considerable effort is already being made by both the UNCCD system and FAO to elaborate SDG 15.3, which would provide a solid basis for the necessary rulebook.

Finally, there is the significant question as to where the initiative would come from for a new global instrument on soil protection based on the Paris Agreement

[33]See Article 3.

[34]This particular element of the Paris Agreement has also been criticised in terms of its likely effectiveness: see Lawrence and Wong (2017).

[35]See Clark (2018).

[36]The Report at 128.

model. As noted earlier, the Report adopts a very pragmatic view that no states at present see themselves sufficiently as potential "winners" to take up an initiative on a new global instrument. This may be so where the aim of the instrument is to impose regulatory obligations with respect to soil protection on participating states, but the same argument is less compelling should the proposed instrument be based on the Paris Agreement model. There also may be a more widely shared "common concern" to establish measures of a global nature that would give greater force and effect to the SDG-inspired goal of achieving LDN by 2030. An initiative of this nature would be very timely if there is to be any real prospect of accomplishing this goal within the suggested time-frame. It is unfortunate that the Report has missed the opportunity to present this option to the German government for its consideration.

It is worth noting that the negotiation of the Paris Agreement began in 2011 with the appointment at the UNCCD Conference of the Parties (COP 17) in Durban of a dedicated, ad hoc subsidiary body (the Ad Hoc Working Group on the Durban Platform for Enhanced Action). The Working group was charged to undertake a process under the UNFCC "to develop a protocol, another legal instrument or an agreed outcome with legal force". Four years later, the hybrid instrument that emerged from the negotiations was considered an agreed outcome with legal force, but it was not a specific protocol or conventional legal instrument. The French government played an important role behind the formal UN processes in promoting this novel approach. The Report might have considered whether the German government could perform a similar role with respect to an international soils instrument based on the model of the Paris Agreement.

A similar process in relation to soils might be initiated in several ways. These include an initiative through the Global Soil Partnership (GSP), which was established in 2012 by FAO with the mandate to "meet the need for a multilateral agreement focusing specifically on soil challenges, and to advocate for sustainable soil and land management at the global level".[37] Alternatively, though perhaps more problematically in the political context, a joint initiative under the UNCCD and CBD might be contemplated.

However, if there is a continued reluctance on the part of states, including Germany, to pursue such an initiative through these or other avenues, it may fall to the non-government sector to do so at first instance. In the case of soil, both the soil science community and the International Union for the Conservation of Nature (IUCN) have a long record in this regard. The opportunity to promote such an initiative presents itself strongly at the moment and it is unfortunate that the Report has failed to recognise this, preferring instead to promote options that focus on improving existing soil governance arrangements. It is to these options that we turn next in this critique.

---

[37]Cited in the Report at 97; it is useful to note here that the FAO (2017) Voluntary Guidelines for Sustainable Soil Management were adopted by the 4th Global Soil Partnership Plenary Assembly (Rome, 25 May 2016). The VGSSM elaborate the principles outlined in the revised World Soil Charter 2014.

4 Improving Existing Governance

The Report extensively discusses existing governance arrangements at the global level relating to soil.[38] In particular, it shows that several existing treaties, institutions and fora that are relevant for soil governance at the international level already address specific soil conditions. These conditions include drylands, wetlands and forest soils; soil components such as soil biodiversity and soil carbon; and soil threats such as soil pollution and soil erosion. The Report suggests "that options to improve international soil governance include the strengthening or expansion of soil related provisions and policies within the existing fora". In this regard, the Report focuses on the UNCCD, CBD, and the role of FAO and the United Nations Environment Program (UNEP),[39] but does not raise the significant role that peak soil-science bodies play in soil governance.[40]

4.1 Treaties

4.1.1 UNCCD

The first treaty addressed by the Report is the UNCCD, where, as noted previously, the suggestion is to "gradually shape the UNCCD more towards the model of the Paris Agreement through strengthening transparency and review".[41] The specific elements of this approach are not explained fully in the Report and it is difficult therefore to assess the feasibility of this recommendation. We have raised the possibility of an alternative approach involving the development of a new international instrument for soil that is styled on the model of the Paris Agreement. We believe this could be a more effective and efficient approach ultimately than the option proposed in the report of "reshaping" the UNCCD through decisions of the Parties, especially given the extensive deficiencies of the UNCCD framework that have been outlined in the Report.

---

[38]The bulk of the report is devoted to a "stocktake of existing international soil governance": see Part 2, pp. 37–125.

[39]See also Boer and Hannam (2015) available at: http://www.austlii.edu.au/au/journals/IntJlRuralLawP/2015/2.pdf; and Boer and Hannam (2003).

[40]E.g., the International Union of Soil Sciences (IUSS) was founded as the International Society of Soil Science (ISSS) on 19th May 1924. The International Union of Soil Sciences (IUSS) is the global union of soil scientists. The objectives of the IUSS are to promote all branches of soil science, and to support all soil scientists across the world in the pursuit of their activities.

[41]The Report at 130.

4.1.2 UNCBD

In relation to the UN CBD, the Report outlines how the mandate of the CBD can provide for soil. It argues that a document could be commissioned by the CBD Secretariat within the CBD's Technical Series, addressing the topic of "sustainable use of soils" under the CBD.[42] It further suggests that, as a minimum, such a document would collate in handbook-style all CBD objectives, principles, rules, decisions etc. with relevance for the sustainable use of soil and describe their respective soil implications.[43] These are sensible proposals that we support.

Regarding the implementation of the CBD's International Biodiversity Strategy, the Report calls for an increase in soil-related activity by the FAO (which co-manages the Strategy) and argues for better integration of soil biodiversity into national biodiversity strategies and action plans, and into sectoral policies and national reports. This critique agrees with this suggestion, but the question remains as to which body would be best placed to manage its implementation. Further suggestions made in the Report which have merit include to integrate the sustainable soil use concept into the next revision of the Strategic Plan for Biodiversity (2021–2030) and to consider the development of a "soil biodiversity" Achi Target[44] when some of these targets expire in 2020.[45]

The Report is cautious with respect to possible changes to the CBD itself to give more explicit recognition to soil biodiversity and does not put forward any specific recommendation in this regard. Integrating an explicit soil focus into the CBD text would require an amendment to the Convention and thus first a decision of the Conference of the Parties (COP) (Art. 29, CBD) and it concludes this would have "limited value added.".[46]

The Report also identifies a far-reaching option to integrate soils into the CBD process through the development of a "Soil Protocol" under the Convention, and suggests this can be considered a medium- to long-term perspective, when all other options have been pursued.[47] This suggestion has merit as a protocol could address

---

[42]The Report at 131; Boer and Hannam (2019) contend "that for the CBD to take on an expanded, more precise role in addressing land degradation and encouraging the sustainable use of land, it would be desirable for technical guidelines to be drafted on sustainable land management, reinforced by an extra protocol to the Convention".

[43]Ibid. The IUCN World Commission on Environmental Law's Specialist Group for Sustainable Use of Soil comprehensively explored in 2009 the potential role for the application of the CBD to achieve sustainable use of soil and drafted a possible protocol under the CBD on this topic. A detailed Commentary also was prepared which comprehensively discussed the legal and scientific background for each Article in the proposed instrument.

[44]https://www.cbd.int/sp/targets/.

[45]The Report at 131. See also at 137, where it is noted that "one proposed indicator for SDG target 15.3 – "proportion of land that is degraded over total land area" – could also be used as an indicator for Aichi Target 5 ("At least halving the rate of loss of all natural habitats"); the same applies for indicators for soil organic carbon as well as for soil organic matter content (for Aichi Target 7, 15)."

[46]Ibid.

[47]The Report at 132; see also, Boer and Hannam (2015), p. 7.

the conservation and sustainable use of soil biodiversity, while promoting sustainable soil management as a strategy to improve terrestrial biodiversity. Unlike the FAO Voluntary Guidelines on Sustainable Soil Management (VGSSM), which are non-binding,[48] a CBD Protocol would constitute a binding international agreement and as such would have substantial political weight. However, as the report also acknowledges, "for this proposal to succeed, a protocol would require a COP decision (Art. 28 CBD) and subsequent ratification by sufficient parties". The first obstacle that would have to be confronted is that, politically, in the short term, it seems unlikely that the CBD parties will engage in developing a new protocol. Considering a Soil Protocol under the CBD, therefore, may be a less effective strategy than the alternative we have suggested of developing a Paris Agreement style of instrument in relation to soil protection, preferably in the near future.

4.1.3 The Paris Agreement

The Report briefly addresses the possible role of the Paris Agreement in sustainable soil management, after examining the role of FAO and UNEP. It notes that "the Agreement has spawned climate initiatives by sub-national authorities and non-state actors which can also add political weight to international soil policy". The Report indicates the nexus between land use and soil management can be used also to further address sustainable soil management within the climate regime but does not develop this idea.

In 2018, the Paris Agreement adopted a framework which includes rules for reporting on and accounting for land use, land-use change and forestry that will replace the existing 1995 United Nations Framework Convention on Climate Change (UNFCCC)[49] framework. The Report outlines that, while it will take some time to shape these rules, the new framework will provide an opportunity to demonstrate efforts to report on climate change management activities.[50] The Report also notes the UNFCCC's "Koronivia" joint work on agriculture, which was mandated in 2017 (Decision C.23), provides an opportunity, through the EU, to shape the UNFCCC's future work on agriculture.[51]

[48]http://www.fao.org/3/a-bl813e.pdf.

[49]UNEP (1992a, b).

[50]The Report at 134.

[51]Ibid. Re Decision C.23, see https://unfccc.int/files/meetings/bonn_nov_2017/application/pdf/cp23_auv_agri.pdf.

4.2 Institutions

4.2.1 FAO

The Report argues that, despite the fact that soils already play a significant role within the FAO, further activities on the part of FAO would strengthen international soil governance.[52] For example, as pointed out in the Report, FAO's Strategic Objectives could be adjusted to better reflect the importance of soils and thereby contribute to the achievement of the FAO's other Strategic Objectives.[53] Further, and based on such a strategic readjustment, a consistency check might be helpful to examine whether there are policies and programmes within FAO (e.g., on agricultural mechanisation, crop production intensification or animal production) that are not fully coherent with the sustainable management of soils (as defined in the VGSSM).[54] Given the strong mandate of the FAO in practical agriculture and food production, especially in dryland and poorer nations of the world, it may be difficult to expect any radical change to a more holistic or ecosystem approach to soil management.

4.2.2 UNEP

The Report suggests that despite a tradition of addressing soils, the United Nations Environment Program (UNEP) has not had a high profile with regard to soil governance in recent years. However, this critique points out that UNEP introduced a comprehensive strategy for soil conservation in 2004.[55] In this regard, it is agreed that the UN Environment Assembly's (UNEA) recent resolution for "Managing Soil Pollution to achieve Sustainable Development"[56] could be used as an entry point to strengthen UNEP's capacities and role with regard to international soil governance. An immediate step within UNEP would be to increase the capacity for implementing the soil pollution resolution and the monitoring of its impact. This would include "not only the envisaged preparation of a global assessment on soil pollution but the

[52]Of significance, the FAO has carriage of the revised World Soil Charter 2014 and the Voluntary Guidelines for Sustainable Soil Management through the Global Soil Partnership.

[53]The Report at 133; including to help eliminate hunger, food insecurity and malnutrition; make agriculture, forestry and fisheries more productive and sustainable; reduce rural poverty; increase the resilience of livelihoods to threats and crises; http://www.fao.org/about/what-we-do/en/.

[54]The Report at 133.

[55]UNEP (2004); it should be noted that the 2004 Strategy outlined the critical issues UNEP saw in environmental assessment, policy guidance and implementation to improve the integration of environmental land and soil aspects across other environmental focal areas, and relevant international, regional and national development processes. The UNEP Strategy promoted the "ecosystem" approach for land management and soil conservation.

[56]UNEA 2 December 2017; https://www.informea.org/en/decision/managing-soil-pollution-achieve-sustainable-development.

provision of technical capacities and earmarked programmes for supporting governments' efforts to strengthen and coordinate national and regional policies to curb soil pollution (through in-house as well as technical and legal assistance etc.)".[57] This critique concurs that, in accordance with the Resolution, this would also include the development of guidelines (jointly or coordinated with FAO and the World Health Organisation) for the prevention and minimization of soil contamination which is a growing world problem.[58]

5 Means of Implementation

5.1 Capacity Building

The Report strongly recommends "the improvement of capacity building for accessing international support, in particular through information and technical assistance". It points out that the means of implementation, including financial support, capacity building and technology development and transfer for developing countries, is a recurring theme in international environmental law and governance". This critique agrees that, besides bilateral development assistance, there are many multilateral channels with significant resources to support activities related to land and soil, notably, the Global Environment Facility (GEF), the CCD's LDN Fund and the Green Climate Fund.[59] Further, it is prudent to point out that the various existing guidelines on the sustainable use of soil could help ensure that adequate provision is made for soil in different types of land management programs, which could be funded through the aforementioned avenues. It is suggested that the Global Soil Partnership (GSP) could possibly take up this role, which could further enhance and extend the application of the GSP and the VGSSM framework.

5.2 Subsidies

The Report recommends the promotion of political commitments by governments, as well as international organisations like the World Bank, to "rationalise, reduce and eliminate those subsidies that are incompatible with sustainable soil management". It

[57] It is noted here that the UNEP (2018) Final Assessment of the Fourth Program for the Development and Periodic Review of Environmental Law (Montevideo IV) at 77, presents an overview of its mandated actions to achieve its objective. This is: "to improve national and international principles and standards and to support efforts under the United Nations Convention to Combat Desertification for the further development of legal approaches for the conservation, restoration and sustainable use of soils".

[58] The Report at 134.

[59] The Report at 134.

points out that this approach could build on the SGDs, which address specific types of unsustainable subsidies.

This review agrees that developing countries should receive financial incentives for voluntarily eliminating subsidies that undermine or run directly counter to soil protection. However, for this to be purposeful, technical assistance for a transition to more sustainable land management also should be provided. The Report goes on to point out "that such assistance needs to be checked against already available finance and other support, as well as development priorities". It is prudent to reform or remove altogether environmentally harmful subsidies. Practices and incentives that have a degrading impact on soil should be removed and revenue-based policies could be applied to protect the soil.[60]

The Report notes that, at the international level, "the SDGs already stipulate that agricultural export subsidies, as well as certain types of subsidies in the fishing sector and 'inefficient' fossil fuel subsidies, should be eliminated". It rightly suggests that this provides a justification for calling for the reduction of subsidies that are detrimental to soil health and which would "counteract the achievement of, among others, SDG 1 (End poverty), 2 (End hunger), 3 (Ensure healthy lives), 14 (Sustainable use of oceans etc.) and 15 (Sustainable use of terrestrial ecosystems)".[61]

6 Enhancing Coordination and Coherence

The Report concludes with the observation that the "current governance of soil at the international level is piecemeal and spread over parts of different mandates such as biodiversity, desertification, food and agriculture".[62] It acknowledges that, "while several institutions address soil, none has a clear or universally accepted mandate to address soil in general". The fact is that the environmental interests of soil are well represented across the world with a number of global and regional soil scientific bodies together with multilateral institutions.[63] However, the final section of the Report suggests that there is scope to improve coordination and coherence in international soil governance through two avenues in particular. These are agreeing on a clearer division of labour between the relevant institutions (in particular, the UNCCD, CBD, Paris Agreement, FAO/GSP and UNEP); and the establishment of a new coordinating forum which should also make provision for the peak global soil institutions. These institutions have many years of experience and involvement in

[60]The Report at 135.

[61]Ibid.

[62]Ibid.

[63]E.g., The International Union of Science, World Association of Soil and Water Conservation, European Society for Soil Conservation, Soil Science Society of America.

soil policy and management and provision should be made for this in future global soil governance.[64]

6.1 Division of Labour

The Report suggests that a clearer division of labour between the institutions "holds significant potential for improving international soil governance and notes that a rudimentary division of labour is already emerging". It observes that there are interesting current developments regarding soil at the international level that could be relevant for a potential future division of labour. For example, "the rivalry between FAO and UNCCD could be resolved if each focused on different but specific issues. At the UNCCD, the SDGs and in particular the LDN target have catalysed action on LDN target setting and national action plans, which advantageously has been specifically supported by the GEF". The FAO has long been active in developing methodological tools and data collection, in particular through the GSP, as well as in developing voluntary standards related to sustainable soil management and land rights.[65] Under the Paris Agreement, there is a new framework for agriculture, and land use is becoming increasingly important as a carbon sink in order to reach the Paris Agreement's climate goals.[66] In 2017, UNEA adopted a ministerial declaration on a 'pollution-free world' as well as the Resolution on soil pollution, which UNEP has yet to follow up on.

The Report clearly points out that while some division of labour appears to be evolving in the wake of the SDGs, it does not seem satisfactory at this point and requires further effort. It is recognized that the relevant institutions addressing soil, as identified above, could do more to work towards a practical division of labour. Importantly, this includes "discussing overlaps and gaps in soil governance with regard to biomes, types of soil and drivers of soil degradation".[67]

The question we pose with respect to this particular recommendation is whether it is likely to prove effective in practice, given the multiplicity of overlapping interests and mandates involved and the associated conflicting priorities. The Report concedes that current efforts to achieve an effective division of labour have been inadequate, but offers little in the way of specific ideas as to how the further effort required can be made to happen. This critique suggests that a soil instrument styled on the Paris Agreement model, under the concept of sustainable soil management, as

[64]See Hurni and Meyer (2002).

[65]Ibid.; it should be noted that the various versions of draft protocols developed by the IUCN World Commission on Environmental Law's Specialist Group on Sustainable Use of Soil make provision for a peak global body on soil, not unlike the concept of the existing GSP, but with a much more extensive role than the GSP.

[66]Paris Agreement Article 5.1 at 6.

[67]The Report at 136.

outlined above, could provide for the establishment of a global peak body on sustainable use of soil. We contend that it could more effectively address the overlaps and gaps in labour by providing a single coordinating forum.[68]

6.2 *Establish a Coordinating Forum*

The Report proposes that a coordinating body could be developed in the medium-term, with a mandate designed to coordinate and promote international soil governance.[69] It suggests that "such a forum could draw on the experience of the Secretariats of international treaties and organisations where they have cooperated in various formalised degrees in the past, through a Joint Liaison Group (JLG) and the Rio Conventions Pavilion".

The example elaborated in the Report is that a "soil coordinating forum" could draw on the high-level approach of the Stockholm, Rotterdam and Basel Conventions which have to some extent institutionalised their cooperation in order to create synergies, e.g. through joint COPs and a joint secretariat. However, it recognises that, "in contrast to these three Conventions, the fora involved in soil are specific types of institutions and this level of formal cooperation may be unsuitable for soil". However, a functioning division of labour regarding soil is unlikely to be achieved through a similar top-down administrative exercise.

Instead, an option is suggested in between the organic development of existing fora and a high-level formalised approach, that is, a "Collaborative Mechanism on Soils" modelled on the "Collaborative Partnership on Forests".[70] A Collaborative Mechanism on Soils could be mandated to coordinate and promote international soil governance as an interagency-forum. It is perceived that this mandate would have more political weight than the existing Inter Agency Advisory Group's mandate, which is limited to developing indicators. Meetings of a Collaborative Mechanism could be back-to-back with meetings of the Global Soil Partnership.[71] It would also provide a supporting mechanism to the extensive work of the global peak soil-science institutions.

This is a worthwhile proposal that deserves further attention, preferably in the more immediate future rather than as a medium to long term option, as suggested in the Report. As suggested above, the proposed Collaborative Mechanism could also provide an appropriate forum to pursue the development of a Paris Agreement type

---

[68]It should be noted that the various versions of draft protocols developed by the IUCN World Commission on Environmental Law's Specialist Group on Sustainable Use of Soil make provision for a peak global body on soil, not unlike the concept of the existing GSP, but with a much more extensive role.

[69]Ibid.

[70]The Report at 137.

[71]The Report at 138.

of global instrument on soil, and it could eventually be enshrined within the proposed instrument.

6.3 Coherence of Soil Information

While there is no lack of general substantive guidance for national soil policies, it is desirable, as the Report suggests, that existing guiding documents "be checked for duplication and coherence and, if necessary, be consolidated, for instance in a non-binding instrument". While there is a very wide range of soil information available for potential review, the Report falls short in describing examples of this material. The UNCCD's Inter-Agency Advisory Group could discuss the issue of duplication and coherence, but it is suggested that this might require an adjustment of its mandate.

A considerable amount of international material exists on soil policy, sustainable land management and LDN that states can draw on for formulating and implementing their national soil policies.[72] While only a limited amount of this material is binding, it provides a basis nevertheless for states to draw on to develop, adopt and implement soil policies.[73] While some specific issues around soil protection might be addressed in a limited way, or might only be addressed to a small extent, the Report recognizes that some issues such as domestic planning and zoning laws are usually not specifically addressed at the international level. In this regard, the Report specifies that consolidated guidance could be modelled on, for example, the "United Nations Forest Instrument" adopted by the General Assembly in 2007.[74] It is agreed that "the added value to existing overarching instruments such as the revised World Soil Charter of 2015 would need to be assessed, taking into account the political and academic effort involved".

7 Conclusions

There are many plausible recommendations made in the Report to improve the global governance of soil, but there are also various institutional, legal and political barriers to achieving this objective. This critique recognizes the complexities involved in achieving Sustainable Development Goal 15.3 with respect to LDN and acknowledges many of the recommendations presented in the Report in this

[72]For example, the FAO Voluntary Guidelines for Sustainable Soil Management have a strong policy element and could readily be turned into policy material.

[73]The Report at 138; see Hannam and Boer (2004) Part IV, 37–80.

[74]Originally adopted as the "Non-legally binding instrument on all types of forests", renamed by UNGA Res 70/199 of 16.2.2016.

regard. Various options are presented to improve the role of the UNCCD but this instrument is limited by its principal focus on dryland ecosystems. Alternatively, the CBD offers greater potential but an amendment of this instrument would be required to provide specifically for soil biodiversity. As an alternative to the enhanced application of these two instruments to improve the global governance of soil, a proposal is presented in this critique for the development of a binding instrument for soil, similar in style to the Paris Agreement.

At the institutional level, the Report makes various recommendations to improve the role of the FAO and the UNEP in soil governance. However, this critique agrees with the position put forward in the Report that the current political economy of the institutions provides significant barriers to a general improvement in soil governance being achieved, including a general lack of cooperation between the institutions. In this regard the proposal advanced in the Report of a "Collaborative Mechanism on Soils", modelled on the "Collaborative Partnership on Forests", is to be welcomed. This mechanism could be mandated to coordinate and promote international soil governance as an interagency-mechanism. However, we suggest that, rather than envisaging this as a medium to long-term initiative, it should be pursued more immediately with a view to possibly enshrining the "Collaborative Mechanism" within a global soil instrument styled on the Paris Agreement.

References

Boer B, Hannam I (2003) Legal aspects of sustainable soils: international and national. Rev Eur Commun Int Environ Law 12(2):149

Boer B, Hannam I (2015) Developing a global soil regime', Special edition 1 Soil governance, International Journal of Rural Law and Policy 1, Available at: http://www.austlii.edu.au/au/journals/IntJlRuralLawP/2015/2.pdf

Boer B, Hannam I (2019) Land degradation. In: Lees E, Vinuales J (eds) Oxford handbook of comparative environmental law. Oxford University Press, Oxford, pp 438–459

Clark K (2018) The Paris agreement: its role in international law and American jurisprudence. Notre Dame J Int Comp Law 8(2):107

Food and Agriculture Organization of the United Nations (2017) Voluntary guidelines for sustainable soil management. Food and Agriculture Organization of the United Nations, Rome

Hannam I, Boer B (2002) Legal and institutional frameworks for sustainable soils: a preliminary report. IUCN, Gland. Xvi + 88p

Hannam I, Boer B (2004) Drafting legislation for sustainable soils: a guide. IUCN, Gland. X + 100pp

Hurni H, Meyer K (eds) (2002) A world soils agenda, discussing international actions for the sustainable use of soils, prepared with the support of an international group of specialists of the IASUS Working Group of the International Union of Soil Sciences (IUSS). Centre for Development and Environment, Berne

IISD, SDG Knowledge Hub (2019) "UNCCD CRIC reviews global assessment of land degradation", 5 February 2019

Institute for Public Policy Research (2019) This is a crisis: facing up to the age of environmental breakdown. Institute for Public Policy Research, London

IPBES (2018) Intergovernmental science-policy platform on biodiversity and ecosystem services, The IPBES assessment report on land degradation and restoration

Keesstra SD et al (2016) The significance of soils and soil science towards realization of the United Nations sustainable development goals. Soil 2:111–128. https://doi.org/10.5194/soil-2-111-2016, 2016 www.soil-journal.net/2/111/2016/
Lawrence P, Wong D (2017) Soft law in the Paris climate agreement: strength or weakness. Rev Eur Comp Int Environ Law 26(3):276
UNCCD (2016) Science-policy brief 02-September 2016
UNEP (1992a) UNEP convention on biological diversity, 31 ILM 822
UNEP (1992b) United Nations framework convention on climate change, 31 ILM (1992) 849
UNEP (1994) United Nations Convention to Combat Desertification (UNCCD) 33 ILM 1328
UNEP (2004) United Nations Environment Program, UNEP's strategy on land use management and soil conservation, A strengthened functional approach, UNEP Policy Series, Nairobi
UNEP (2018) Final assessment of the fourth program for the development and periodic review of environmental law (Montevideo IV), Evaluation Office of UN Environment
UNEP (2019) Measuring progress: towards achieving the environmental dimension of the SDGs

The Global Soil Partnership: Tackling Global Soil Threats Through Collective Action

Natalia Rodríguez Eugenio

1 The Global Soil Partnership

Soil is a finite natural resource. On a human time-scale it is non-renewable. However, despite the essential role that soil plays in the life of people, as a basis for food security and their provision of key ecosystem services, including climate change adaptation and mitigation, there is increasing degradation of soil resources due to inappropriate practices, burgeoning population pressures and inadequate governance over this essential resource. The increasing degree and extent of soil degradation processes due to mismanagement and land use changes are threatening this resource. Urgent action is needed to reverse this trend if we are to assure the necessary food production for future generations, mitigation of climate change, provision of clean groundwater, and halting the loss of biodiversity.

The renewed recognition of the central role of soil resources as the basis for life on Earth has led to numerous regional and international projects, initiatives and actions over the past decade. Despite these numerous emergent activities, soil resources are still seen as a second-tier priority. There has been no international governance body to advocate for and coordinate initiatives to ensure that knowledge and recognition of soils were appropriately represented in global change dialogues

The authors wish to thank Fiona Bottigliero, Camilla Gomes da Silva, Giulia Stanco, Yuxin Tong, Debra Turner, Isabelle Verbeke, Kostiantyn Viatkin, and Yusuf Yigini, who contributed data and editorial review of the chapter.

N. Rodríguez Eugenio (✉)
Global Soil Partnership, Food and Agriculture Organization of the United Nations, Rome, Italy
e-mail: Natalia.rodriguezeugenio@fao.org

H. Ginzky et al. (eds.), *International Yearbook of Soil Law and Policy 2019*,
International Yearbook of Soil Law and Policy 2019,
https://doi.org/10.1007/978-3-030-52317-6_11

and decision-making processes until the inception of the Global Soil Partnership (GSP).

Aware of the existence of this gap in international soil governance and in order to reverse the growing trend of soil degradation through the promotion of sustainable soil management (SSM),[1] the Committee on Agriculture (COAG) of the Food and Agriculture Organization of the United Nations (FAO), at its Twenty-third Session (May 2012), endorsed the establishment of the GSP as a voluntary body within the FAO structure. The GSP was automatically composed of all FAO member countries, but was open to any other public or private bodies, universities and research centres, NGOs and farmers' or civil society associations wishing to join the partnership.

The mandate of the GSP is to improve governance of the limited soil resources of the planet in order to guarantee healthy and productive soils for a food secure world, as well as support other essential ecosystem services, in accordance with the sovereign right of each State over its natural resources.

The GSP has adopted three interrelated approaches to achieve its mandate and to address current global soil issues: (1) policy advocacy and public awareness-raising at all levels, (2) development of technical tools that allow countries to generate their own soil information, and (3) programmatic actions focused on the implementation of SSM practices at field level.

During the first phase (2012–2014), the GSP focussed activities on building a strong network with multi-stakeholder partners and positioning soils, as well as the importance of SSM,[2] in the 2030 Sustainable Agenda being developed by the United Nations.[3] The first step was the establishment of eight regional soil partnerships (RSPs): in Africa; Asia; Central America, the Caribbean and Mexico; Europe, including the Eurasian subregional soil partnership; Near East and North Africa; North America; the Pacific; and South America, composed of the member states of the Food and Agriculture Organisation of the United Nations (FAO). Most of the 194 FAO member countries have appointed a focal point to the GSP, usually civil servants of their country's ministry of agriculture with a strong working knowledge and insight on soil management. Of the FAO member countries not officially represented in the GSP, many are informally represented by staff of national institutions that are part of the GSP's technical networks. Each region has developed its respective plan of action after an open consultation, in which national focal points to the GSP, FAO regional and country offices, soil scientist and other relevant stakeholders contributed, to define regional priorities, while maintaining a link with global priorities, as detailed in the Programme Document of the GSP for the period 2014–2018.[4] The RSPs facilitate synergies between countries with similar soil health threats and provide opportunities to share successful SSM practices across diverse regions.

[1]Montanarella and Vargas (2012).

[2]FAO (2014a).

[3]UNGA (2015).

[4]FAO (2014b).

In addition, the Intergovernmental Technical Panel on Soils (ITPS), consisting of 27 soil scientists, was established to provide scientific and technical advice to all GSP members on global soil issues, including overseeing the preparation of the first-ever report on the Status of the World's Soil Resources.[5] This report was the result of a collaboration between a large number of experts and contributed to the inclusion of soil-related targets and indicators in the Sustainable Development Goals (SDGs).[6,7] The ITPS has a 3-year mandate, and the current panel (2018–2021) is the third since its inception.

During the second working period (2015–2017), the GSP focussed considerable attention on the formulation of normative documents and technical guidelines, including the revised World Soil Charter,[8] and the Voluntary Guidelines for Sustainable Soil Management (VGSSM),[9] the Farmer's Compost Handbook[10] and the Handbook for Saline Soil Management.[11] Other main targets of the GSP during this period were the strengthening of international cooperation and the promotion of SSM[12] building on the momentum generated by the 2015 International Year of Soils, which was endorsed by the United Nations General Assembly.[13]

In the current working period (2018–2020), core efforts are focussed on addressing the global threats to soils identified in the Status of World's Soil Resources (SWSR) report[14] and on the implementation of the VGSSM as a way to halt soil degradation. Detailed information on current activities is provided in Sect. 3.

To implement the priorities for action[15] in the most efficient way, the GSP's priorities are articulated around the implementation plans of the five pillars of action: Pillar 1: Promote sustainable management of soil resources for soil protection, conservation and sustainable productivity;[16] Pillar 2: Encourage investment, technical cooperation, policy, education, awareness and extension in soil;[17] Pillar 3: Promote targeted soil research and development focusing on identified gaps, priorities and synergies with related productive, environmental and social development

[5]FAO, ITPS (2015a, b).

[6]Bouma and Montanarella (2016).

[7]Tóth et al. (2018).

[8]FAO (2015a).

[9]FAO (2017a).

[10]FAO (2015b).

[11]FAO, Lomonosov Moscow State University (2018).

[12]Rojas and Caon (2016).

[13]United Nations General Assembly (2014).

[14]FAO, ITPS (2015a, b).

[15]GSP (2019a).

[16]http://www.fao.org/3/CA2217EN/ca2217en.pdf.

[17]http://www.fao.org/3/a-az899e.pdf.

actions;[18] Pillar 4: Information and Data;[19] and, Pillar 5: Harmonisation of methods, measurements and indicators for the sustainable management and protection of soil resources.[20] The implementation plans for the five pillars have been adopted at different times and include measures for a period of 5 years. The activities proposed at the global level are also reflected at the regional level, in accordance with the priorities defined by the RSPs.

2 The World's Soils Under Threat

It is well-recognised that since the second half of the twentieth century, soil degradation is increasing at a global scale.[21,22] The land expansion of urban and industrial infrastructures into agricultural and natural areas that are essential for providing ecosystem services, including sufficient food, forests, fuel, fibre and raw materials for a growing population, is a great challenge to humanity. This reality, together with unsustainable consumption patterns, put enormous pressure on natural resources, whose resilience or adaptive capacity is limited.

Eight threats were identified as affecting European soils: erosion, soil organic matter decline, salinisation, compaction, floods/landslides, contamination, sealing and the decline in soil biodiversity.[23] However, at the global level, two additional threats were identified: soil nutrient imbalance and soil sodification.[24] These threats vary in terms of extent, intensity and trends depending on geographical context; nevertheless, they all need to be addressed in order to achieve sustainable development and food security.

These ten soil threats have often been addressed separately and considered as separate degradation processes. However, they should be considered as symptoms of a broader and more systemic problem, caused by overexploitation of soil resources associated with the current system of production and consumption of goods and food. In an attempt to halt the advance of these threats and reverse existing degradation, the GSP proposes the adoption of sustainable soil management practices.

[18]http://www.fao.org/3/a-az920e.pdf.

[19]http://www.fao.org/3/a-bl102e.pdf.

[20]http://www.fao.org/3/a-az922e.pdf.

[21]Oldeman et al. (1991).

[22]Montanarella et al. (2016).

[23]European Commission (2006).

[24]FAO, ITPS (2015a, b).

3 Assisting Countries in Raising Awareness and Promoting Policies for Soil Protection

One of the main priorities of the GSP is to raise awareness about soils and to foster soil protection policies. Since the adoption by the United Nations General Assembly (UNGA) of the International Year of Soils and the World Soil Day[25] in 2015, the GSP has reached millions of people with all the awareness-raising activities that have taken place for their celebrations. The GSP has acted as a unified and authoritative voice for soils to better coordinate efforts and pool limited resources. The worldwide celebrations are a unique opportunity to convey the message of the importance of soil for food security, healthy ecosystems and human well-being.

These communication activities have had a strong impact at the national level by raising awareness of society as a whole, contributing to support the GSP's mission of improving soil governance and promoting SSM to ensure healthy and productive soils and support essential ecosystem services. Numerous politicians and policy-makers have joined the voices in favor of soil protection, and have contributed to the strengthening of legislation in their respective countries.

In order to support national efforts to improve soil governance, to achieve the political component of Pillar 2, and to ensure that policy-makers and governments are more aware of the legal instruments at their disposal to protect this precious resource, the GSP is working on the preparation of a legal platform called SoiLEX.

SoiLEX is grounded in the already existing FAO database (FAOLEX),[26] which is one of the world's largest electronic collections of national laws, regulations and policies on food, agriculture and natural resources management. The aim of SoiLEX is to provide a selection of the most relevant legal instruments for soil protection and conservation, clustered by the type of degradation process they address or by the tools that they provide for SSM. SoiLEX will facilitate the analysis of the existing policies and regulations, and the assessment of their effectiveness for controlling soil degradation around the world.

According to an ongoing analysis of national soil-related legal instruments included in SoiLEX, legal instruments dealing with soil protection and conservation are often nonexistent or scattered through many other legislation. There are several dozen countries in the world that have enacted soil protection and conservation acts, such as Argentina,[27] Ghana,[28] New Zealand,[29] and Georgia.[30] However, in most countries, legal instruments that protect and conserve soils are included in broader legislation, such as environmental, water or waste management acts.

---

[25]UNGA (2014).

[26]FAO (2019a, b).

[27]Presidencia de la Nación Argentina (1981).

[28]The Republic of Ghana (1957).

[29]Ministry for the Environment, Government of New Zealand (1941).

[30]FAOLEX Database (2014).

International and national experts will analyse the instruments included in SoiLEX, and their effective implementation will be verified with field surveys and assessments, supported by national soil monitoring systems, which will allow the identification of the most successful cases. The most successful legislation can provide ideas to countries lacking such instruments to develop new legal instruments, taking into account the socioeconomic, administrative and legal particularities of each country.

In addition, to support and facilitate national efforts on soil protection and conservation, the GSP has developed two normative documents, the revised World Soil Charter,[31] adopted by the FAO member countries in the 1981 FAO Conference, and the VGSSM,[32] endorsed by the 155th session of the FAO Council in 2016. By presenting generally accepted, practically proven and scientifically-based principles to promote SSM, the VGSSM provides a framework for action against soil degradation.[33] They are an accessible and clear reference for a wide range of stakeholders involved in soil management, including government officials and policy makers, who should translate them into concrete actions and practices to support farmers with SSM.[34]

The GSP advocates for the implementation of the VGSSM at the national and regional level. In addition to the organisation of global symposia on the main soil threats, the organisation of national multi-stakeholder workshops is encouraged. Workshops aim to shape the implementation of the VGSSM around dominant human activities, climatic conditions and local and indigenous traditions.

Additionally, a protocol to assess whether soils are being sustainably managed as per the VGSSM is currently being developed by the ITPS and the pillar 1 working group. By addressing the ten main soil threats identified in the SWSR report and discussed in the VGSSM, and by providing a set of indicators to choose from, the protocol will allow land users to determine if soils are being sustainably managed or not, and if their conditions are improving, declining or remaining stable. The availability of quantifiable indicators is also important for other stakeholders, such as policymakers or extension services technicians, as it will allow them to make better and informed decisions based on the current use of soil resources. After an open consultation process to be launched during the second semester of 2019, the indicators will be defined and included in the protocol.

To facilitate the operationalisation of the VGSSM, technical manuals on SSM practices focussing on addressing the ten main soil threats as identified in the SWSR report are being developed by the ITPS and an ad-hoc expert working group. These technical manuals are intended to compile state-of-the-art technical knowledge and offer tools to adapt them to regional and local conditions. The first technical guidelines being developed concern the management of soils for soil organic carbon

[31]FAO (2015a).

[32]FAO (2017a).

[33]Baritz et al. (2018).

[34]Baritz et al. (2018).

(SOC) protection and sequestration at the national and local scales, and the prevention, management and remediation of soil pollution, described in detail in Sects. 4.1 and 4.2, respectively. The principles and measures mentioned in the VGSSM are defined in detail and with practical examples in these technical manuals.

However, despite all the technical knowledge that is being generated, the success of the VGSSM in preventing and reversing soil degradation depends also on stronger political action. The inclusion of the principles of the revised World Soil Charter and SSM practices in national and local legislation would facilitate the mobilisation of financial resources and effective action against soil degradation.

4 Addressing Soil Threats by Improving Soil Information Worldwide

FAO and UNESCO led the preparation of a world-wide collaborative work to obtain a global soil map at a 1:5,000,000 scale,[35] which was done over 20 years, and finally published in 1974. Since 1995, collaborative efforts have been made between FAO, UN Environment, the Joint Research Centre of the European Commission (JRC), and ISRIC (International Soil Reference Information Centre)—World Soil Information to make regional updates of the FAO/UNESCO soil map of the world, notably under the SOTER (Soil and Terrain Database) program. A SOTER database with global coverage was never achieved, but SOTER databases were developed for various regions, countries and continents.[36] Recent efforts focus on using digital soil mapping techniques (mapping soil properties continuously rather than soil associations with their related soil properties). In 2009, coordinated by the Digital Soil Mapping Working Group of the International Union of Soil Sciences (IUSS) and led by academic and research centres in all continents, the GlobalSoilMap Project was created as a global consortium of soil scientist with the aim of updating the global soil information using state-of-the-art and emerging technologies.[37] The GlobalSoilMap.net has produced maps of soil properties for several countries, such as France,[38] Nigeria,[39] and for Sub-Saharan Africa,[40] but once again, a global product has not been achieved. In 2014, ISRIC developed a Global Soil Information service called "SoilGrids", which consists of spatial predictions of soil properties and classes using limited soil profile data.[41] This global product has been developed using a top-down approach and has been reported to over or underestimate some soil

[35]FAO, UNESCO (1974).

[36]ISRIC (2016).

[37]Arrouays et al. (2014).

[38]Mulder et al. (2016).

[39]Akpa et al. (2016).

[40]Hengl et al. (2015).

[41]Hengl et al. (2017).

properties.[42] The pedometrics community supports the need for developing global products following a bottom-up approach to ensure accuracy and reduce uncertainties.[43]

The Global Soil Partnership is responding to the previously mentioned global gap on soil information and data, and promoting soil data collection to update existing information. FAO and the GSP have ongoing projects for the development of national soil information systems (NSIS) using new and legacy data in Afganistan, Cambodia, Lesotho, Turkey, Ukraine and Sudan, which will eventually be part of a federated Global Soil Information System (GloSIS), referring to a distributed or shared information system, which will enable data institutions to own their data and be able to control their access.[44] By establishing and strengthening the NSIS, countries are empowered to better manage their soils and to adequately monitor and report on the status and trends of soils in their territories.

As mentioned above, the scientific community accepts that new soil information and data should focus on generating information on soil properties rather than on soil classification maps. The GSP Pillar 4 working group is, therefore, focusing on building national capacities on soil properties mapping for the preparation of different thematic products that are explained in detail in the following sections.

In addition, in order to generate soil information comparable around the globe, the working group formed by the national contacts for the Global Soil Laboratory Network (GLOSOLAN) and its regional branches (RESOLANs), part of the GSP Pillar 5, is working on the preparation of harmonised soil analysis methods, measurements and indicators at regional and global levels. The aim of GLOSOLAN is to facilitate communication and knowledge exchange. This is of utmost importance, for example, to utilise soil information for policy development and the building of observation systems.

In 2016, the decision-making body of the GSP, the Plenary Assembly, approved the holding of annual symposia that address the most relevant soil threats, to gather the most recent scientific and technical knowledge to support the development of a global agenda for action. Work on soil threats was prioritised according to their urgency, their impact on the environment and human well-being, and their relevance to the global agenda. For the time being, five soil threats are being addressed by the GSP and the RSPs, namely soil organic carbon loss, soil pollution, soil erosion, soil biodiversity loss and soil salinisation. The plan is to continue to address all soil threats and to complete the updating of soil information at the national, regional and global levels.

The concrete actions to implement the VGSSM and to improve the availability and quality of soil information and data, addressing each of the threats to soil, which is being carried out by the GSP, are outlined in Sects. 4.1–4.5 below.

---

[42] Tifafi et al. (2018).

[43] Arrouays et al. (2017).

[44] FAO (2018a).

4.1 *Loss of Soil Organic Carbon*

Addressing the loss of soil organic carbon (SOC) was selected by the ITPS as the first priority action area due to its impact on soil productivity and climate change.[45] In addition, SOC has a major impact on other soil threats, such as soil erosion, nutrient imbalance and soil pollution. SOC has been internationally recognised to play a key role in soil health, biodiversity conservation and climate change mitigation. It was therefore included as one of the metrics to report on the SDG indicator 15.3.1 *Proportion of degraded land,*[46] corresponding to SDG target 15.3: *"By 2030, combat desertification, restore degraded land and soil, including land affected by desertification, drought and floods, and strive to achieve a land-degradation neutral world"*. The global reporting process on this target is coordinated by the United Nations Convention to Combat Desertification (UNCCD).[47] FAO has provided technical guidance in the development of methodologies and tools to assess land degradation and is supporting countries in regular data collection and analysis through capacity building programs in the use of FAOSTAT and FAO's products for enhanced analysis in member countries.[48]

Because soils are the largest terrestrial carbon reservoir,[49] SOC sequestration has recently been promoted as a solution to climate change.[50] However, the carbon cycle in soils is a complex process of transformations and fluxes of carbon between atmosphere, organisms (plants and plant material, macrofauna and microorganisms including bacteria and fungi) and soils, which is influenced by several factors including land management and climatic conditions.[51] Well managed and healthy soils should keep a balance between inputs and outputs of carbon. Mismanaged and degraded soils can, on the contrary, be a powerful source of CO_2. In addition, not every soil has the same potential for storing carbon in stable forms.[52] Soils' ability to accumulate and retain carbon does not follow an infinite increase curve, but stabilises at different levels depending on intrinsic soil properties and management practices.[53]

National reporting on the SDG 15.3.1 indicator for SOC requires robust and spatially exhaustive information on the SOC stocks in the topsoil (0–30 cm soil layer). Recognising the need to improve national capacities for obtaining such information, the GSP has launched a Global Soil Organic Carbon map (GSOCmap) programme. Within this programme, the GSP has developed three key products:

---

[45]ITPS (2016).

[46]UNSTAT (2018).

[47]UNCCD (2018).

[48]FAO (2017b).

[49]Blum (2005).

[50]Smith (2012).

[51]Stockmann et al. (2013).

[52]Six et al. (2002).

[53]Beare et al. (2014).

(1) the GSOCmap guidelines,[54] which provide definitions and specifications for the GSOCmap products; (2) the technical manual 'GSOC Mapping Cookbook',[55] based on the latest advancements in digital soil mapping; and, (3) the GSP Data Policy,[56] which ensures that existing ownership rights to shared soil data are respected. A worldwide programme of training workshops was launched to improve national capacities in SOC mapping. Thus, national experts were able to produce SOC maps for their countries according to IPCC/UNCCD specifications. National maps were subsequently submitted to the GSP Secretariat and compiled into the GSOCmap.

Covering around 70% of the world with national data, the GSOCmap is an essential source of information for countries to report on the SDG 15.3.1 indicator on SOC stocks. Until the creation of the GSOCmap, national information was scattered across many research centres, universities and government agencies, there was no national harmonised value for soil organic carbon stocks and in many cases, governments did not know how to access and process raw data to obtain a meaningful value for reporting. GSOCmap relies on national SOC maps, which were produced by national experts using national soil data. Therefore, national GSOCmap contributions are the first building block for soil carbon monitoring and reporting worldwide.

All current activities on SOC of the GSP are articulated under the framework for implementation of the outcome document, "Unlocking the potential of soil organic carbon",[57] a summary of the discussion and recommendations agreed during the Global Symposium on Soil Organic Carbon (GSOC17), co-organised with the Intergovernmental Panel on Climate Change (IPCC), UNCCD and the World Meteorological Organization (WMO). Due to the urgency to halt SOC losses and enhance SOC worldwide, the GSP started immediately after the symposium with the implementation of the eight recommendations included in the outcome document of the GSOC17.[58]

As a way to improve cooperation, information exchange and the efficiency of global efforts, the outcome document was presented at the 13th meeting of the Conference of the Parties (COP13) of the UNCCD in Ordos, China, in 2017,[59] and at the COP23 of the United Nations Framework Convention on Climate Change (UNFCCC), held in Bonn in November 2017.[60] Both COPs acknowledged the significance of the agriculture sector in adapting to and mitigating climate change and the primary role that soil carbon, soil health and soil fertility will play in

---

[54]Pillar 4 Working Group (2017).

[55]FAO (2018b).

[56]FAO (2017c).

[57]FAO (2017d).

[58]Vargas Rojas et al. (2019) in Ginzky et al. (2019).

[59]UNCCD (2017), p. 71.

[60]UNFCCC (2018).

achieving the Conventions' objectives, as reflected in the UNFCCC's Koronivia joint work on agriculture (KJWA).[61]

According to the second recommendation of the GSOC17 outcome document, a working group for the preparation of the Guidelines for measuring, mapping, monitoring and reporting on SOC was established in 2018, with the participation of more than 100 experts from all over the world. These guidelines aim to provide methodologies and tools for policymakers and governmental technicians to gather and generate information on SOC. The guidelines are planned to be published at the end of 2020.

Since 2017, an ad-hoc working group of experts has been working on the preparation of a technical manual on SOC management, which summarises the best practices in different land uses and soil types to maintain and/or increase the amount of soil organic carbon. A novelty since its conception is the inclusion of a chapter focused on carbon-rich soils (peatlands, wetlands and black soils). The manual is still under preparation and it is foreseen to be published at the end of 2019 and correspond to the implementation of the three recommendations under theme 3 of the GSOC17 outcome document.

Regarding the sustainable management of these carbon-rich soils, and in accordance with the discussions and decisions of the GSOC17, the GSP launched the International Network of Black Soil (INBS) on 21 March 2017. The objective of the network is to support and facilitate joint efforts towards sustainable management of black soil resources for food security, agricultural sustainability and climate change adaptation and mitigation. Twenty-seven countries rich in black soils have officially joined the INBS.

The FAO GSOCmap[62] estimates that the upper-layer of soil contains a global SOC stock amounting to 680 Pg C.[63] In the literature and international discussions, it has been indicated that the majority of soils around the world are far from reaching their carbon sequestration limit[64] and there is, therefore, great potential for carbon sequestration.[65] Soils could contribute to offsetting annual anthropogenic greenhouse gas emissions.[66,67] The IPCC has also remarked on the co benefits of SOC sequestration, such as the improvement of soil health and increased biodiversity.[68] However, many voices from the scientific community call for caution in considering soil organic carbon sequestration as a viable long-term solution to climate change because of the lack of supporting data on the real potential of soils and the

[61]FAO (2018c).

[62]FAO, ITPS (2017).

[63]1 Pt (Petagram) = 1 Billion Metric Tonnes.

[64]Kane (2015).

[65]Rumpel et al. (2018).

[66]4p1000 (2015).

[67]FAO (2017e).

[68]IPCC (2018), paragraph C.3.5, p. 19.

permanence of sequestered carbon,[69] and the diversion from the critical need to reduce greenhouse gas emissions.[70,71]

In order to provide a solid scientific response and help policy makers and the international community better understand the potential of soil organic carbon sequestration in the fight against climate change, the GSP is working on a global map of the potential for soil organic carbon sequestration (GSOC seq) using the GSOCmap as an input. Building on the strengthening of national capacities and networks of specialists created during the preparation of the GSOCmap, and given the success of the "bottom-up" approach (based on the best available data), this map will be prepared in collaboration with national experts. Through the implementation of widely used and validated soil organic matter simulation models, it will also be developed in parallel with a validation process. The GSOCseq potential map is aimed at contributing to the strengthening of public policies related to sustainable management of SOC and its relationship with degradation, food security and climate change mitigation and adaptation. This product is also expected to support evidence-based decisions on sustainable land and soil management, help build capacity at the national and regional levels, and generate long-term global data sets.

4.2 *Soil Pollution and Contamination*

Soil pollution refers to the presence in the soil of a chemical or substance out of place and/or present at a higher than normal concentration that has adverse effects on any non-targeted organism.[72] Soil pollution poses a serious risk to human health through direct contact (dermal exposure or inhalation of polluted soil particles) or indirectly, by consuming plants or animals that have accumulated large amounts of soil pollutants.

In addition, soil pollution jeopardises the achievement of the 2030 Agenda for Sustainable Development.[73] Pollutants present in the soil can enter the food chain, reducing food security[74] (SDG 2) by both reducing crop yield[75,76] and food safety.[77] As it has been mentioned, soil pollution poses a serious risk to human health[78] (SDG 3). Soils act as a filter and buffer of contaminants, preventing their entry into surface

[69]He et al. (2016).

[70]Amundson and Biardeau (2018).

[71]Kon Kam King et al. (2018).

[72]Rodríguez Eugenio et al. (2018).

[73]FAO (2018d).

[74]Ahmad and Ashraf (2012).

[75]Hossain et al. (2005).

[76]Zhu et al. (2005).

[77]Vasseur and Cossu-Leguille (2006).

[78]Oliver and Gregory (2015).

and groundwater bodies,[79] but when soil capacity is exceeded, contaminants are leached or transported by runoff, reducing the quality of water bodies (SDG 6). When N-fertilisers are applied in excess in soils, the different converted forms of nitrogen are released or leached from soils: nitrates can reach drinking water causing severe health impacts in humans[80] and nitrous oxide (N_2O) is lost as a gas, which has a potent greenhouse effect, exacerbating global climate change (SDG 13). In addition, soil pollutants are toxic not only for humans but also for soil-dwelling organisms, reducing biodiversity below and above ground[81] (SDG 15).

Since soil pollution is difficult to be perceived visually, unlike other degradation processes, it has been neglected for a long time, despite the threat it poses to human well-being and development. Fortunately, concern is growing in all regions and the issue has been integrated into major international debates. A clear example of the growing importance of soil pollution at the international level was its inclusion in the debate of the United Nations Environment Assembly (UNEA), at its third session, in December 2017, which addressed the theme "Towards a pollution-free world". The adopted resolution 3/6 "*Managing soil pollution to achieve sustainable development*"[82] calls on countries to address soil pollution in an integrated manner and with a special focus on prevention. It also instigates UN agencies to work together to strengthen capacities and to develop and improve available information on sources and impacts of soil pollution.

FAO and the GSP, as interested parties, took up the challenge, understanding that soil pollution is a severe threat that has long been ignored by the scientific community and at the political level. In response to that call, the Global Symposium on Soil Pollution (GSOP18)[83] was organised to respond to the mandate and gather the latest information about soil pollution. To join forces and address the wide diversity of soil pollution sources and impacts, the GSP invited UNEP, the World Health Organisation (WHO) and the Basel, Rotterdam and Stockholm chemical Conventions (BRS Conventions) to co-organise the symposium and to cooperate in the following efforts to prevent, reduce and manage soil pollution. As in the previous edition of the global symposium, the outcome document; "*Be the solution to soil pollution*",[84] sets an agenda for the next decade.

During the GSOP18, more than 500 participants from 100 countries discussed the major threats to human health, the environment and food safety posed by soil pollution. The GSOP18 outcome document summarises the eight key recommendations that will be implemented by FAO according to global needs (Table 1).

The first step in the implementation of the outcome document was the preparation of the International Code of Conduct for the Sustainable Use and Management of

[79]Hatfield et al. (2017).
[80]Cameron et al. (2013).
[81]Thiele-Bruhn et al. (2012).
[82]UNEP (2017).
[83]GSP (2018).
[84]FAO (2018d).

Table 1 Recommendations from GSOP18

Recommendation 1	To support the development and implementation of tools and guidelines that would support the prevention and remediation of soil pollution, such as the International Code of Conduct for the Sustainable Use and Management of Fertilizers
Recommendation 2	To include soil pollution assessment and minimisation measures in the Soil Doctors Programme in order to support land users to maintain healthy soils under local conditions for long-term benefits
Recommendation 3	To advocate for the implementation of existing guidelines, such as the Codex Alimentarius, the Revised World Soil Charter, the Voluntary Guidelines for Sustainable Soil Management, the Code of Conduct on Pesticide Management, the Global Action Plan on Antimicrobial Resistance amongst others, to ensure safe food in the face of emerging soil pollutants
Recommendation 4	To establish an expert and multi-stakeholder working group to develop feasible and regionally contextualised guidelines for assessing, mapping, monitoring and reporting on soil pollution
Recommendation 5	To establish a working group to develop guidelines for the management of polluted soils, including a database of good practices for addressing soil pollution (management and remediation)
Recommendation 6	To implement capacity development building and training activities covering the full cycle of soil pollution: from its assessment to its remediation—including the strengthening of facilities for data analysis and management
Recommendation 7	To implement the activities of the Global Soil Laboratory Network (GLOSOLAN), including harmonised methods to identify and measure soil contaminants
Recommendation 8	To implement a global assessment of the status of global soil pollution using a country-driven process in line with the UNEA3 resolution "Managing soil pollution to achieve sustainable development" and to promote the establishment of national soil information systems that include data/information on contaminants

Fertilizers, following an open consultation, to support and implement the VGSSM. The FAO Conference has recently endorsed the Fertilizer Code.[85] The Fertilizer Code aims to assist countries to address the multiple and complex issues related to the responsible use and management of fertilisers in agriculture at the farm, ecosystem and national level. It also aims to address issues of a global perspective in terms of contributing to the SDGs. These include ensuring sustainable food production systems, ending hunger and malnutrition, enhancing food security and nutrition, reducing pollution, and improving food safety. Thus, the Fertilizer Code addresses the judicious use and management of fertilisers to prevent inappropriate use, underuse and overuse.

The Soil Doctors Global Programme is the GSP's farmer-to-farmer training system that aims to build the capacity of smallholder farmers on the practice of SSM and, by doing so, support governmental agencies and organisations working on agricultural extension at the field level (promoting broader impacts and a reduction

[85]FAO (2019b).

of costs). The programme also aims to educate farmers on soil science principles for SSM practices and aims to achieve this by providing them with a set of tools composed of educational materials and a soil testing methods (STM) manual for preliminary soil analysis. According to the second recommendation of the outcome document, soil pollution has been included in the educational materials by providing farmers with simple measures that will prevent pollution in their fields, such as waste management and integrated fertility and pest management.

Two multi-stakeholder expert working groups were created after an open call. About 100 scientists, professionals, private sector and government representatives are developing technical guidelines on how to measure, map, monitor and report soil pollution, as well as how to manage and remediate soil pollution using the best available technologies. The launch of the guidelines is planned to take place in mid-2021.

Soil pollution has an immense diversity of possible sources, from agrochemicals to chemicals used in our daily lives, industrial and urban wastes, mining, wars and military activities, transportation, etc. These source activities have different degrees of importance in terms of surface affected in different regions and countries, as well as intensity levels. For example, the use of agrochemicals in one region may be more extensive than in another, but the amount of persistent chemicals may be greater in the latter, presenting worse problems in terms of human health impacts and food safety. In order to understand the global occurrence of soil pollution and management, and prevention trends in different regions of the world, a global assessment is being conducted by the GSP.

A questionnaire covering various aspects related to soil pollution, ranging from legal instruments to human health and soil-monitoring systems, was prepared by the GSP[86] in cooperation with UNEP and WHO and circulated among the GSP national focal points and other partners. This questionnaire is expected to gather information about how local and national governments perceive soil pollution. The questionnaire has also been made available to the public through the GSP website and disseminated through the GSP communication channels. The information will be assessed, completed and contrasted with publicly available findings in national and international reports and scientific publications to construct a final picture of the soil pollution situation at the global level. The information on soil pollution and contamination included in the Status of the World's Soil Resources report[87] will be considered as a baseline for establishing trends in different regions.

A technical report and a policy brief will be generated at the end of the consultation process, and will be presented at the fifth session of the UNEA in 2021. Both documents will summarise the situation at global and regional level, including the main knowledge and policy gaps, the major constraints the countries face in tackling soil pollution and recommendations on the way forward to address the problem effectively at all levels.

[86]GSP (2019b).

[87]FAO, ITPS (2015a, b).

4.3 Soil Erosion

Although soil erosion is a natural process that is part of soil and landscape formation and evolution, it is significantly accelerated by human activities such as the removal of vegetative cover, down-slope tillage, overgrazing, and shifting cultivation associated with agriculture.[88] Climate change, continuous mechanical tillage of agricultural land, some land use changes, deforestation, lack of erosion control amongst others, are leading drivers of accelerated soil erosion with consequences of extreme events such as landslides, increased emissions of greenhouse gases and soil organic carbon losses.[89,90,91,92] Soil erosion negatively affects agricultural food production, water quality, and environment in general. On farmlands, soil erosion reduces infiltration capacity, moisture availability, drainage capacity, plant rooting depth, and loss of soil nutrients. The displaced soil particles from eroded sites cause sedimentation and pollution of surface water storage, blockage of waterways, and destruction of infrastructures.

An average global loss of 0.3% of annual crop yields occurs as a result of soil erosion,[93] which if continued might lead to a reduction of 10% in crop yields by 2050. According to the SWSR report, soil erosion is the major threat to soil health worldwide, even though there are major differences in the conditions and trends for soil erosion between different regions of the world.[94] In addition, soil erosion constitutes an indicator for the Sustainable Development Goal 2.4.1.[95,96]

The Global Symposium on Soil Erosion (GSER19) has been organised by FAO and the GSP's ITPS, together with the Science-Policy Interface (SPI) of the UNCCD and the Joint FAO/IAEA Programme of Nuclear Techniques in Food and Agriculture. The GSER19 gathered a large range of stakeholders, from experts on soil erosion, NGOs, and farmers, to economists, policy-makers and governmental bodies, to establish a common platform to present and discuss the latest information on the status of interventions and innovations in the field of soil erosion and related land management.

As a result of this symposium, a joint work plan has been developed with the participants, which includes the main measures that countries could take to effectively address and halt soil erosion. Two major outputs will be prepared: (1) a technical manual with the methodology for the preparation of harmonised national

[88] Borrelli et al. (2017).
[89] Routschek et al. (2014).
[90] Lal (2014).
[91] Borselli et al. (2006).
[92] Zhang et al. (2018).
[93] FAO, ITPS (2015a, b).
[94] FAO, ITPS (2015b), pp. 66–67.
[95] FAO (2018e), p. 201.
[96] FAO (2018e).

maps of soil erosion that will ultimately feed the global map on soil erosion; (2) a global economic study on the effects of soil erosion.

During the Symposium, the report *Soil and nutrients loss in Malawi: an economic assessment*[97] was launched. This report is a clear example of the critical importance that soils have on a country's economy. The loss of fertile soils by erosion results in a decrease in agricultural productivity, increased expenditure on fertilisers, and a general decline in profitability of crop production. Therefore, a 10% increase in soil loss would produce monetary losses of about 0.26% of Malawi's GDP and 0.42% of the total agricultural production value. Higher soil loss values lead to larger impacts. Degradation processes like this jeopardise food security and socio-economic development and welfare, mainly in agriculturally dependent countries.

4.4 Loss of Soil Biodiversity

Soils host a wide diversity of organisms that play fundamental roles in driving many ecological processes that sustain the functioning of terrestrial ecosystems, such as soil formation, nutrient and water cycling, climate regulation, production of food, medicines and fibre, disease and pest control.[98] The Convention on Biological Diversity (CBD) defines soil biodiversity as "the variation in soil life, from genes to communities, and the ecological complexes of which they are part, that is from soil micro-habitats to landscapes".[99] Soils are subject to numerous pressures, as are the organisms that inhabit them. However, only a few scientific publications have addressed the relationship between the threats to soil health and the loss of soil biodiversity.[100]

At its sixth meeting, the COP of CBD proposed an International Initiative for the Conservation and Sustainable Use of Soil Biodiversity as a crosscutting initiative within the programme of work on agricultural biodiversity, to emphasise the full range of ecosystem services provided by soil biodiversity.[101] FAO was invited to coordinate this initiative, so the GSP and its ITPS, in an effort to promote soil biodiversity, decided to organise the Global Symposium on Soil Biodiversity in 2020, in collaboration with the Global Soil Biodiversity Initiative (GSBI) and the UN Convention on Biological Diversity (CBD). During the symposium, a consultation will be held on the way forward for the implementation of the International Initiative for the Conservation and Sustainable Use of Soil Biodiversity. In addition, the GSP is preparing a Global Assessment of Soil Biodiversity following a bottom-up approach similar to the one used for the global assessment on soil pollution.

---

[97] Asfaw et al. (2018).

[98] Orgiazzi et al. (2016a).

[99] CBD (2012).

[100] Orgiazzi et al. (2016b).

[101] CBD (2002), paragraph 13.

Global, regional and country assessments on the status of knowledge on soil biodiversity would help to prioritise actions to close the knowledge gaps and hence more effectively inform stakeholders of the consequences of their actions on land use, climate change, biodiversity, and ecosystem services.

4.5 Soil Salinisation

Soil salinisation is the accumulation of salts in the soil. The accumulated salts include sodium, potassium, magnesium and calcium, chloride, sulphate, carbonate and bicarbonate. Salinisation can be primary or natural, caused by the accumulation of salts from parent materials, groundwater, or long-term accumulation of salts contained in rainfall, or it can be human-induced, caused by human interventions such as inappropriate irrigation practices, e.g., with salt-rich irrigation water and/or insufficient drainage. Sodification can be associated with salinisation and is the accumulation of sodium and/or sodium salts in the solid and/or liquid phases of the soil. Degradation due to salinity and sodicity varies geographically with climate, agricultural activities, irrigation methods and land management policies and is mainly restricted to irrigated farming systems.[102]

Salinisation and sodification mainly affect arid and semi-arid ecosystems,[103] and in some regions have become worrying issues that are affecting food security and becoming an economic disaster,[104] as in the case of Australia[105] or Central Asia.[106] Due to the extent and severity of the problem, international organisations have been addressing soil salinisation for decades,[107,108] but no recent statistics are available on the global extent of salt-affected soils.

Under the GSP, several activities have been carried out to address secondary or human-induced salinisation, mainly in the Eurasian region, where in recent years an international training workshop has been organised on salinity management[109] and small research projects aimed at mitigating and adapting to soil salinity have been funded. Currently, a working group formed by the ITPS and external experts is preparing the technical specifications to develop an updated global map of salt-affected soils using a methodology similar to the GSOC map, a bottom-up approach led by the countries. This global map will be supported by a document on best available practices for the management of salt-affected soils, which will build on

---

[102]FAO, ITPS (2015a).

[103]Thomas and Middleton (1993).

[104]World Bank (2009).

[105]Vengosh (2003).

[106]Khitrov et al. (2008).

[107]UNEP, FAO, ISRIC (1990).

[108]Martinez Beltran and Licona Manzur (2005).

[109]GSP (2017).

existing documents, including case studies from Central Asia,[110] and be complemented by relevant practices in other regions.

5 The Way Forward

This chapter has distinctly demonstrated the need for the work and activities performed by the GSP as a way to promote and increase the sustainable management of global soils. It is becoming increasingly evident that soil resources, which are finite, play a crucial role in global food and nutrition security, and climate change adaption and mitigation. The GSP will continue its efforts in keeping soils high on the global agenda and seize the enabling momentum that has emerged since the Partnership was established.

The GSP actions described in this chapter focussed on promoting better knowledge of the state of our global soil resources and creating strong networks of experts to provide solid scientific knowledge to forge the basis for action at global, regional and local levels.

For the first time at the international level, the GSP has created a bottom-up working structure, empowering national technicians and governments to be able to value, promote and protect their soil resources. The consolidation of the Global Soil Information System and its national nodes is one of the main challenges facing the GSP. There is a need to improve the available soil information in order for them to be used for informed decision-making.

The GSP should play a leading role to ensure a better commitment from member countries. Indeed, countries need to encourage and sponsor the active participation of their official focal points at both international and regional meetings and allocate resources to implement activities at the national and local level. The establishment of the National Soil Partnership (NSP) should be fostered to improve coordination among different actors within countries, create a shared space to develop joint programs, facilitate distribution of tasks and allow resource savings. The successful NSP establishments from countries such as Nicaragua, Portugal, Italy, Mongolia, Malawi, Thailand, Ukraine, Brazil and Syria can serve as an example to follow for the countries willing to proceed with its establishment within the GSP framework. The GSP will continue its efforts to facilitate the translation in various languages of the Voluntary Guidelines for Sustainable Soil Management.

The other soil threats, as highlighted in the Status of the World's Soil Resources and the Voluntary Guidelines for Sustainable Soil Management, will continue to be addressed following the structure created so far. The latter has proven to be effective in mobilising not only technical but also economic resources, and bringing together different stakeholders affected by these threats to discuss and propose concrete solutions and actions. The symposium planned for 2022 will address soil fertility

[110]FAO, Lomonosov Moscow State University (2018).

and its importance for a nutritious food supply by tackling important topics such as nutrition-sensitive agriculture, and how soils are part of the solution in overcoming malnutrition and micronutrient deficiencies. It is expected that the symposium will bring together all those experts and organisations working on related topics so that an integrated response can be given to such an important matter in the achievement of the SDGs.

The GSP also needs to move towards a greater presence on the ground. The implementation of the Global Soil Doctors Programme, a farmer-to-farmer training programme, will favour the transfer and adaptation of knowledge, which the GSP expert networks are generating, to farmers, field technicians and other land users. Technical support will also continue to be provided through continuous training actions.

Communication activities will continue to support the GSP mission to improve soil governance, promote Sustainable Soil Management (SSM) and raise awareness on the need for healthy and productive soils. Since 2014 and thanks to the leading role of the GSP, World Soil Day is now an official UN observance and efforts will persist to engage a wider audience. The Glinka World Soil Prize will be further disseminated among individuals or organisations committed to solving acute national, regional or local problems of soil degradation. In addition, the World Soil Day organisers will be encouraged to apply for the yearly World Soil Day award that aims to prize exceptional efforts in the organisation of high impact WSD celebrations.

Despite the recognition of the important work done by the GSP, the partnership faces setbacks on a daily basis, both economic and technical, that hinder the implementation of the activities presented in this chapter and upcoming activities. The fact that the GSP is a voluntary partnership indicates that FAO member countries are under no obligation to financially contribute, and implies that in most cases expert contributions are also voluntary and therefore unpaid, threatening its long-term continuity.

References

4p1000 (2015) Welcome to the "4 per 1000" initiative | 4p1000 [WWW Document]. URL https://www.4p1000.org/. Accessed 16 Apr 2019

Ahmad MSA, Ashraf M (2012) Essential roles and hazardous effects of nickel in plants. In: Whitacre DM (ed) Reviews of environmental contamination and toxicology. Springer New York, New York, pp 125–167. https://doi.org/10.1007/978-1-4614-0668-6_6

Akpa SIC, Odeh IOA, Bishop TFA, Hartemink AE, Amapu IY (2016) Total soil organic carbon and carbon sequestration potential in Nigeria. Geoderma 271:202–215. https://doi.org/10.1016/j.geoderma.2016.02.021

Amundson R, Biardeau L (2018) Opinion: soil carbon sequestration is an elusive climate mitigation tool. Proc Natl Acad Sci U S A 115:11652–11656. https://doi.org/10.1073/pnas.1815901115

Arrouays D, McKenzie N, Hempel J, de Forges AR, McBratney AB, McKenzie N, Hempel J, de Forges AR, McBratney AB (2014) GlobalSoilMap : basis of the global spatial soil information system. CRC Press, Boca Raton. https://doi.org/10.1201/b16500

Arrouays D, Lagacherie P, Hartemink AE (2017) Digital soil mapping across the globe. Geoderma Reg 9: 1–4. https://doi.org/10.1016/j.geodrs.2017.03.002

Asfaw S, Orecchia C, Pallante G, Palma A (2018) Soil and nutrients loss in Malawi: an economic assessment. Food and Agriculture Organisation of the United Nations, UNDP-UNEP poverty-environment initiative, and the Ministry of Agriculture, Irrigation and Water Development, Malawi

Baritz R, Wiese L, Verbeke I, Vargas R (2018) Voluntary guidelines for sustainable soil management: global action for healthy soils. In: Ginzky H, Dooley E, Heuser IL, Kasimbazi E, Markus T, Qin T (eds) International yearbook of soil law and policy 2017, international yearbook of soil law and policy. Springer International Publishing, Cham

Beare MH, McNeill SJ, Curtin D, Parfitt RL, Jones HS, Dodd MB, Sharp J (2014) Estimating the organic carbon stabilisation capacity and saturation deficit of soils: a New Zealand case study. Biogeochemistry 120:71–87. https://doi.org/10.1007/s10533-014-9982-1

Blum WEH (2005) Functions of soil for society and the environment. Rev Environ Sci Biotechnol 4:75–79. https://doi.org/10.1007/s11157-005-2236-x

Borrelli P, Robinson DA, Fleischer LR, Lugato E, Ballabio C, Alewell C, Meusburger K, Modugno S, Schütt B, Ferro V, Bagarello V, Oost KV, Montanarella L, Panagos P (2017) An assessment of the global impact of 21st century land use change on soil erosion. Nat Commun 8:2013. https://doi.org/10.1038/s41467-017-02142-7

Borselli L, Torri D, Øygarden L, De Alba S, Martínez-Casasnovas JA, Bazzoffi P, Jakab G (2006) Land levelling. In: Boardman J, Poesen J (eds) Soil erosion in Europe. John Wiley & Sons, Ltd., Chichester, pp 643–658. https://doi.org/10.1002/0470859202.ch46

Bouma J, Montanarella L (2016) Facing policy challenges with inter- and transdisciplinary soil research focused on the UN sustainable development goals. Soil 2:135–145. https://doi.org/10.5194/soil-2-135-2016

Cameron KC, Di HJ, Moir JL (2013) Nitrogen losses from the soil/plant system: a review: nitrogen losses. Ann Appl Biol 162:145–173. https://doi.org/10.1111/aab.12014

CBD (2002) COP 6 decision VI/5 agricultural biological diversity [WWW Document]. URL https://www.cbd.int/decision/cop/default.shtml?id=7179. Accessed 10 July 2019

CBD (2012) Introduction to the convention on biological diversity [WWW Document]. URL https://www.cbd.int/agro/soil.shtml. Accessed 10 July 2019

European Commission (2006) Communication from the Commission to the Council, the European Parliament, the European Economic and Social Committee and the Committee of the Regions – Thematic Strategy for Soil Protection, COM/2006/0231 final

FAO (2014a) Post 2015 and SDGs. Nourishing people, nurturing the planet

FAO (2014b) Healthy soils facility of the global soil partnership. Programme Document 2014–2018

FAO (2015a) Revised World soil charter. FAO, Rome

FAO (2015b) Farmer's compost handbook. Experiences in Latin America. Food and Agriculture Organisation of the United Nations, Rome

FAO (2017a) Voluntary guidelines for sustainable soil management. FAO, Rome

FAO (2017b) FAO and the SDGs indicators: measuring up to the 2030 agenda for sustainable development. FAO, Rome

FAO (2017c) GSP soil data policy. FAO, Rome

FAO (2017d) Unlocking the potential of soil organic carbon. Outcome document. Food and Agriculture Organisation of the United Nations, Rome

FAO (2017e) Recarbonization of global soils: a dynamic response to offset global emissions. Food and Agriculture Organisation of the United Nations, Rome

FAO (2018a) Report of the fourth workshop of the International Network of Soil Information Institutions (INSII). FAO, Rome

FAO (2018b) Soil organic carbon mapping cookbook, 2nd edn. FAO, Rome

FAO (2018c) The Koronivia joint work on agriculture and the convention bodies: an overview. FAO, Rome

FAO (2018d) Be the solution to soil pollution. Outcome document of the global symposium on soil pollution. Food and Agriculture Organisation of the United Nations, Rome
FAO (2018e) SDG indicator 2.4.1. Proportion of agricultural area under productive and sustainable agriculture. Methodological note. FAO, Rome
FAO (2019a) FAOLEX database. Food and Agriculture Organisation of the United Nations, Rome. [WWW Document]. URL http://www.fao.org/faolex/en/. Accessed 16 Apr 2019
FAO (2019b) The international code of conduct for the sustainable use and management of fertilizers. FAO, Rome
FAO, ITPS (2015a) Status of the world's soil resources: main report. Food and Agriculture Organisation of the United Nations, Rome
FAO, ITPS (2015b) Status of the world's soil resources. Technical summary. Food and Agriculture Organisation of the United Nations, Rome
FAO, ITPS (2017) Global soil organic carbon map v.1.2.0
FAO, Lomonosov Moscow State University (2018) Handbook for saline soil management
FAO, UNESCO (1974) FAO – UNESCO Soil map of the world, 1: 5 000 000
FAOLEX Database (2014) Resolution No 415 of 2013 of Georgian Government on charter for determining level of fertility of soil and charter for soil conservation and fertility monitoring [WWW Document]. URL http://www.fao.org/faolex/results/details/en/c/LEX-FAOC167682/. Accessed 10 July 2019
Ginzky H, Dooley E, Heuser IL, Kasimbazi E, Markus T, Qin T (eds) (2019) International yearbook of soil law and policy 2018, international yearbook of soil law and policy. Springer International Publishing, Cham
GSP (2017) Training workshop on soil salinity management in Kharkiv, Ukraine | Global Soil Partnership | Food and Agriculture Organisation of the United Nations [WWW Document]. URL http://www.fao.org/global-soil-partnership/resources/events/detail/en/c/1032709/. Accessed 16 Apr 2019
GSP (2018) Global symposium on soil pollution | Food and Agriculture Organisation of the United Nations [WWW Document]. URL http://www.fao.org/about/meetings/global-symposium-on-soil-pollution/en/. Accessed 16 Apr 2019
GSP (2019a) Why the partnership? | Global Soil Partnership | Food and Agriculture Organisation of the United Nations [WWW Document]. Glob. Soil Partnersh. URL http://www.fao.org/global-soil-partnership/about/why-the-partnership/en/. Accessed 16 Apr 2019
GSP (2019b) Fill out the global survey on soil pollution | Global Soil Partnership | Food and Agriculture Organisation of the United Nations [WWW Document]. URL http://www.fao.org/global-soil-partnership/resources/highlights/detail/en/c/1184829/. Accessed 16 Apr 2019
Hatfield JL, Sauer TJ, Cruse RM (2017) Soil. In: Advances in agronomy. Elsevier, Amsterdam, pp 1–46. https://doi.org/10.1016/bs.agron.2017.02.001
He Y, Trumbore SE, Torn MS, Harden JW, Vaughn LJS, Allison SD, Randerson JT (2016) Radiocarbon constraints imply reduced carbon uptake by soils during the 21st century. Science 353:1419–1424. https://doi.org/10.1126/science.aad4273
Hengl T, Heuvelink GBM, Kempen B, Leenaars JGB, Walsh MG, Shepherd KD, Sila A, MacMillan RA, de Jesus JM, Tamene L, Tondoh JE (2015) Mapping soil properties of Africa at 250 m resolution: random forests significantly improve current predictions. PLoS ONE 10:e0125814
Hengl T, de Jesus JM, Heuvelink GBM, Gonzalez MR, Kilibarda M, Blagotić A, Shangguan W, Wright MN, Geng X, Bauer-Marschallinger B, Guevara MA, Vargas R, MacMillan RA, Batjes NH, Leenaars JGB, Ribeiro E, Wheeler I, Mantel S, Kempen B (2017) SoilGrids250m: global gridded soil information based on machine learning. PLoS One 12:e0169748
Hossain MF, White SK, Elahi SF, Sultana N, Choudhury MHK, Alam QK, Rother JA, Gaunt JL (2005) The efficiency of nitrogen fertiliser for rice in Bangladeshi farmers' fields. Field Crops Res 93:94–107. https://doi.org/10.1016/j.fcr.2004.09.017
IPCC (2018) Global warming of 1.5 °C. An IPCC special report on the impacts of global warming of 1.5 °C above pre-industrial levels and related global greenhouse gas emissions pathways, in the context of strengthening the global response to the threat of climate change, sustainable

development, and efforts to erradicate poverty. Summary for policy makers. (No. ISBN 978-92-9169-151-7). Intergovernmental Panel on Climate Change, Switzerland

ISRIC (2016) Soil and Terrain (SOTER) database programme | ISRIC [WWW Document]. URL https://www.isric.org/projects/soil-and-terrain-soter-database-programme. Accessed 10 July 2019

ITPS (2016) Report of the Fifth Working Session of the Intergovernmental Technical Panel on Soils

Kane D (2015) Carbon sequestration potential on agricultural lands: a review of current science and available practices. National Sustainable Agriculture Coalition Breakthrough Strategies and Solutions, LLC

Khitrov NB, Kalinina NV, Krylova ES, Rukovich DI, Vilchevskaya EV (2008) The degree of soil salinity. In: Interactive agricultural ecological atlas of Russia and neighboring countries. Economic plants and their diseases, pests and weeds

Kon Kam King J, Granjou C, Fournil J, Cecillon L (2018) Soil sciences and the French 4 per 1000 initiative—the promises of underground carbon. Energy Res Soc Sci 45:144–152. https://doi.org/10.1016/j.erss.2018.06.024. Special Issue on the Problems of Methods in Climate and Energy Research

Lal R (2014) Soil conservation and ecosystem services. Int Soil Water Conserv Res 2:36–47. https://doi.org/10.1016/S2095-6339(15)30021-6

Martinez Beltran J, Licona Manzur C (2005) Overview of salinity problems in the world and FAO strategies to address the problem. Int. Salin. Forum Manag. Saline Soils Water Sci. Technol. Soc. Issues Oral Present. Abstr. Conv. Cent. Riverside Calif. USA 25–28 April 2005 311-314 311–314

Ministry for the Environment, Government of New Zealand (1941) Soil Conservation and Rivers Control Act 1941

Montanarella L, Vargas R (2012) Global governance of soil resources as a necessary condition for sustainable development. Curr Opin Environ Sustain Terrestrial Syst 4:559–564. https://doi.org/10.1016/j.cosust.2012.06.007

Montanarella L, Pennock DJ, McKenzie N, Badraoui M, Chude V, Baptista I, Mamo T, Yemefack M, Singh Aulakh M, Yagi K, Young Hong S, Vijarnsorn P, Zhang G-L, Arrouays D, Black H, Krasilnikov P, Sobocká J, Alegre J, Henriquez CR, de Lourdes Mendonça-Santos M, Taboada M, Espinosa-Victoria D, AlShankiti A, AlaviPanah SK, Elsheikh EAEM, Hempel J, Camps Arbestain M, Nachtergaele F, Vargas R (2016) World's soils are under threat. Soil 2:79–82. https://doi.org/10.5194/soil-2-79-2016

Mulder VL, Lacoste M, Richer-de-Forges AC, Martin MP, Arrouays D (2016) National versus global modelling the 3D distribution of soil organic carbon in mainland France. Geoderma 263:16–34. https://doi.org/10.1016/j.geoderma.2015.08.035

Oldeman LR, Hakkeling RTA, Sombroek WG (1991) World map of the status of human-induced soil degradation : an explanatory note, 2nd. rev. ed. (No. 9789066720466). ISRIC [etc.], Wageningen [etc.]

Oliver MA, Gregory PJ (2015) Soil, food security and human health: a review: Soil, food security and human health. Eur J Soil Sci 66:257–276. https://doi.org/10.1111/ejss.12216

Orgiazzi A, Bardgett RD, Barrios E, Behan-Pelletier V, Briones MJI, Chotte J-L, De Deyn GB, Eggleton P, Fierer N, Fraser T, Hedlund K, Jeffery S, Johnson NC, Jones A, Kandeler E, Kaneko N, Lavelle P, Lemanceau P, Miko L, Montanarella L, Moreira FMS, Ramirez KS, Scheu S, Singh BK, Six J, van der Putten WH, Wall DH, European Commission, Joint Research Centre (2016a) Global soil biodiversity atlas. Publications Office, Luxembourg

Orgiazzi A, Panagos P, Yigini Y, Dunbar MB, Gardi C, Montanarella L, Ballabio C (2016b) A knowledge-based approach to estimating the magnitude and spatial patterns of potential threats to soil biodiversity. Sci Total Environ 545–546:11–20. https://doi.org/10.1016/j.scitotenv.2015.12.092

Pillar 4 Working Group (2017) GSP Guidelines for sharing national data/information to compile a Global Soil Organic Carbon (GSOC) map

Presidencia de la Nación Argentina (1981) LEY N° 22.428. Régimen legal para el fomento de la acción privada y pública tendiente a la conservación y recuperación de la capacidad productiva de los suelos
Rodríguez Eugenio N, McLaughlin MJ, Pennock D, Global Soil Partnership (2018) Soil pollution: a hidden reality. Food and Agriculture Organisation of the United Nations, Rome
Rojas RV, Caon L (2016) The international year of soils revisited: promoting sustainable soil management beyond 2015. Environ Earth Sci 75:1184. https://doi.org/10.1007/s12665-016-5891-z
Routschek A, Schmidt J, Kreienkamp F (2014) Impact of climate change on soil erosion — a high-resolution projection on catchment scale until 2100 in Saxony/Germany. CATENA 121:99–109. https://doi.org/10.1016/j.catena.2014.04.019
Rumpel C, Lehmann J, Chabbi A (2018) '4 per 1,000' initiative will boost soil carbon for climate and food security. Nature 553:27. https://doi.org/10.1038/d41586-017-09010-w
Six J, Conant RT, Paul EA, Paustian K (2002) Review stabilisation mechanisms of soil organic matter: implications for C-saturation of soils
Smith P (2012) Soils and climate change. Curr Opin Environ Sustain Terrestrial Syst 4:539–544. https://doi.org/10.1016/j.cosust.2012.06.005
Stockmann U, Adams MA, Crawford JW, Field DJ, Henakaarchchi N, Jenkins M, Minasny B, McBratney AB, De Courcelles VDR, Singh K, Wheeler I, Abbott L, Angers DA, Baldock J, Bird M, Brookes PC, Chenu C, Jastrow JD, Lal R, Lehmann J, O'Donnell AG, Parton WJ, Whitehead D, Zimmermann M (2013) The knowns, known unknowns and unknowns of sequestration of soil organic carbon. Agric Ecosyst Environ 164:80–99. https://doi.org/10.1016/j.agee.2012.10.001
The Republic of Ghana (1957) Land Planning and Soil Conservation (amendment) Act, 1957
Thiele-Bruhn S, Bloem J, de Vries FT, Kalbitz K, Wagg C (2012) Linking soil biodiversity and agricultural soil management. Curr Opin Environ Sustain 4:523–528. https://doi.org/10.1016/j.cosust.2012.06.004
Thomas DSG, Middleton NJ (1993) Salinisation: new perspectives on a major desertification issue. J Arid Environ 24:95–105. https://doi.org/10.1006/jare.1993.1008
Tifafi M, Guenet B, Hatté C (2018) Large differences in global and regional total soil carbon stock estimates based on SoilGrids, HWSD, and NCSCD: intercomparison and evaluation based on field data from USA, England, Wales, and France. Glob Biogeochem Cycles 32:42–56. https://doi.org/10.1002/2017GB005678
Tóth G, Hermann T, da Silva MR, Montanarella L (2018) Monitoring soil for sustainable development and land degradation neutrality. Environ Monit Assess 190:57. https://doi.org/10.1007/s10661-017-6415-3
UNCCD (2017) Report of the Conference of the Parties on its thirteenth session, held in Ordos, China, from 6 to 16 September 2017. Part two: Action taken by the Conference of the Parties at its thirteenth session. United Nations Convention to Combat Deserfication, Bonn, Germany
UNCCD (2018) Default data: methods and interpretation. A guidance document for 2018 UNCCD reporting. United Nations Convention to Combat Desertification, Bonn, Germany
UNEP (2017) Resolution 3/6. Managing soil pollution to achieve sustainable development, UNEP/EA.3/Res.6
UNEP, FAO, ISRIC (1990) Global Assessment of Human-induced Soil Degradation (GLASOD) | ISRIC [WWW Document]. URL https://www.isric.org/projects/global-assessment-human-induced-soil-degradation-glasod. Accessed 12 Apr 2019
UNFCCC (2018) Report of the Conference of the Parties on its twenty-third session, held in Bonn from 6 to 18 November 2017. United Nations Framework Convention on Climate Change, Bonn, Germany
UNGA (2014) Resolution adopted by the General Assembly on 20 December 2013. 68/232. World Soil Day and International Year of Soils
UNGA (2015) Resolution adopted by the General Assembly on 25 September 2015. 70/1. Transforming our world: the 2030 agenda for sustainable development

United Nations General Assembly (2014) Resolution 68/232. World soil day and international year of soils

UNSTAT (2018) Metadata-Indicator 15.3.1: proportion of land that is degraded over total land area

Vargas Rojas R, Cuevas-Corona R, Yigini Y, Tong Y, Bazza Z, Wiese L (2019) Unlocking the potential of soil organic carbon: a feasible way forward. In: International yearbook of soil law and policy 2018. Springer International Publishing, pp 373–395. ISBN: 978-3-030-00757-7

Vasseur P, Cossu-Leguille C (2006) Linking molecular interactions to consequent effects of persistent organic pollutants (POPs) upon populations. Chemosphere 62:1033–1042. https://doi.org/10.1016/j.chemosphere.2005.05.043

Vengosh A (2003) Salinisation and saline environments. In: Treatise on geochemistry. Elsevier, Amsterdam, pp 1–35. https://doi.org/10.1016/B0-08-043751-6/09051-4

World Bank (2009) The World Bank Annual Report 2009. Year in review. The World Bank, Washington

Zhang H, Fan J, Cao W, Harris W, Li Y, Chi W, Wang S (2018) Response of wind erosion dynamics to climate change and human activity in Inner Mongolia, China during 1990 to 2015. Sci Total Environ 639:1038–1050. https://doi.org/10.1016/j.scitotenv.2018.05.082

Zhu JH, Li XL, Christie P, Li JL (2005) Environmental implications of low nitrogen use efficiency in excessively fertilized hot pepper (Capsicum frutescens L.) cropping systems. Agric Ecosyst Environ 111:70–80. https://doi.org/10.1016/j.agee.2005.04.025

Part III
Regional/National Reports

Soil Law in Mexico

José Juan González Márquez and Ana María Pacheco Ruiz

1 Introduction

This chapter briefly surveys the Mexican soil constitutional and legal system. The study includes the analysis of the soil property regime, the nation's power to impose on private property, those modalities dictated by the public interest, based on the third paragraph of Article 27 of the Political Constitution of the United Mexican States (1917) and that gives private property a statutory character, and the regime of public property of those soils that, by constitutional mandate, have to remain into the inalienable and imprescriptible domain of the nation without prejudice that they can be used and exploited by particulars through concessions.

2 Constitutional Basis of Soil Law

Article 27 of the Political Constitution of United Mexican States, as regarded by Serna de la Garza (2007) provides the basis for legislation on natural resources use and conservation, including the soil.[1] Firstly, the first paragraph of article 27 as it

[1] See: Serna de la Garza (2007), pp. 473–494.

J. J. González Márquez (✉)
Mexican Institute for Environmental Law Research, Mexico City, Mexico
e-mail: jjgonzalez@gonzalezasociados.com

A. M. Pacheco Ruiz
Metropolitan Autonomous Unioversity - Campus Azcapotzalco, Faculty of Law, Ciudad de México, Mexico

H. Ginzky et al. (eds.), *International Yearbook of Soil Law and Policy 2019*,
International Yearbook of Soil Law and Policy 2019,
https://doi.org/10.1007/978-3-030-52317-6_12

was passed in 1917 recognized that the original property of the lands and waters of the Mexican territory was vested in the nation and it had the power of transferring their domain to private persons to create private property on soil and waters. Secondly, in line with paragraph 2 of article 27, the State can use its power of eminent domain to compel a private property owner to relinquish ownership and dominion of any property to the state, but it only may be taken for reasons of public use and subject to payment of compensation. Thirdly, paragraph 3 of article 27 empowers the nation to impose on private property such restrictions as the public interest may demand, as well as to regulate, for social benefit, the use and exploitation of those natural resources, which are susceptible of appropriation.

Finally, paragraphs 4 and 5 of article 27 provide a foundation for the rectory of the State on the use, exploitation and conservation of certain natural resources considered strategic. In accordance with the mentioned constitutional provisions, waters and natural resources into the subsoil are considered as exclusive property of the nation and, therefore, can only be exploited by individuals through the granting of concessions and contracts in case of hydrocarbons. However, the domain of the nation on the resources mentioned is inalienable and not subject to the statute of prescription.

As Glenn (2010) analyses, in the Mexican legal system as well as in most of the legal regimens that belong to the civil law tradition, the soil is typically a natural resource of private property, with the exemption of those cases in which it remains the original property of the nation and it has not been transmitted to private persons or when it belongs to the direct domain of the nation.[2]

Thus, in accordance with the Constitution, in Mexico, we can distinguish four categories of soil property: original property of the soil, that is vested in the nation; private property of the soil, which is constituted by transmitting the original national domain of the soil to private persons; public property of the soil, that includes those soils that the law reserves to the nation and cannot be transmitted in any case to private persons; and finally *ejidal*[3] and communal property of the agricultural soils, which are considered social property of land.[4] The following sections analyse these different categories of soil property with special emphasis on private property of soil.

---

[2]See: Glenn (2010), p. 99.

[3]In Mexico, an ejido is an area of communal land used for agriculture, on which community members individually farm designated parcels and collectively maintain communal holdings. The system of ejidos was based on an understanding of the Aztec Calpulli and the medieval Spanish ejido.

[4]In the Mexican legal regime, 'soil' and 'land' are synonymous.

3 Public Ownership of Soil

Public property of soil is basically ruled by the General Law on the National Assets.[5] According to this law, there are areas of the national territory considered as property of the nation, inalienable and imprescriptible—not subject to the statute of prescription—, such as the Federal Maritime and Riverside Zones, beaches and lands reclaimed from the sea.[6] The federal maritime zone is defined as 'the band of twenty metres of firm land, passable and contiguous to beaches or, when applicable, to riverbanks, from their mouth in the sea to one hundred metres upstream.[7] The riverside is defined as the band of 10 m aside the rivers measured from the maximal level of the waters.[8]

According to article 6, paragraph II, of the General Law on National Assets, beaches are also owned and managed by the nation. Beaches are defined as the land between the lowest annual ebb of the tide and the highest flow of the tide (article 7, paragraph IV, General Law on the National Assets).

Soils reclaimed from the sea are gained lands created by the action of the sea,[9] according to article 70 of the General Law on the National Assets, such gained soils are owned by the nation, but may be conveyed to the owner of the adjacent beachfront property only after such lands are disincorporated from the public domain by executive decree.[10] Such gained soils may also be allocated to the use and enjoyment of the state governments without being disincorporated, however, such allocations do not transfer property rights to the local government.[11]

Private persons can only enjoy the federal maritime and river zones, beaches and lands reclaimed from the sea through concessions granted by the federal government, but, as the definition of concession by López (1995) states, such concessions do not transfer property rights.[12]

[5]Diario Oficial de la Federación (May 20, 2004).

[6]Article 13, General Law on the National Assets.

[7]See Articles 3, 14, 20 and 40, Regulation for the Use and Enjoyment of the Territorial Sea, Navigable Waterways, Beaches, Federal Maritime Zone and Lands Reclaimed from the Sea. See also national official standard 'NORMA OFICIAL MEXICANA NOM-146-SEMARNAT-2005' issued by the Department of Environment and Natural Resources in 2005.

[8]See article 3, paragraph XLVII General Waters Law.

[9]See national standard NORMA OFICIAL MEXICANA NOM-146-SEMARNAT-2005.

[10]See Article 6, paragraph IX, General Law on the National Assets.

[11]See Article 70 of the General Law on the National Assets.

[12]See López Ramón (1995), p. 1341.

4 Social Ownership of Soil

The agrarian property of the soil is subject to the modalities imposed by the Agrarian Law,[13] the General Environmental Law and the Rural Development Law.[14] Among these modalities to agrarian soil use imposed by the Agrarian Law are those aimed at:

- Providing, under the terms of the regulatory law, for the collective organisation and exploitation of ejidos and communities;
- Developing small rural property; and promoting agriculture.

The Agrarian Law regulates matters related to the use of soil for agricultural purposes. This law regulates the creation and operation of various forms of agricultural communities, such as *ejidos* (communal village farmlands).

Additionally, the General Law for Ecological Balance and Environmental Protection of 1988 (the Environmental Law) establishes the general criteria and policies regulating soil use for the purpose of preserving and restoring the ecological balance. Article 98 sets forth the general principles for soil use regulation and provides that the soil is to be used in a manner compatible with its essential purpose, so that its physical integrity and productive capacity are not altered. Practices that induce soil erosion, degradation, or other adverse environmental impacts must be avoided; restoration projects in areas of severe erosion or deforestation must be devised; and erosion preventing activities are to be implemented during the construction of public or private infrastructures. These principles must be considered by the federal government when adopting support programmes for agricultural activities, including investment, financing, and technical assistance.

Article 103 of the General Environmental Law also requires that soil conservation and restoration practises necessary to prevent soil and environmental deterioration be implemented in connection with all farming activities. In areas of serious desertification and erosion, the Ministry of the Environment, in cooperation with other federal, state, and local authorities, may design and implement special programmes regulating land uses and the exploitation of natural resources, with the aim of restoring the ecological balance. Notice that, previously to their implementation, such programmes must be published in the *Federal Official Journal (Diario Oficial de la Federación)*.

The Rural Development Act is intended to promote Mexico's sustainable rural development and an appropriate environment. The law applies to *ejidos* and rural communities and associations, as well as to individuals and entities that carry out rural activities.

The Act seeks to achieve its objective by promoting a project of social and economic transformation, encouraging productive activities and social development of the rural sector, recognizing the fragility of the rural sector in order to achieve the improvement of populations living conditions, respecting the conservation and

---

[13]Diario Oficial de la Federación (February26,1992).

[14]Diario Oficial de la Federación (December 7, 2001).

sustainable use of natural resources and diversifying the productive activities of the countryside to increase productivity, profitability and employment of the rural population.

5 Private Ownership of Soil

Private property of soil is typically ruled by federal and local civil codes. However, as Booth (2002) states, private property of soil is also subject to a series of modalities and limitations imposed by a number of administrative laws based on the power granted to the nation by the first part of third paragraph of article 27 of the Constitution.[15] As a consequence, private property is far from an absolute right in Mexico.[16]

Based on the power granted to the nation by the third paragraph of article 27 of the Constitution, various administrative laws complement and modify the classic rules of the civil codes to give content to the idea of *the social function of property*[17] that the Constitution postulates and that makes the right to private property in Mexico an eminently statutory property.

5.1 Modalities on Private Property and Soil Uses

Third paragraph of article 27 of the Mexican Constitution vests the nation with the power to define soil uses. The mentioned paragraph points out: 'The Nation shall at all time the right to impose on private property such restrictions as the public interest may demand.' These limitations on the property right are based on the public interest and could have one of the following objectives:

- to make an equitable distribution of public wealth,
- to take care of its conservation,
- to achieve the balanced development of the country,
- to improve the quality of life of the rural and urban population.

---

[15]See: Booth (2002), p. 73.

[16]See: Mirow (2011), pp. 1183–1217.

[17]According to Duguit (1920), the social function of property means that property is no longer exclusively a right of individuals but that it plays a social function, this is recognized in the third paragraph of article 27 of the Mexican Constitution, which holds that: The Nation shall at all time have the right to impose on private property such restrictions as the public interest may demand, as well as to regulate, for social benefit, the utilisation of those natural resources which re susceptible of appropriation, in order to make an equitable distribution of public wealth, to conserve them, to achieve a balanced development of the country and to improve the living conditions of rural and urban populations.

Article 27 does not indicate what the soil uses should be, but it does refer that, to give fulfilment to the faculty of the nation to impose modalities on the private property (including soils and waters) the necessary measures will be dictated for:

- Ordering human settlements.
- Establishing adequate provisions, uses, reserves and destinies of lands, waters and forests.
- Preserving and restoring ecological balance.
- Dividing large rural estates
- Providing, under the terms of the regulatory law, the collective organisation and exploitation of *ejidos* and communities.
- Developing small rural property.
- Promoting agriculture, livestock, forestry and other economic activities in rural areas.
- Avoiding the destruction of natural elements and damage that property may suffer to the detriment of society.

Secondary laws, based on these constitutional principles, establish soil uses. Among the laws that define soil uses, are the Agrarian Law (AL) 1992, the General Law of Sustainable Rural Development (Rural Development Law) 2001, the General Law of Sustainable Forest Development (Forest Development Law) 2018, the Environmental Law 1988, and the Human Settlements, Land Use Planning and Urban Development of 2016 (General Urban Law).

5.2 Soil Uses Defined by Zoning and Land Use Planning

Soil uses are regulated by both, the Environmental Law and General Urban Law. The Environmental Law regulates the ecological ordinance of the territory whereas the General Urban Law governs urban planning.

The Environmental Law defines the ecological zoning plan of Mexican territory as 'the instrument of environmental policy that regulates or induces land use with the aim of protecting the environment and preserving and to exploiting natural resources in a sustainable manner by analysing the deterioration trends and the potentialities for exploiting the land';[18] whereas according to the General Urban Law the territorial ordering of human settlements is 'A public policy whose purpose is the rational use and occupation of the territory as well as to establish the spatial basis of socio-economic development strategies and environmental preservation'.[19]

The Environmental Law recognizes four levels of environmental planning:

- Program of General Ecological Ordinance of the Territory (PGEOT),

[18]Article 3, paragraph XXIV of the Environmental Law.

[19]Article 3, paragraph XXVI of the General Urban Law.

- Regional programs for ecological ordinance,
- Local programs for ecological ordinance, and
- Maritime zone ecological ordinance programs.

The PGEOT is issued by the Federal Executive. In accordance with article 20 of the Environmental Law, it aims to establish:

- The ecological regionalisation of the national territory and of the zones on which the nation exercises sovereignty and jurisdiction, based on the diagnosis of the characteristics, availability and demand of natural resources, as well as of the productive activities that develop in them and of the location and status of existing human settlements; and
- The ecological guidelines and strategies for the preservation, protection, restoration and sustainable use of natural resources, as well as for the localisation of productive activities and human settlements.

Under such a legal basis, the PGOET was issued in 2012. It divides the national territory into 145 biophysical units and establishes ecological guidelines and strategies to applicable to each one of them.

In accordance with article 20, Bis-1 and 20 Bis-2, the regional programs for ecological ordinance are aimed at ordering the soil use in regions that cover all or part of the territory of a local state. These programmes have to be issued by local state governments. However, in the following cases participation of federal government is mandatory:

- When an ecological region is located within the territory of two or more federative entities, the Federal Government, the Government of the respective States and Municipalities and, in its case, the Government of Mexico City, within the scope of their competencies, may formulate a regional ecological management program. To this aim, the federation shall celebrate the appropriate coordination agreements or conventions with the involved local governments.
- When a regional ecological management program includes a protected natural area, under the federal jurisdiction, or part of it, the program must be drafted and approved jointly by the Ministry and the governments of the States, Mexico City and Municipalities in which it is located. As the case may be.

In accordance with article 20, *bis*-4 and 20 *bis*-5 of the Environmental Law, local programs for ecological ordinance shall be elaborated by the municipal authorities and must have the following objectives:

- To determine the different ecological units that are located in the area or region concerned, describing their physical, biotic and socio-economic attributes, as well as the diagnosis of their environmental conditions, and the technologies used by the inhabitants of the area.
- To regulate, outside of the urban areas, the uses of the soil with the purpose of protecting the environment and preserving, restoring and making sustainable use of the respective natural resources, mainly in the accomplishment of productive activities and the location of human settlements; and

- To establish ecological regulation criteria for the protection, preservation, restoration and sustainable use of natural resources within the urban areas, in order to be considered in the corresponding urban development plans or programs.

Finally, the Maritime zone ecological ordinance programs are formulated and issued by the Federal Executive and in accordance with article 20 *bis*-7 of the Environmental Law the maritime eco-management programs must contain at least:

- The precise delimitation of the area that will comprise the program;
- The determination of the ecological zones based on the characteristics, availability and demand of the natural resources included in them, as well as the type of productive activities that they develop; and
- Guidelines, strategies and other provisions for the preservation, protection, restoration and sustainable use of natural resources, as well as the carrying out of productive activities and other works or activities that may affect the respective ecosystems.

Similarly, the General Urban Law establishes the following levels or planning:[20]

- the National Urban Development Program,
- local programmes for urban development,
- programs for zoning metropolitan areas,
- municipal plans and programs for urban development,
- programs for urban development in population centres, and
- other plans and programs derived from the afore mentioned and considered by local laws.

Given that neither the Environmental Law nor the General Urban Law clearly establishes the limit of their scope of territorial application, it can be argued that both laws are responsible for ordering the same territory, that is to say, the territory of the country as a whole. Although each of these mechanisms can be interpreted as ordering the national territory from two different points of view, ecological ordering does so from the point of view of productive activities, while urban planning does so from the point of view of human settlements.

However, according to article 22 of the General Urban law, the territorial ordering of human settlements is limited to ordering the territory of population centres. When referring to the National System of Planning of Territorial Ordering, Urban and Metropolitan Development, this legal provision states: 'The planning, regulation and evaluation of the Territorial Ordering of Human Settlements and Urban

[20]The National Urban Development Program is an instrument to guide the urban development and the territorial ordering of Mexico, that determines the policy, the strategy and the actions of the urban development, as well as the bases to expedite the programs of the subsequent application fields. These local, municipal and more specific programs must be created in accordance with the National Program, which must also be in harmony with the National Development Plan, a document created by the President at the beginning of each administration and is the Federal Executive's guiding document in which the national objectives, strategies and priorities of the country's integral and sustainable development are specified.

Development of the Population Centres are part of the National Democratic Planning System, as a global, sectorial and regional policy that contributes to the achievement of the Objectives of the National Development Plan, federal programs and state and municipal plans'.[21]

On the other hand, the analysis of the Environmental Law allows us to assert that the ecological ordinance of territory has supremacy on the urban ordering. The above position is based on the following reasons:

- The PGEOT establishes guidelines and ecological strategies for locating productive activities and human settlements (Article 20, section I).
- Regional Programs for Ecological Ordinance determine ecological regulation criteria for productive activities and the location of human settlements (Article 20a 3, section II).
- Programs or Ecological Ordinance establish the ecological regulation criteria to be considered by the corresponding urban development plans or programs (Article 20*bis* 4, section III).
- The urban planning programmes must be submitted to the environmental impact assessment procedure (Article 32 of Environmental Act).
- Urban development plans or programmes must take into account the guidelines and strategies contained in the ecological planning programmes of the territory (Article 23, section I)

5.3 *Forest Use of Soil*

The Forest Development Law (2018) indicates in its article 7 fraction LXXI that *forest soils* are those covered by forest vegetation, while in its fraction LXXII provide that *preferably forest soils* are those that have been covered by forest vegetation, but currently are not covered by it, however, due to their climate conditions, soil and topography are more suitable for forest use than for other alternative uses, excluding those already urbanised.

In order to allocate *forest soil* or *preferably forest soil* for a different use, it is mandatory to obtain an *authorisation for the soil use change*. In consequence, forest soil will continue being forest soil until the *authorisation for the soil use change* is issued. This is recognized by Article 135 of the Regulations of the General Law on Sustainable Forestry Development of 2019 (Forest Regulations), which points out:

> Article 135: forest lands shall continue to be considered as such even if they lose their forest cover due to illicit actions, plagues, diseases, fires, landslides, hurricanes or any other cause.

[21] See Article 22 of the General Urban Law.

The identification of forest soil is carried out through the forest zoning referred to in article 7, sections LXXXIV and 49 of the Forest Development Law. According to the first article, forest zoning is 'the instrument of planning in which the forest and preferably forest soils are identified, grouped and ordered within hydrographic basins, under criteria of conservation, restoration and sustainable management.'

In accordance with article 49 of the Act, the National Forestry Commission (NFC) is responsible for elaborating forest zoning, based on the National Inventory of Forest and Soils and on the Ecological Ordering's Programmes.

The analysis of the Forest Development Law revels that changing soil use in forest soils is in principle prohibited. According to article 68, section I; 69, section I and 93 of the mentioned legal body, it is evident that the Ministry of the Environment and Natural Resources can only authorise the change of soil use by exception.

In this direction, article 93 states that:

> The Ministry shall authorise the change of soil use in forest soils, by exception, with the prior technical opinion of the members of the State Forestry Council in question and based on justifying technical studies that demonstrate that biodiversity of affected biodiversity prevails and that erosion, deterioration of water quality or a decrease in water catchment is mitigated into the affected areas by the removal of forest vegetation.

On the other hand, articles 68, section I and 69, section I above are reiterative when expressing:

> ARTICLE 68—corresponds to the Ministry to issue the following acts and authorisations.
>
> I. Authorisation of soil use change in forest soils, by exemption;
>
> ARTICLE 69—corresponds to the Ministry to grant the following authorisations:
>
> I. Change of soil use in forest soils, by exception

In such a way that the Ministry of the Environment and Natural Resources is entitled, upon request of the interested party, to lift the prohibition established by the cited precepts. This faculty is not a regulated but discretionary faculty, since the Law does not indicate in which cases or under which circumstances the individual's request must be positively resolved. On the other hand, except in the case contemplated by the third paragraph of article 97, the law does not establish in which cases the authority must deny the request for change of soil use, the referred precept states:

> No authorisation may be granted to change the use of soil in a burned area without 20 years having passed, and it is reliably accredited to the Ministry that the affected forest vegetation regenerated, through the mechanisms established for this purpose in the regulation of this law.

However, the above does not mean that the law grants the Ministry of the Environment and Natural Resources an arbitrary faculty to decide on the soil use change, rather that it must base and adequately motivate its resolution. Moreover, the very first paragraph of article 93 in its second part states that the authority must decide a request for soil use change taking into account:

- The opinion of the members of the State Forestry Council; and
- The Justifying Technical Studies that the applicant must present, as stated in article 138 of the Regulation.

In such a way that the authority when basing and motivating the resolution issued on the application for change of soil use must necessarily refer to Article 93 to deny the application for change of soil use.

This means that the discretion that the law grants to the authority is limited in both its positive and negative aspects. That is to say, the authority could only deny the lifting of the prohibition when it is based on and motivates that any of the following three assumptions are accredited:

- The updating of the hypothesis foreseen in the third paragraph of article 97 of the law.
- The analysis of the justifying technical studies demonstrate that biodiversity is compromised, and that soil erosion, deterioration of water quality or a decrease of water catchment cannot be mitigated in the affected areas.
- The State Forest Council's opinion is not favourable to the request.

On the positive side, the authority's discretion is limited by the Eighth Transitory Article of the 2013 Constitutional Reform Decree on energy matters, as well as by article 96 of the Hydrocarbons Law (2014) and 71 of the Electricity Industry Law (2014), which indicates that the use of land for energy sector activities is preferable over any other use. Similarly, the Miner Law (1992) also states that the miner use of the land is preferable over any other, as well as the Forest Development Law. In both cases, the authority must give the land use change even though the State Council were negative or the technical studies inadequate.

5.4 Environmental Use of Soil

The Environmental Law also deals with allocating soil for environmental use through the declaration of natural protected areas.

According to Pavón and González, natural protected areas are places that have not been significative altered by human activities or that require preservation or restoration of their natural environments, biodiversity, ecosystems, surrounding areas, or those areas used for entertainment, culture, and national identity.[22] Under federal

[22]Pavón and González Márquez (2007).

jurisdiction, there are different kinds of natural protected areas such as biosphere reserves, national parks, natural monuments, natural resources protected areas, flora and fauna protected areas, and sanctuaries. However, this definition should be complemented by the provisions of the second paragraph of article 44 and article 57 of the Environmental Law. The first of them—modified in 1996—states that 'owners, possessors or holders of other rights over lands, waters and forests included within protected natural areas shall be subject to the modalities that, in accordance with this Law, and the other provisions contained in the management program and in the corresponding ecological planning programs.' The second article states that protected natural areas are established by Executive Order, issued by the federal or local government.

The natural protected areas define the different uses that can be given to the soil on which they are established. The original text of the Environmental Act of 1988 indicated that in the case of biosphere reserves the limitations to the property right that were established in the respective Declaration Decree should be different for the core area than for the buffer zone of the natural protected areas.

Soil use restrictions in the core area should be greater because in that area it is the natural phenomenon that is intended to protect, while in the buffer zone it could withstand greater uses because it is an area that protects the core area from the external impact. However, a reform made in 2005 also established for the rest of the natural protected areas different zone divisions with diverse use soil restrictions.

Declarations of natural protected areas do not expropriate land ownership but deprive the owner of a series of prerogatives inherent in the right of property. In this regard, article 63 of the Environmental Law states that 'the protected natural areas established by the Federal Executive may include, partially or totally, properties subject to any property regime.' However, the second paragraph of article 44 states that 'owners, possessors or holders of other rights over land, water and forests included within natural protected areas shall be subject to the modalities that, in accordance with this Law, establish the decrees declaring such areas, as well as the other provisions contained in the management programme and in the applicable environmental management programmes'.[23]

On the other hand, in principle, landowners where natural protected areas are decreed lose the right to manage it. This is clear from the reading of article 47 of the Environmental Law, which states:

> In the establishment, administration and management of the protected natural areas referred to in the preceding article, the Ministry (of the Environment) shall promote the participation of its inhabitants, owners or holders, local governments, indigenous peoples, and other public and private social organisations with the aim of promoting the integral development of the community and ensuring the protection and preservation of ecosystems and their biodiversity.

[23]Pavón and González Márquez (2007).

Even though natural protected areas are controlled by the federal government, we see a new trend in environmental policy in which their management has been delegated to the states and municipalities. Local governments may create state parks and reserves in relevant areas at state level. In a like manner, it is the responsibility of the municipalities to establish ecological preservation areas in population centres pursuant to the provisions established by their local legislation. The Ministry of Environment and Natural Resources is responsible for promoting the participation of citizens, landowners, holders, local governments, indigenous people, and social, public, and private organisations in the establishment, administration, and management of natural protected areas. This is in order to encourage integral development of the community and to guarantee the protection and preservation of the ecosystems and their biodiversity. These natural protected areas combine different development systems with land use regulations and natural preservation strategies. The Environmental Law considers the reacquisition and designation of private property as natural protected areas as a mere limitation on property rights under paragraph 3 of article 27 of the Constitution; however, in reality, these intrusions represent the use of the government's eminent domain power established by paragraph one of the same article and should be subject to compensation.

6 Soil Pollution

Soil pollution is inappropriately ruled by the current environmental legislation because it is associated almost exclusively with the disposal of waste even when it results from various causes.

The Environmental Law contains a chapter called 'prevention and control of soil pollution' which is composed of articles 134–144. Almost all of these legal provisions are programmatic and only, article 137 states that it is subject to the authorisation by municipalities or by the federal district, the operation of the collection, storage, transportation, accommodation, reuse, treatment, and final disposal of municipal solid waste systems.

In the same vein, article 142 of the mentioned legal body prohibits import of waste for its spillage, deposit, confinement, storage, incineration, or any treatment for its destruction in the national territory or in the areas in which the nation exercises sovereignty or jurisdiction. However, these two provisions only refer to soil pollution indirectly.

Finally, according to article 141 of the Environmental Act, the Ministry of Environment, in coordination with the Ministries of Economy and Health, shall issue the technical standards[24] for the manufacture and use of containers and

[24]These standards are called: Mexican official standard (Norma Oficial Mexicana): the technical regulation of mandatory observance issued by the competent authorities, which establishes rules,

packages for all kinds of products whose materials make possible the reduction of solid waste generation.

Other provisions of this chapter are much more programmatic and refer to the criteria for the prevention and control of soil contamination, the issuance of technical standards in this area and the process of entering into coordination agreements with the states and the municipalities.

By contrast, in a different chapter, article 152*bis* specifically refers to soil pollution and establishes a parameter of soil remediation by stating: 'when the generation, management or final disposal of hazardous materials and waste produces soil pollution, those responsible for these operations must carry out the necessary actions to recover and restore the conditions of the soil, so that it can be destined to any of the activities foreseen in the urban development program or of ecological ordinance plan that results applicable for the specific site or area'.

The Waste Act of 2003 also contains a chapter on soil pollution called 'liability for contamination and remediation of sites', which includes articles 68–79. The chapter governs five issues: the obligation to repair the damages caused by soil pollution; authorisation for the transfer of contaminated sites; abandonment of polluted sites; identification, inventory, registration, and categorisation of sites polluted with hazardous wastes and declarations of remediation.

Transfer of polluted sites is in principle banned by article 71 of the Waste Law, but this prohibition can be lifted by means of the authorisation granted by the Ministry of Environment.

In the case of sites polluted by hazardous waste that are abandoned, article 73 of the Waste Law provides that the Ministry of Environment, in coordination with state and municipal governments, may formulate and execute the corresponding remediation programmes and make effective the guarantees provided by those who abandoned the site in question.

Likewise, under article 75 of the Waste Act, the Ministry of the Environment and local governments are obliged to carry out actions of identification, inventory, registration, and categorisation of sites contaminated with hazardous waste, to proceed to its remediation.

Finally, according to article 74 of the Waste Law, the Remediation Declarations may establish modalities to the property right, but the Law does not indicate to whom it corresponds to issue such declarations.

---

specifications, attributes, guidelines, characteristics or prescriptions applicable to a product, process, installation, system, activity, service or method of production or operation, as well as those related to terminology, symbology, packaging, marking or labelling and those referring to its compliance or application.

7 Institutional Arrangement for Soil at Federal and State Level

In Mexico, soil is generally ruled by local civil codes. In terms of land use, generally the municipalities are in charge of granting the land-use permits. However, in environmental matters there are concurrent competences between the states and the federation, since the federal government can impose different modalities to the use of land. Local and federal authorities have the power to enforce environmental administrative laws and federal laws. At the federal level, there are two ministries that have specific faculties in the matter, the Ministry of the Environment and the Ministry of Agrarian, Territorial and Urban Development.

Article 32 bis of the Organic Law of the Federal Public Administration (1976) grants the Ministry of the Environment the power to:

- Propose to the Federal Executive the establishment of protected natural areas, and promote the participation of federal or local authorities for their administration and monitoring.
- Organize and manage natural protected areas, and supervise the conservation, protection and surveillance of these areas.
- Exercise possession and ownership of the nation on beaches, federal maritime land zone and land reclaimed from the sea.
- Grant contracts, concessions, licenses, permits, authorisations, assignments, and to recognize rights, in matters of water, forest, beaches, federal maritime terrestrial zone and lands gained to the sea.

The Ministry of Agrarian, Territorial and Urban Development is empowered by article 41 of the same law to:

- Promote, in coordination with state and municipal authorities, the planning and ordering of the national territory for maximum use,
- Foresee at the national level the needs of soil for urban development and housing, considering the availability of water determined by the Ministry of the Environment, and to regulate, in coordination with federal and municipal governments, the mechanisms to satisfy those needs.
- Develop, support and implement programs to meet the needs of urban soil and the establishment of provisions and territorial reserves for the proper development of population centres, in coordination with different authorities
- Exercise the right of expropriation for reasons of public utility in cases not assigned to another authority.

8 Conclusions

Soil property in Mexico is, in principle, ruled by private law. However, in accordance with the Mexican Constitution, public law could modify the traditional rules of civil codes. Administrative laws establish the different uses of the soil and limit the prerogatives of the soil owner with the aim of protecting the environment and the public interest.

In that sense, in Mexico, property of soil is governed by the principles of *social function of property* and environmental function of property. As a result, the extension of the right of private property is defined by the law.

References

Booth P (2002) From property rights to public control. The quest for public interest in the control of urban development. Town Plan Rev 73(2):153–170. RU, Liverpool University Press

Duguit L (1920) Las transformaciones generales del derecho privado desde el Código de Napoleón. Librería española y extranjera, Madrid

Glenn HP (2010) Legal traditions of the world. Oxford University Press, New York, p 99

López Ramón F (1995) Concesión. In: Enciclopedia Jurídica Básica. Civitas, Madrid, p 1341

Mirow MC (2011) Origins of the social function of property in Chile. Fordham Law Rev 80 (3):1183–1217

Pavón G, González Márquez JJ (2007) Land use planning in Mexico: as framed by social development and environmental policies. In: Chalifour NJ, Kameri-Mbote P, Lye LH, Nolon JR (eds) Land use law for sustainable development. Cambridge University Press, Cambridge

Serna de la Garza JM (2007) El régimen constitucional de la propiedad en México. In: Serna de la Garza JM (ed) Derecho comparado Asia-México. Culturas y sistemas jurídicos comparados. UNAM, México, pp 473–494

Laws

Environmental Law 1988

General Law on the National Assets 2004

General Waters Law 1992

General Urban Law 2016

Official Journal of the Federation

Organic Law of the Federal Public Administration 1976

Political Constitution of United Mexican States 1917

Brumadinho's Disaster, Mine Tailing Dams and the Environmental Licencing in Brazil: Preventing Risks to Human Wellbeing, Soils, and the Environment

Paula Galbiatti Silveira, Marina Demaria Venâncio, and José Rubens Morato Leite

1 Introduction

Land degradation is a significant challenge that affects not only soil quality but also food security, biodiversity, and the climate. Among human activities that impact soil the most, mining is considered one of the most degrading. In the context of a global socio-ecological crisis, represented by the transgression of planetary boundaries and climatic changes, the 'business as usual' scenario in the management of natural resources has led to escalating impacts on the environment and avoidable mining disasters. Their consequences illustrate that society and the environment often pay the costs of successful modernization, while governments and corporations remain silent – in a true scenario of unaccountability and global risks.

In Brazil, for instance, mining activities have accounted for an important part of the country's economy since colonial times, because of its abundant reserves of important minerals. The country is the second biggest exporter of iron in the world, and one of the largest exporters of tin, lithium, niobium, bauxite, and manganese. One of the biggest producers of steel.[1] However, the combination of medium and large-scale mining, corruption, and poor environmental management, has led to

[1]For more information check the website of the Brazilian Mining Institute (Instituto Brasileiro de Mineiração 2019).

P. G. Silveira
Federal University of Santa Catarina, Santa Catarina, Brazil

University of Bremen, Bremen, Germany
e-mail: galbiatt@uni-bremen.de

M. D. Venâncio (✉) · J. R. M. Leite
Federal University of Santa Catarina, Santa Catarina, Brazil
e-mail: marina.venancio@posgrad.ufsc.br; morato.l@ufsc.br

H. Ginzky et al. (eds.), *International Yearbook of Soil Law and Policy 2019*, International Yearbook of Soil Law and Policy 2019,
https://doi.org/10.1007/978-3-030-52317-6_13

major socio-environmental impacts, culminating in two recent disasters, namely the collapse of two mine tailings dams in Mariana (2015) and Brumadinho (2019)—both owned by the same company, Vale. These cases are among the biggest socio-environmental disasters in the history of the country and raised several questions regarding enforcement of Brazilian environmental legislation, as well as negligence and corruption in the public and private sectors.

Within this context, this chapter aims at addressing the issue of Brumadinho's disaster and mine tailing dams, highlighting the role and importance of the environmental licencing process in Brazil as a preventive instrument in risk assessment. To achieve that, this research is divided into three sections. First, it discusses the elements of the socio-ecological crisis, contextualizing the issue of mining disasters to the theory of the world risk society. Secondly, it summarizes the main issues concerning Brumadinho's dam collapse. Lastly, it outlines the most relevant aspects of the Brazilian environmental licencing process and the best available technology (BAT) as a technical clause, stressing their essential role in the prevention of major mining disasters.

2 Elements of the Socio-Ecological Crisis

According to the Global Footprint Network,[2] humanity has been operating in overshoot and superseding Earth's capacity to regenerate renewable resources. Currently, 1.7 planets are needed to make current consumption trends of natural resources sustainable.[3] This alarming number illustrates the severity of the socio-ecological crisis, which is deeply rooted in a complex and rather unsustainable development path that often neglects planetary boundaries.

Planetary boundaries can be understood as thresholds within which humanity can operate safely, and whose transgression may lead to undesirable and potentially catastrophic scenarios.[4] Given the nine boundaries identified by Rockström and colleagues,[5] it is estimated that we have already transgressed three of them, namely the interference with the global nitrogen cycle, the rate of biodiversity loss, and climate change. This means that the business as usual scenario is leading ecological systems—the precondition to any kind of human activities—to exhaustion and destabilization, which has been bourne out by major environmental studies.[6]

Concerning land degradation, changes in soil systems are also included in the planetary boundaries. Furthermore, soil degradation is a global challenge that leads to food insecurity and biodiversity loss, intensified by global climate changes.

[2]Global Footprint Network (2019).

[3]Global Footprint Network (2019).

[4]Rockström et al. (2009a).

[5]Rockström et al. (2009a).

[6]Rockström et al. (2009b).

According to the *Assessment Report on Land Degradation and Restoration*, produced by the Intergovernmental Science-Policy Platform on Biodiversity and Ecosystem Services (IPBES),[7] land degradation "[...] is a pervasive, systemic phenomenon occurring in all parts of the world, is an urgent priority in order to protect the biodiversity and ecosystem services that are vital to all life on Earth and to ensure human well-being". The Report also affirms that land degradation impacts 3.2 billion people and represents an economic loss in the order of 10% of the annual global gross product.[8]

As stated above, the connection between soil degradation and biodiversity loss is clear and alarming. For instance, the preliminary summary for policymakers of the Global Assessment Report on Biodiversity and Ecosystem Services, the largest global assessment concluded since the Millennium Ecosystem Assessment, notes that "[h]uman actions threaten more species with global extinction now than ever before".[9] Also, according to the study led by IPBES, roughly 1 million species are threatened with extinction, which represents an average of 25% of assessed animal and plant species.[10]

Likewise, while climate change might exacerbate the impacts of land degradation and might limit options for addressing it, land degradation is also a major contributor to climate changes.[11] In an equally alarming tone, the latest report from the Intergovernmental Panel on Climate Change (IPCC)[12] on the global warming of 1.5 °C stated that urgent, coordinated and unprecedented action is needed to tackle harmful climatic changes. According to the IPPC, the current pledges made under the Paris Agreement are not sufficient to meet its targets, and "[p]athways limiting global warming to 1.5 °C with no or limited overshoot would require rapid and far-reaching transitions in energy, land, urban, infrastructure (including transport and buildings), and industrial systems".[13]

Nonetheless, despite the sense of urgency in IPCC's message, there was a major tension in the 24th Conference of the Parties (COP) to the United Nations Framework Convention on Climate Change (UNFCCC) regarding whether or not to "welcome" the report (raised by the United States, Russia, Saudi Arabia, and Kuwait). Instead, the failure to reach a consensus resulted in the Conference welcoming its "timely completion", falling short in properly integrating its findings and relevance into the outcome document. This illustrates the complexity of addressing the major challenges in a truly anthropogenic era.

As the sixth edition of the Global Environment Outlook (GEO-6) assessment highlights, "unsustainable human activities globally have degraded the Earth's

7IPBES (2018), p. 2.

8IPBES (2018), p. 2.

9IPBES (2019).

10IPBES (2019).

11IPBES (2018).

12IPCC (2018).

13IPCC (2018), p. 40; IPCC (2014), p. 17.

ecosystems, endangering the ecological foundations of society".[14] GEO-6 points out that rapid urbanization, population pressure, technological innovation, and economic development are major drivers of environmental change, and are "strongly intertwined, complex, and spread widely and unevenly across the world".[15]

As a result, the sociologist Beck[16] argues that we are in the making of a reflexive, also known as second, modernization, in which there is the "continuity of the principles and discontinuity of basic institutions of nation-state modernity": a dialectics of modernity. Societies have become increasingly capitalistic and modern, and are giving rise to new potentials opposing the industrial modernity.[17]

In this context, we may refer to a 'world risk society', that is, a stage "of modern society that no longer merely casts off traditional ways of life but rather wrestles with the side effects of successful modernization - with precarious biographies and inscrutable threats that affect everybody and against which nobody can adequately insure".[18] According to Beck,[19] the theory of the world risk society, states that "the global anticipation of global dangers and catastrophes rock the foundations of modern societies".

It is important to note that 'risk' presumes decision, modernization, and is caused by humans, whereas 'danger' is caused by nature.[20] The global risks in the reflexive modernization are delocalized, incalculable, difficult to insure against and to compensate.[21]

> In radicalized modernity, the new risks are hence manufactured or fabricated uncertainties and dangers, because the range of potential catastrophes and uncertainties grows with technological and scientific progress and more industrialization, more cars and more wealth also cause more environmental problems. [...] The new risks are, in other words, manufactured uncertainties and dangers: modernity is faced with its own destructive potential of social and technological development without having adopted adequate answers.[22]

In the case of mining activities and its manufactured risks, for instance, a potential disaster would not only affect the place of the incident but would also have rather long-lasting transboundary effects, posing several challenges in terms of insurance, compensation, and liability. Regarding the latter, one of the characteristics of the world risk society is the so-called 'organized irresponsibility', which can be defined as a state in which no one takes responsibility for the results of successful modernization. According to Beck,[23] in the reflexive modernization,

---

[14]UN Environment (2019a), p. 4.

[15]UN Environment (2019a), p. 6.

[16]Beck (2008), p. 2.

[17]Beck (2008), p. 2.

[18]Beck (2012), p. 8.

[19]Beck (2008), p. 4.

[20]Beck (2008), p. 5.

[21]Beck (2008), p. 5.

[22]Beck (2008), p. 5.

[23]Beck (1995), p. 64.

> [I]t is the application of prevalent norms that guarantees the non-attributability of systemic hazards: hazards are writ small as risks, compared away and legally and scientifically normalized into improbable 'residual risks', making possible the stigmatization of protest as outbreaks of 'irrationality'. Those who uphold maximum pollution levels turn white into black, danger into normality, by act of government. Whoever waves the banners of rigorous causal proof while demanding that the injured parties do the same, not only demands the unachievable, as science has meanwhile adequately confirmed, but thereby also holds aloft a shining shield to keep rising, collectively conditioned hazards out of the reach of politics or attribution to individuals.

Within this context and as a symptom of the socio-ecological crisis, the case of disasters involving tailing dams, their licencing and flawed regulation, becomes an emblematic example of *organized irresponsibility*, in which the normalization of risks leads to catastrophic scenarios with no 'culprit'. It is also a symbol of this momentum defined as reflexive modernization, calling for new governance instruments, proper answers, and an overall new attitude from policymakers, decision makers and society.

3 Mining Tailing Dams and the Brumadinho Dam Failure in Brazil

From the perspective of the socio-ecological crisis and the world risk society, we can now further address the Brumadinho disaster. Since the twentieth century, the exploitation of resources (such as metals and minerals) has increased remarkably.[24] In terms of mining, the activity constitutes "the most energy-intensive of all industries".[25] Among its negative social impacts, we can highlight the devastation of the environment, people's displacement, great consumption of natural resources (such as energy and water), and severe pollution of groundwater and soils.[26] Additionally, "[m]ining activities are often associated with deforestation, the dredging of huge pits, the disposal of waste rocks, acids, and ores, and the possible formation of AMD [acid mine drainage]".[27]

As stated by the UN Environment,[28] with reference to declining stocks and increased demand for metals,

> [w]hen declining ore grades are combined with the larger project scales needed to extract enough ore to supply market demand, greater risks threaten the natural environment. More land is cleared, or simply removed and shipped away, as mountain-top removal illustrates. Larger volumes of mine waste accumulate, with heavy metals and reactive agents

---

[24]UN Environment (2019b), p. 91.

[25]Weiss et al. (2016), p. 93.

[26]Weiss et al. (2016), p. 93.

[27]Weiss et al. (2016), p. 93.

[28]UN Environment (2019b), p. 92.

recombining into noxious compounds. Water pollution risks, especially from acid and metalliferous drainage, increase. Threats to biodiversity become more complex [. . .].

In a context of globalized fabricated risks, we can also highlight the risks of *mining disasters*, especially those involving mining tailing dams, which are the focus of this chapter. In consonance with the United Nations Office for Disaster Risk Reduction (UNDRR), disaster can be defined as a severe "disruption of the functioning of a community or a society at any scale due to hazardous events interacting with conditions of exposure, vulnerability and capacity, leading to [. . .] human, material, economic and environmental losses and [or] impacts".[29] For instance, a disaster occurs when an event causes a serious disturbance, associated with conditions such as poverty, leading to major impacts. This was the case of the collapse of the Fundão and Satarém tailing dams in 2015, in the Municipality of Mariana (Minas Gerais), which released 62 millions of cubic meters of mud and iron ore tailings into the region's ecosystem, completely flooding the subdistrict of Bento Rodrigues, affecting roughly 6 million people.[30]

Concerning these structures, "tailings" are a high-volume waste produced by the beneficiation of ore, that is, "the residue of an ore that remains after it has been milled and the desired metals have been extracted".[31] They are often disposed through "the use of a wet tailings impoundment facility", such as a dam.[32] Around the world, thousands of deaths can be attributed to tailing dam's failures, with more than 120 incidents registered since 1961.[33] This depicts that the issue is not recent and should be properly addressed by law.

In short, mining disasters due to tailing dams can result in short- and long-term impacts on livelihoods, e.g. access to basic services, health, ecosystems,[34] and soils. These consequences have the potential to affect regions located hundreds of kilometres away from the incident,[35] and are transboundary. Again, we argue that such disasters are emblematic examples of the reflexive modernization and the world risk society, whose global risks are not restricted to a single region, but rather omnipresent.

More specifically in Brazil, there are around 24,092 dams registered by monitoring bodies, among which 790 are tailing dams.[36] According to the National Water Agency (ANA or *Agência Nacional de Águas*), at least 3387 dams have a high 'associated potential damage' (DPA, acronym in Portuguese) or 'risk category' (CRI, acronym in Portuguese), including five mine tailing dams.[37]

[29]UNDRR (2017).

[30]Mesquita (2017).

[31]Environmental Law Alliance Worldwide (2010), p. 6.

[32]Environmental Law Alliance Worldwide (2010).

[33]WISE Uranium Project (2019).

[34]Fundação Oswaldo Cruz (2019), p. 10.

[35]Fundação Oswaldo Cruz (2019), p. 10.

[36]Agência Nacional de Águas (2018).

[37]Agência Nacional de Águas (2019).

The Brazilian State of Minas Gerais—where the municipality of Brumadinho is located—has in total 754 dams, among which 317 are destined to store mine tailings.[38] The lack of planning from the government, control measures, and adequate technical procedures in the planning and developing of mines, has led to socio-environmental conflicts in the State of Minas Gerais.[39] In this State alone, seven mine tailings have breached over the last 15 years, including the dams of Fernandinho (1986), Macacos (2001), Cataguases (2003), Rio Pomba/Cataguases (2007), Herculano (2014), Fundão (2015), Santarém (2015),[40] and Córrego de Feijão (2019). The latter has become known as "the Brumadinho disaster" and resulted in roughly 238 deaths and 32 reports of missing people.

According to the Emergency Mission's Report about Brumadinho, done by the National Council on Human Rights (CNDH, or *Conselho Nacional dos Direitos Humanos*) just after the disaster, on the 25th of January 2019, around 12:28 p.m., the Córrego do Feijão's mine tailings dam breached.[41] The dam was constructed in 1976, using the method of upstream, with 86 m of height and 720 m of length.[42] The tailings used to occupy an area of 249.5 thousand square meters and a volume of 11.7 million cubic meters.[43] The breach leaked approximately 12 million cubic meters of iron ore tailings directly into the *Ferro-Carvão* river,[44] and later reached Paraopeba river, one of the biggest tributaries of São Francisco river. It hence 'washed away' operational equipment (like trains, cars, and machines) and the administrative centre of Vale company, killing hundreds of workers who were in the field or having lunch.[45]

Among the impacts caused by the disaster, we can also highlight the destruction of habitats and killing of fauna in the affected rivers; water contamination; suspension of water supply to rural and urban activities; silting of rivers and reservoirs for energy production; burial of lagoons and water springs; destruction of vegetation; alteration in the water flow; soil contamination, and many others.[46]

Furthermore, according to a technical note released by the Osvaldo Cruz Foundation,[47] the Brumadinho disaster can lead to an increase in the incidence of pre-existing regional illnesses (such as yellow fever and schistosomiasis), mental health issues (such as depression and anxiety), and vector-borne diseases. Likewise, it can aggravate pre-existing chronic illnesses, such as cardiovascular diseases,

[38] Alves (2015).

[39] Leite (2016), pp. 382–383.

[40] Alves (2015).

[41] CNDH (2019).

[42] CNDH (2019).

[43] CNDH (2019).

[44] Fundação Oswaldo Cruz (2019), p. 1.

[45] CNDH (2019).

[46] CNDH (2019).

[47] Fundação Oswaldo Cruz (2019).

hypertension, renal failure, and diabetes.[48] These impacts on health, as stated by the report, will require monitoring over the many years, increased health surveillance, including diagnostic and treatment of non-transmissible illnesses preventive measures against vector-borne diseases (e.g. yellow fever and dengue), the reconstruction (and establishment) of sewage systems, and water quality surveillance.[49] Several other impacts in local communities were identified, affirming a recurrent violation of human rights in dam's management in Brazil.[50]

Owned by Samarco, a joint venture between the English-Australian BHP and the Brazilian Vale, the failure of the Ferro-Carvão dam embodies and illustrates a series of structural problems in the Brazilian environmental planning and management of tailing dams. The advanced environmental legislation in the country has not adequate to prevent such disasters. For instance, a set of environmental licences and an Environmental Impact Assessment (EIA or *Estudo de Impacto Ambiental*) are required for the installation of mining activities. However, despite these preventive instruments, the powerful mining lobby and the lack of a more structured licencing system with well-equipped environmental monitoring agencies, has been preventing the legislation from being adequately implemented and monitored.

For example, in the years between Mariana and Brumadinho, the Brazilian government dedicated fewer resources to the monitoring of dams and weakened the legislative framework of environmental licencing.[51] In Minas Gerais, where both tailing dams used to be located, a bill (later approved by the House of Representatives) established fewer or lighter requirements to granting environmental licences.[52]

The situation now is even more concerning, with an increasing trend in Brazilian politics towards the weakening of environmental policies. In this context, in an unprecedented meeting, eight former Brazilian ministers of environment convened in São Paulo in early May 2019 to address the current political scenario of the country, after the election of current president Jair Bolsonaro. The group stressed their concerns with the ongoing process of dismantlement of relevant policy structures, which has been affecting the Ministry's capacity of formulating and implementing environmental policies—including those aimed at tackling deforestation, climate change, and corruption.[53] The document released in the meeting also condemned the government's attempts to weaken the environmental licencing process, referring to the case of Brumadinho as a tragic example of poor environmental management.[54]

[48]Fundação Oswaldo Cruz (2019), p. 11.

[49]Fundação Oswaldo Cruz (2019), p. 11.

[50]CNDH (2019).

[51]CNDH (2019).

[52]CNDH (2019).

[53]Ricupero et al. (2019).

[54]Ricupero et al. (2019).

4 Environmental Regulation of Mining Activities in Brazil

As noted above, Brumadinho is an emblematic case of organized irresponsibility, in the sense that environmental regulations and environmental bodies responsible for monitoring mining activities have not been able to prevent such a disaster from occurring. In relation to the legal aspects of mining activities in Brazil, the environmental licencing of potentially harmful activities and the EIA[55] can be considered the most important tools to prevent, to monitor, and to mitigate risks. The importance of environmental licencing was even stressed by UN Human Rights Experts, who criticized the recent 'deregulation' of environmental protection laws and requested the Brazilian government not to licence any installations of new tailing dams until the safety of the existing structures is attested.[56] Within this context, they called upon the government to give priority to safety evaluations of existing dams and rectify current licencing and safety inspections processes, in order to prevent such disasters from occurring.[57]

Having said that, we can now delve more specifically into the co-related Brazilian legislation. The Brazilian Constitution, whose environmental core is found in article 225, prescribes a subjective right to an ecologically balanced environment. In this same article, paragraph 1 establishes duties to protect the environment, including, in item IV, the need to conduct previous EIA to the installation of works or activities that might potentially cause significant environmental degradation. It also prescribes, in item V, the obligation to control the production, commercialization, and use of techniques, methods and substances that might risk life, quality of life, and the environment. Furthermore, the constitutional text addresses environmental mining impacts explicitly, stating in paragraph 2 that whoever explores mineral resources is obliged to recover the degraded environment according to technical solutions required by the competent public body.

Among the instruments to fulfil the constitutional requirements for environmental protection, environmental licencing is regarded as one of the most efficient tools to control and to manage those environmental risks. It is mainly a preventive instrument, to assess potential risks and to help the State in the decision-making of activities that use natural resources or that might impact the environment. It is also a precautionary instrument, through the use of the BAT, and it supports possible mitigation and/or compensation of damages, as well as the evaluation of costs and benefits of projects.

[55]In Brazil, according to CONAMA Resolution n. 1/1986, the instrument is named Environmental Impact Assessment (EIA), even though environmental impacts also consider the ones that might affect social and economic activities. Pursuant to article 6, I, c, socioeconomic aspects shall be considered in the environmental diagnosis of the influence area of the project.

[56]Office of the United Nations High Commissioner for Human Rights (2019).

[57]Office of the United Nations High Commissioner for Human Rights (2019).

Generally speaking, environmental licencing in Brazilian law is an administrative process[58] through which the competent public environmental body grants a licence to allow the geographical location, installation, construction, expansion, modification and operation of projects and activities that might use natural resources, considered effective or potentially pollutant, or that might cause environmental degradation.[59]

Within the process of environmental licencing, an EIA can be required before the installation of a project or activity that might cause significant environmental degradation. The main laws and resolutions regulating the matter consist of (1) Law No 6,938/1981; (2) National Environmental Council CONAMA (National Environmental Council or *Conselho Nacional do Meio Ambiente*) Resolution No 01/1986; (3) CONAMA resolution No 237/1997; and (4) Supplementary Law (LC) No 140/2011, which regulates the distribution of competences between the federative entities.

4.1 Environmental Licencing and the Environmental Impact Assessment of Mining Activities in Brazil

In Brazil, there is a general regulation for the environmental licencing of potentially degrading activities, including mining. It is the most important instrument to prevent disasters, to control risks, and to protect the environment, weighing up environmental interests and other interests, such as the need to extract minerals. The LPNMA or Law No 6,938/1981 has inaugurated environmental law in Brazil as it is currently known. This Law is the basis for Brazilian environmental law once it establishes the

[58]There is an ongoing discussion in the literature about the nature of the environmental licencing in Brazil, whether it is a procedure or a process. Laws and resolutions define it at times as a procedure (see article 1, I, CONAMA Resolution No 237/1997, and article 2, I, LC No 140/2011) and sometimes as a process. This paper understands it, however, based on Farias, as a process. This differentiation is important, as the classification of environmental licencing as procedure or process leads to a minor or greater transparency and public participation. To be considered as an administrative process, it includes the obligation to make it public, to observe the right to adversary system and full defense, to fundament the decisions, and to open to public participation access to it. Its importance can be seen, as explained by Farias, in hydropower projects, in which the future and possible displacement of whole communities might be decided within an environmental licencing process. Therefore, one of the most important effects to understand the environmental licencing as a process is to strengthen social control, because the affected communities and all those who are interested might have access to the process and to be part of it. However, in practice, there is a wide margin for discretion within the process and the environmental bodies might make easier or more difficult depending on the interests related to the project. In those cases, the environmental licencing is normally questioned in the Judiciary. See Farias (2019).

[59]Legal concept established in article 1, I, and 2, CONAMA Resolution No 237/1997, and by article 2. I, LC No 140/2011. Free translation. In the following sections and chapters, the citation of laws and norms are based on a free translation by the authors from the original texts, except when expressively written otherwise.

objectives of the environmental national policies, defines environment, degradation, and pollution, creates the National Environmental System (SISNAMA, acronym in Portuguese) and the CONAMA, and defines instruments for environmental protection.

Among its instruments, article 9, IV, establishes the environmental licencing and revision of potentially or effectively pollutant activities. It is important to highlight the broad definition given by the law to the concepts of 'environmental degradation' and 'pollution'. In accordance with article 3, II, LPNMA, 'environmental quality degradation' is defined as an adverse alteration of the environment's characteristics. From this concept, 'pollution' is defined by article 3, III, LPNMA, as the degradation of the environmental quality resulting from activities that directly or indirectly (1) harms people's health, safety, and well-being; (2) creates adverse conditions to social and economic activities; (3) affects negatively the biota; (4) affects environment's aesthetic or sanitary conditions; and (5) release materials or energy in disagreement with established environmental standards.

Annex VIII of the law exemplifies potentially or effectively pollutant activities, in a total of 20 (twenty), which means that an environmental licence is legally required for them, without discretion by the environmental body, explicitly including mining activities in the first item.[60] The regulation of the environmental licencing process can be found in the CONAMA Resolution No 237/1997. According to articles 8 and 18, there are three types of licence, which can be granted isolated and successively, in a three-phase process, by the competent environmental body:

1. Preliminary Licence (LP): granted in the preliminary phase (project planning), approving the project's location, conception, attesting its environmental viability, and establishing the basic requirements and conditions. When applicable, an EIA is a pre-requisite for this licence. The validity of this licence shall be no longer than 5 (five) years;
2. Installation Licence (LI): authorizes project or activity's installation in accordance with the approved plans, programmes, and projects. The validity of this licence shall be no longer than 6 (six) years;
3. Operation Licence (LO): authorizes the project or activity operation, after verifying that all the conditions established in the previous licences were fulfilled. The validity of this licence shall be between 4 (four) and 10 (ten) years.

According to Farias,[61] the preliminary licence is the most important one, as it is the licence upon which the project's conception is built, environmental impacts are mapped, mitigation and compensation measures are established, and the project is discussed with the community. Therefore, it must function as a foundation for the whole project. After the LP is granted, the entrepreneur elaborates the executive project, structuring a more detailed project with the technical measures to protect the

[60]Annex 1 of CONAMA Resolution n. 237/1997 also establishes projects and activities that shall be submitted to the environmental licencing process, with a similar structure as the one mentioned.

[61]Farias (2019), pp. 73–74.

environment.[62] Lastly, after all the previous licences requirements were fulfilled and the environmental body inspects the installed project attesting it, the LO might be granted.[63]

In summary, fulfilling the requirements for a preliminary licence is always a precondition for the issuance of the next licence. Nonetheless, receiving one of the licences does not automatically grant its holder the right to receive the following. Each step has its requirements, including that all the conditions and requirements from the previous licence were complied with.[64]

One of the major criticisms of the environmental licencing process, from the private sector, is the excess of bureaucracies and its slowness.[65] Therefore, the law establishes maximum deadlines for the environmental body to decide if the licence is to be granted or not. Article 14, CONAMA Resolution n. 237/1997 prescribes a maximum deadline of six months, from the licence request, and a maximum of twelve months for those in which an EIA/RIMA and/or a public audience is included. Different deadlines might be established according to technical specificities of the project, under article 19, §1, Decree n. 99.274/1990.

It is important to highlight that the Law does not allow the so-called "tacit licence".[66] It means that even if the environmental body does not accomplish the maximum deadlines established in the law and the delay is not to be caused by the entrepreneur, the licence will not be automatically granted.

In relation to the EIA, it is a process of evaluating the possible environmental impacts of a specific project. The United Nations Environmental Program (UNEP)

---

[62]Farias (2019), p. 81.

[63]Farias (2019), p. 82.

[64]Farias (2019), p. 72.

[65]Those are valid critiques, as most of the environmental bodies in Brazil are not well equipped with financial and human resources to fulfil their competences with quality in due time. Therefore, many environmental licencing processes that should be easy and for small impact projects end up being too expensive and taking so much time making the activity unfeasible. On the other hand, for large-scale, complex or medium to high impact projects, the procedures within the environmental process should not be seen as an obstacle, but an important preventive environmental instrument to analyse the impacts and for society to discuss them. Most of those critiques, in sum, have been done in order to violate the law and many have been the attempts in the National Congress to soften environmental protection rules, especially the environmental licencing process.

[66]STJ has consolidated its stand on the matter that the licence must be previous and preventive and cannot be understood as tacit due to delay or inertia by the environmental bodies. Until an express decision, there is an *iuriset de iure* presumption of non-licence. A contrary understanding would be against the constitution by inverting the preventive logic established by it and by punishing the innocent (present and future society). See STJ. REsp 1728334/RJ. Relator Ministro Herman Benjamin. SegundaTurma. J: 05/06/2018. P: DJe 05/12/2018. https://ww2.stj.jus.br/processo/revista/inteiroteor/?num_registro=201703077091&dt_publicacao=05/12/2018. Accessed 8 May 2019. In another case, the Court grounded its decision by stating that it is not acceptable that a state's omission or an inexistent legal act can be a free-pass or can allow the use or abuse of natural resources. See also STJ. REsp 1245149/MS. Relator Ministro Herman Benjamin. SegundaTurma. J: 09/10/2012. P: DJe 13/06/2013. RSTJ v. 238, p. 20. https://ww2.stj.jus.br/processo/revista/inteiroteor/?num_registro=201100383719&dt_publicacao=13/06/2013. Accessed 8 May 2019.

defines EIA "as a tool used to identify the environmental, social and economic impacts of a project prior to decision-making", predicting environmental impacts at an early stage.[67] In addition, principle 17 of the Rio Declaration on Environment and Development states that "[e]nvironmental impact assessment, as a national instrument, shall be undertaken for proposed activities that are likely to have a significant adverse impact on the environment and are subject to a decision of a competent national authority".[68]

As stated above, EIA is considered a type of environmental evaluation[69] (also an instrument of the LPNMA, article 9, III) and is a constitutional requirement for projects or activities that might potentially cause significant environmental degradation, also required in article 3, CONAMA Resolution No 237/1997, and is regulated by the CONAMA Resolution No 01/1986.

The EIA enables the environmental body to evaluate the possible impacts of a proposed project before a permission is granted, and its accompanying report (so-called RIMA) is a resume of the EIA in accessible language to publicize it and to allow society to participate in the process. It is only required when a project or activity might cause a *significant* impact, which means that not every project or activity submitted to an environmental licencing must present an EIA/RIMA.

Legal concepts and guidance in relation to the meaning of a *significant* impact can be found in article 2 of the CONAMA Resolution No 01/1986, which establishes, in an illustrative listing, the activities that depend on the elaboration of the EIA/RIMA to request an environmental licence, such as mining activities. If the impact is not considered significant, other environmental studies might be presented instead of EIA/RIMA.

The EIA is a complex and extensive study and must contain, according to article 5, I, II, III, and IV, CONAMA Resolution No 01/1986, the following general guidelines: (1) all the technological and locational alternatives to the project, confronting them with the possibility of not executing it; (2) systematic identification and evaluation of the possible environmental impacts to be caused during instalment and operation; (3) definition of the limits of the geographic area to be direct or indirect affected by the impacts, the so-called influence area of the project, which shall also consider its water basin; (4) consideration of the compatibility between the project and governmental plans and programmes proposed and in implementation in the influence area; and (5) additional requirements to be fixed by the competent environmental body, according to the characteristics of the project or of the area.

Lastly, according to article 11, §2, the competent environmental body might promote a 'public hearing', in order to inform society about the project and its

[67]Convention on Biological Diversity (2019).

[68]United Nations (1992).

[69]Principle 17 of the Rio Declaration on Environment and Development, adopted during the United Nations Conference on Environment and Development, in 1992, prescribes that "Environmental impact assessment, as a national instrument, shall be undertaken for proposed activities that are likely to have a significant adverse impact on the environment and are subject to a decision of a competent national authority". See United Nations (1992).

environmental impacts. The public hearing is regulated by the CONAMA Resolution No 09/1987, and the idea behind them is to show all interested parties the content of the EIA/RIMA, giving an opportunity to address critics and suggestions.

Article 2 of the Resolution establishes that the competent environmental body shall conduct a public hearing when deemed necessary, or when requested by the Public Prosecution Service, by a civil entity, or by a minimum of 50 (fifty) citizens. In case that a public hearing is requested but not promoted, the licence is considered invalid. The hearing must be in an accessible place and, when the subject is complex or when the location is complicated, there might be more than one audience within the same RIMA.

The hearing's minutes will inform, together with the RIMA, the decision-making process of the environmental body to grant a licence or not. The hearing is, therefore, not binding. It means that if the affected population or people interested in the project are against or in favour of it, it does not bind the authority to approve or not the project. In this sense, it is, unfortunately, the only possibility explicitly established by the law to include public participation, it means that only the complex projects, the ones that might cause significant environmental impact (and therefore those obliged by law to present and EIA/RIMA) might request a public hearing. According to Farias,[70] the environmental licencing is an administrative process that allows local communities, non-governmental organisations or anyone interested to participate or to follow the process, especially after disasters such as Brumadinho have occurred.

The hearings are important to provide not only public participation but also transparency and public control over projects and activities that might cause environmental impacts. As the EIA/RIMA only takes place within the environmental licencing process, the licencing competent authority evaluates it, in accordance with its own environmental reports and also the ones that civil society and other interested environmental councils (for instance, the ones managing a conservation unity, in case it might be impacted by the project) might produce. During this evaluation, the competent authority analyses the positive and negative effects that a project might cause and decides whether the alternative chosen in the EIA/RIMA is the best one or other alternative, including the possibility of not implementing the project.

In summary, mining activities in Brazil are considered in general as activities that might cause significant environmental degradation and, therefore, are submitted to an environmental licencing process and to an EIA. In addition to these general guidelines, it is important to highlight that in 2017, the Law No 13,575 changed some regulations for mining activities, substituting the former competent body and the former Mining Code (Law Decree No 227/1967) and creating the Mining National Agency (ANM or *Agência Nacional de Mineração*), which happened after Mariana's disaster. Its primary focus is on the new regulatory agency and on the financial compensation for the exploration of mining resources.

[70]Farias (2019), pp. 170–172.

4.2 The Need to Adopt the Best Available Technology (BAT) in the Environmental Licencing of Mining Activities

The adoption of the BAT for the management of tailing dams is an important preventive and precautionary instrument to avoid disasters, such as Mariana and Brumadinho, and to control risks and guide the restoration of degraded soil and environment. As stated previously, reports conducted after those disasters affirmed that the method of construction of upstream tailing dams was not the best and that their safety should be proved before more dams could be licenced. This confirms the need for conducting the BAT studies in general and for tailing dams in Brazil, to inform the decision-makers during the environmental licencing process about the risks that such techniques might cause.

In Brazilian environmental law, the duty to adopt the best available technologies can be found in the cited article 225 of the Brazilian Constitution, in its reference to the State's duty of controlling techniques that might risk life, quality of life and environment, and also to the requirement of conducting an EIA for installations that might significantly impact the environment.

The EIA requirement in the Federal Constitution implicitly recognizes the mandatory use of BAT, since it would not be useful to carry out the studies and to analyse the impacts if it were not to decide for the minor ones.[71] The system would lack rationality in demanding such studies if there were not, as a reflex, three practical implications: the mandatory analysis of possible options for the undertaking; the impact of such options on environmental protection; and the adoption of BATs to make compatible the use and access to environmental assets and their protection.[72]

It derives from the precautionary principle, whose primary function is risk management and positive and anticipatory action against risk, in order not to cause harm to the environment. The technical clause of the BAT aims to have constant advancement and search for technologies that reduce the impact of human activity on the environment.

In view of the inexistence of zero risks, i.e. that all human activity is degrading, it does not mean that there should be a relaxation of environmental standards for the good of these impacting activities, but rather the firm rule of reducing environmental degradation to the lowest level technically and scientifically possible.

In order to oblige entrepreneurs to adopt the BAT to reduce to a minimum the negative impacts of these activities and maintain installations within legally determined environmental standards, the State exercises forms of control in environmental licencing and in the evaluation of environmental impacts.[73] In this context, the BAT is intrinsically related to environmental licencing and is a fundamental part of the permits and the contents of the environmental studies that inform them. This

---

[71]Loubet (2014).

[72]Loubet (2014).

[73]Bechara (2009).

happens, because environmental licencing is currently, in Brazilian law, the main instrument for conditioning the use and access to environmental assets and the prevention and precaution of damage, imposing control and management over environmental risks.

Moreover, environmental licencing also provides the public bodies with documents, information and data about the activities developed in the country. In addition, it includes planning and development of activities, which is why the BAT should be adopted from the outset, both to ensure ongoing concern for environmental quality and to minimize costs, incorporating the technologies at the beginning of the project. In fact, once the environmental licence has been granted, the entrepreneur must follow the terms under which the licence was granted and if it is verified, in the exercise of State monitoring, that the technology currently used causes damage, there is an obligation to adopt another less polluting one and also that there is recovery of the degraded area and possible compensation for damages.

It is important to point out that there is no acquired right to pollute. This is because the nature of the right to the ecologically balanced environment, once fundamental and diffuse, does not grant the entrepreneur acquired right to, through the development of his activity, attack nature, causing damage to present and future generations. According to Brazilian Superior Court of Justice (STJ or *Superior Tribunal de Justiça)* case law, there is no acquired right to pollute or to degrade the environment, and time is incapable of solving environmental illegalities of a permanent nature, since some of the protected subjects—future generations—lack voice and representatives who speak or omit themselves on their behalf.[74]

Within the environmental licencing and EIA regulations, the BAT can be found specifically in the need to analyse all the technological and locational alternatives to the project, according to CONAMA Resolution No 001/1986. The first evident application of the BAT is almost expressed when the cited Resolution determines that the entrepreneur must present all the technological and locational alternatives of the project, confronting them with its non-execution. It is not enough, therefore, to survey the technological alternatives, but it should also confront them, evaluate them, analyse their impacts, checking which are the most polluting, which are their types and how to eliminate or to reduce them.[75]

Furthermore, the assessment should require an integrated analysis of the project. It should also present an objective methodology for inventory and classification of environmental impacts and efficiencies, as well as preventative technologies. It is also important to emphasize that the assessment cannot focus only on the specific project that is in the licencing process, but on the synergetic impacts of it with those already installed in the area, the level of pollution already existing and even the perspective of new developments to be developed or already with licencing

[74]See, for instance, REsp 948.921/SP, SegundaTurma, Rel. Min. Herman Benjamin, j. 23/10/2007, DJe 11/11/2009; and REsp 1172553/PR, Rel. Min. Arnaldo Esteves Lima, Primeira Turma, j. 27/05/2014, DJe 04/06/2014.

[75]Loubet (2014).

applications, which may influence the maximum limit of pollutants that can be issued in the plant.[76]

If the analysis of the environmental impacts caused by the project and its alternatives is essential, the technological alternatives and a comparative study should be covered. It must also be pointed out that all the evaluation of the technological alternatives and their impacts must have connection and classification with the diagnosis of the environment, phase in which the assessment must analyse the tolerability to the impacts that will be studied, technology by technology and its interaction with the impacts arising from other anthropic activities already existing in the area.

In summary, the adoption of the BAT as a mandatory requirement in the environmental licencing process has the potential to help prevent disasters, such as Brumadinho, and guide public bodies to monitor them. An example of such obligation can be found in the BAT Reference Document (BREF) for the management of tailings and waste-rock in mining activities in the European Union, for the management of ores that might significantly impact the environment. It also aims to raise awareness of the extractive sector to the use of the BATs and to manage risks in activities with high impacts, such as mining.

5 Final Remarks

All in all, mining activities in Brazil have major social and environmental impacts, including soil degradation and contamination of large areas, followed by deforestation, biodiversity loss and, after Mariana and Brumadinho disasters, even the "death" of rivers. Local communities and traditional people are the ones who suffer the most from these activities and are forced to deal with externalities that could have been prevented.

Given that, the environmental licencing process and the adoption of the BAT are crucial in preventing such disasters from occurring. The environmental licencing system, if well-structured, has the potential to provide essential information to decision-makers and monitoring agencies, assisting them in making informed decisions and in managing and surveilling mine tailings dams properly. A better structure for the environmental licencing would include independent staff to analyse the projects and better resources to monitor mining activities correctly.

However, the lack of political willingness to implement the environmental legislation has pressured legislators to promote changes in the licencing system, as an ultimate attempt to open protected areas to mining activities. The recent attempts to change the environmental licencing system in the country lobbied by the private sector is confirmed by several reports to be a major cause of environmental

[76]Loubet (2014).

degradation in the country, weakening the environmental legislation, environmental monitoring and fostering risks and disasters.[77]

Brumadinho's disaster, considered the major one in the country, is an example of organized irresponsibility in Brazil, and illustrates the current moment of weakening environmental legislation, impeding environmental management and monitoring to fulfil constitutional and legal obligations. The adoption of BAT[78] and the promotion of a better environmental licencing system have the potential to prevent such disasters and to implement more environmental quality, and therefore need to be urgently addressed, fostering preventive and precautionary approaches and increasing monitoring of activities that might significantly impact the environment.

References

Agência Nacional de Águas (2018) Relatório de segurança de barragens. Agência Nacional de Águas, Brasília

Agência Nacional de Águas (2019) Barragens com fiscalização prioritária. http://www3.ana.gov.br/portal/ANA/rioparaopeba. Accessed 22 Apr 2019

Alves HR (2015) O rompimento de barragens no Brasil e no mundo: Desastres mistos ou tecnológicos? Dom Helder Câmara, Belo Horizonte

Bechara E (2009) Licenciamento ambiental na Lei do Sistema Nacional das Unidades de Conservação (SNUC). Atlas, São Paulo

Beck U (1995) Ecological politics in an age of risk. Polity Press, Malden

Beck U (2008) World at risk: the new task of critical theory. Dev Soc:1–21

Beck U (2012) World at risk. Polity Press, Cambridge

CNDH (2019) Relatório da Missão Emergencial a Brumadinho/MG após o rompimento da Barragem da Vale S/A. CNDH, Brasília

Convention on Biological Diversity (2019) What is impact assessment? https://www.cbd.int/impact/whatis.shtml. Accessed 9 May 2019

Environmental Law Alliance Worldwide (2010) Guidebook for evaluating mining project EIAs. Environmental Law Alliance Worldwide, Eugene

Farias T (2019) Licenciamento ambiental: Aspectos teóricos e práticos. Fórum, Belo Horizonte

Fundação Oswaldo Cruz (2019) Avaliação dos impactos sobre a saúde do desastre da mineração da Vale (Brumadinho, MG). Technical note. Fundação Oswaldo Cruz, Rio de Janeiro

Global Footprint Network (2019) Earth overshoot day. https://www.overshootday.org/newsroom/past-earth-overshoot-days/. Accessed 22 Apr 2019

[77]There are several legislative proposals to change the environmental licence system in Brazil. PL 3729/2004, for instance, establishes the possibility of exempting several activities from licence, such as roads and agribusiness (main deforestation drivers in the Amazon); it limits public participation in the process; it establishes shorter evaluation deadlines; and excludes the manifestation of several governmental bodies, such as FUNAI (indigenous protection foundation), IPHAN (institute for historical heritage and artistic heritage), and ICMBio (institute responsible for federal conservation unities).

[78]A legislative proposal, PL 2785/2019, which aims at defining general norms for environmental licence of mining activities, states in article 1, §3, that the licencing authority shall demand the use of BAT in all phases of the mining activity.

Instituto Brasileiro de Mineração (2019) Instituto Brasileiro de Mineração. http://www.ibram.org.br/. Accessed 22 May 2019
Intergovernamental Science-Policy Platform on Biodiversity and Ecosystem Services (2018) Summary for policymakers of the assessment report on land degradation and restoration. IPBES, Bonn
Intergovernamental Science-Policy Platform on Biodiversity and Ecosystem Services (2019) Summary for policymakers of the global assessment report on biodiversity and ecosystem services of the Intergovernmental Science-Policy Platform on Biodiversity and Ecosystem Services. Paris
IPCC (2014) Climate change 2014 synthesis report: summary for policymakers. IPCC, Geneva
IPCC (2018) Global warming of 1.5°C. Technical Report, IPCC
Intergovernamental Science-Policy Platform on Biodiversity and Ecosystem Services (2019) Summary for policymakers of the global assessment report on biodiversity and ecosystem services of the Intergovernmental Science-Policy Platform on Biodiversity and Ecosystem Services. Paris
Leite VR (2016) A mineração em Minas Gerais: Uma análise de sua expansão e os impactos ambientais e sociais causados por décadas de exploração. Sociedade & Natureza:375–384
Loubet L (2014) Licenciamento ambiental: A obrigatoriedade da adoção das melhores técnicas disponíveis (MTD). Del Rey, Belo Horizonte
Mesquita RV (2017) Os números da tragédia um ano depois. Planeta
Office of the United Nations High Commissioner for Human Rights (2019) Brazil: UN experts call for probe into deadly dam collapse. Office of the United Nations High Commissioner for Human Rights. https://www.ohchr.org/en/NewsEvents/Pages/DisplayNews.aspx?NewsID=24128&LangID=E. Accessed 13 May 2019
Ricupero R et al (2019) Comunicado dos ex-ministros de estado do meio ambiente. Instituto de Estudos Avançados da Universidade de São Paulo, São Paulo
Rockström J et al (2009a) A safe operating space for humanity. Nature 461:472–475
Rockström J et al (2009b) Planetary boundaries: exploring the safe operating space for humanity. Ecol Soc 14:1–35
UN Environment (2019a) Global Environmental Outlook - GEO 6: Summary for policy makers. Cambridge University Press, Nairobi
UN Environment (2019b) Global Environment Outlook – GEO-6: Healthy planet, healthy people. Cambridge University Press, Cambridge
UNDRR (2017) Terminology. https://www.unisdr.org/we/inform/terminology. Accessed 29 Apr 2019.
United Nations (1992) Report of the United Nations Conference on Environment and Development. Rio de Janeiro
Weiss F, Leuzinger M, Zurbrügg C, Eggen R (2016) Chemical pollution in low and middle-income countries. Swiss Federal Institute of Aquatic Science and Technology, Dübendorf
WISE Uranium Project (2019) Chronology of major tailings dam failures. http://www.wise-uranium.org/mdaf.html. Accessed 10 May 2019

The Healthy Soil Program

An Analysis of California's Effort to Build Soil and Mitigate Climate Change

Danika Desai

1 Background

California's agricultural abundance is legendary. In terms of both value and crop diversity, California is the leading agricultural producer in the United States.[1] The counties within the San Joaquin Valley produce more food than any other comparably sized region in the world.[2] No other state, or combination of states, matches California's productivity per hectare.[3] This productivity is due to multiple factors, including a temperate climate, a highly-engineered water distribution system, few pests, and California's rich soils.[4]

Indeed, early agricultural growers recognised the excellent quality of California Central Valley's soils and the quality of crops they could produce.[5] A 19th century California vineyard owner described the quality of his soil in almost poetic terms: "the deep stratum of heavy, marly, sub-soil, overlaid by rich, black loam, with surface water enough to maintain a moist condition of the sub-soil without saturation—the vegetation being influenced by the warm summers of the San Joaquin Valley, tempered at that point by the inward flow of moist air which follows tide

Attorney, Earthjustice (San Francisco, California). The views expressed in the article are those of the author and no endorsement by Earthjustice, its clients, or staff is intended.

[1]CDFA (2017a).

[2]Gunasekara (2013), p. 28.

[3]Rees (2016), p. 26.

[4]Olmstead and Rhode (2017), p. 7.

[5]*Id.*

D. Desai (✉)
Oakland, CA, USA
e-mail: desai2018@lawnet.ucla.edu

H. Ginzky et al. (eds.), *International Yearbook of Soil Law and Policy 2019*,
International Yearbook of Soil Law and Policy 2019,
https://doi.org/10.1007/978-3-030-52317-6_14

water to Stockton. . . .Perfect maturity of large crops is attained. . .under these conditions, and the composition of the soil insures the qualities sought by connoisseurs."[6] The promise of agricultural abundance was so great in California that in 1884, a California court protected farmland at the expense of the gold industry in one of the state's first environmental judicial opinions.[7]

Even with ideal conditions for agriculture, however, the state's soils—like much of the arable land around the world—face serious challenges. More than half of California's irrigated cropland is affected by salinisation,[8] which is the accumulation of salts in soils due to over irrigation. Soil salinisation reduces crop yields in California by eight million tonnes a year.[9] If salinisation continues at its current rate until 2030, it could cost the state between 1 and 1.5 billion dollars.[10] The misuse of agricultural soils also causes water pollution; namely, 419,000 tonnes of nitrogen leach into California's groundwater each year, 88% of it from fertiliser and manure applications to agricultural soils.[11] Groundwater in the Tulare Lake Basin and Salinas Valley regularly exceeds state and federal standards for nitrate levels in drinking water, raising public health concerns.[12]

Despite these serious problems, governmental programs focused on building healthy soils are few. Most of the soil conservation programs that currently exist were born out of the Dustbowl in the 1930s.[13] Today, a new environmental crisis has reinitiated interest in healthy soils: climate change. For several decades now, scientists have documented soil's important role in the planet's carbon cycle.[14] Soil has the potential to be a vast carbon sink.[15] Irrespective of its ability to sequester carbon, however, soil is precious in its own right. Soil is literally humanity's food-growing medium, and humanity's survival depends on arable soils. Ironically, the degradation of soil is one of humanity's oldest environmental problems—a problem that seems to replay itself in civilisation after civilisation. As soil scientist David Montgomery put it: "pronounced episodes of soil erosion [are] associated with the rise and subsequent decline of civilizations in the Middle East, Greece, Rome, and Mesoamerica, as well as other regions around the globe."[16] Despite soil's well documented role in the rise and decline of civilisations, soil protection measures in the United States occur only incidentally to other crises. The latest such crisis is climate change.

---

[6]West (1890), p. 526.

[7]*Woodruff v. North Bloomfield Gravel Min. Co.*, 18 F. 753, 756 (C.D. Cal. 1884).

[8]CDFA (2009).

[9]Welle and Mauter (2017), p. 12.

[10]Water Education Foundation (2019).

[11]Tomich et al. (2015), p. 7.

[12]*Id.* at 10.

[13]NRCS (2019).

[14]Morgan et al. (2010), p. 6A.

[15]*Id.*

[16]Montgomery (2007), p. 13270.

California's Healthy Soil Initiative was born out of this growing awareness of the connection between healthy soils and climate change mitigation. In 2016, California's governor Jerry Brown appropriated $7.5 million for the Healthy Soils Initiative from California Climate Investments, proceeds from California's greenhouse gas cap and trade program.[17] California has an ambitious goal to reduce greenhouse gas emissions in the state to1990 levels by 2020, and then by another 80% by 2050.[18] To achieve this, the California Air Resources Board (CARB) created a scoping plan, last updated in 2017. The scoping plan outlines how every sector in California, including agriculture, can reduce greenhouse gas emissions.[19] Agriculture is responsible for 8% of California's total greenhouse gas emissions. Of that, 60% is methane from livestock,[20] mostly from the animals themselves and manure management at dairies and is thus not directly related to healthy soils. Despite agriculture's relatively small contribution to the state's greenhouse gas emissions, soil features in California's scoping plan because soil has the potential to sequester additional carbon beyond current agricultural greenhouse gas emissions; the ultimate goal for California is to turn working and natural lands from sources of carbon in the atmosphere to carbon sinks.[21]

2 The Healthy Soils Program: Structure and Function

The main component of the Healthy Soils Initiative is the Healthy Soils Program (HSP), administered by the Office of Environmental Farming and Innovation within the California Department of Food and Agriculture (CDFA). The Office of Environmental Farming was created by the Cannella Environmental Farming Act of 1995, which tasked CDFA with establishing an environmental farming program that "shall provide incentives to farmers whose practices promote the well-being of ecosystems, air quality, and wildlife and their habitat."[22] The HSP is an example of such a program.

The HSP has three goals: (1) to incentivize California growers and ranchers to implement agricultural management practices that sequester carbon, reduce atmospheric greenhouse gases, and improve soil health; (2) to fund demonstration projects that conduct research and/or showcase conservation management practices that mitigate greenhouse gas emissions and improve soil health; and (3) to create a platform to promote conservation practices throughout the state.[23]

[17]CDFA (2019a).

[18]CARB (2018a).

[19]CARB (2017).

[20]*Id.* at 85.

[21]CARB (2019).

[22]Cal. Food & Ag. Cod. § 566 (Deering 2019).

[23]CDFA (2018a).

To achieve these goals, the HSP has two component parts: the HSP Soils Incentive Program and the HSP Demonstration Projects.[24] The Incentive Program directly funds growers to implement conservation practices.[25] The Demonstration Projects research and showcase conservation practices to encourage other growers to adopt them.[26]

2.1 *The HSP Incentives Program*

The Incentive Program "provide[s] financial incentives to California growers and ranchers for implementation of agricultural management practices that sequester carbon, reduce atmospheric greenhouse gases (GHGs), and improve soil health."[27] Incentive projects can be funded up to $75,000.[28] Growers and ranchers can apply to implement various conservation practices from a pre-approved list, which includes, among others, cover cropping, forage and biomass planting, alley cropping, and prescribed grazing.[29] Most of the conservation practices are also listed by the U.S. Natural Resources Conservation Service (NRCS), a permanent agency within the U.S. Department of Agriculture, as conservation practices eligible for financial and technical assistance, but California also funds compost applications, which is not a standard practice under the NRCS.[30] In addition to the normal conservation aims and benefits, the Incentive Program also requires that project participants measure greenhouse gas reductions using a "Quantification Methodology" developed by the California Air Resources Board.[31] Individual growers and ranchers are eligible to apply for Incentive Program funding.[32]

---

[24] *Id.*

[25] CDFA (2019b).

[26] CDFA (2019c).

[27] CDFA (2019b).

[28] CDFA (2018a).

[29] *Id.* at 6–7. Eligible practices include cover cropping, conservation crop rotation, mulching, nutrient management, residue and tillage management, strip cropping, compost applications, establishing conservation cover, buffer strips, field borders, filter strips, grassed waterways, herbaceous wind barriers, ribarian herbaceous cover, vegetative barriers, forage an biomass planting, alley cropping, hedgerow planting, multi-story cropping, riparian forest buffers, tree-shrub establishment, windbreak establishment, prescribed grazing, range planting, and silvopasture.

[30] *Id.*

[31] *Id.* at 11. CARB is the California Air Resources Board, the agency in charge of ensuring California reaches its climate change goals.

[32] *Id.* at 4.

2.2 HSP Demonstration Projects

Demonstration Projects must implement conservation practices and conduct outreach to other growers to showcase how the project enhances soil conservation and sustainability.[33] The goal of Demonstration Projects is to "create a platform promoting widespread adoption of conservation management practices throughout the state."[34] CDFA requires that all Demonstration Projects showcase the project to at least 120 growers over the 3-year period of the project.[35] Individual ranchers and growers are not eligible for Demonstration Project funding unless they are working in collaboration with "Not-for-profit entities, University Cooperative Extensions, Federal and University Experiment Stations, Resource Conservation Districts (RCDs), Federal and California Recognized Native American Indian Tribes."[36] CDFA requires partnerships with established organizations because the goal of Demonstration Projects is to conduct outreach to growers. Often, universities and other established institutions have better infrastructure and expertise to conduct this outreach compared to individual growers.

There are two kinds of Demonstration Projects eligible for funding: Type A and Type B.[37] Type A projects have more stringent data collection requirements. Type A project participants must "collect data on field measurements of GHG emissions, and . . . collect co-benefit data including benefits to soil health and environmental water and air quality data to address knowledge gaps regarding implementation of specific practices."[38] Like the Incentive Program, Demonstration Projects must implement conservation practices from a pre-approved list.[39] The list for Type A projects, however, contains more eligible practices than the list for Type B projects or Incentive Program projects.[40] Specifically, Type A projects can implement experimental practices that do not yet have peer-reviewed literature demonstrating their effectiveness in either carbon sequestration or soil health benefits. Type A projects thus serve as a mechanism to collect data and fill knowledge gaps. Eligible Type A practices include all the practices allowed for the Incentive Program described above, as well as new practices such as microbial inoculation with compost tea,

[33] CDFA (2019c).

[34] CDFA (2018b).

[35] *Id.* at 13.

[36] *Id.* at 5.

[37] CDFA (2019c).

[38] CDFA (2018b).

[39] *Id.*

[40] *Id.*

vermicompost application, and whole orchard recycling.[41] Type A projects can be awarded up to $250,000.[42]

Type B projects must implement conservation practices from a pre-approved list and then conduct outreach to other growers showcasing how these practices improve soil conservation and sustainability.[43] Type B projects can be awarded up to $100,000.[44] Unlike Type A projects, Type B projects do not have the additional measurement requirements.[45] As such, the list of eligible practices for Type B projects is limited to practices that have peer-reviewed literature demonstrating effectiveness in mitigating climate change and improving soil health.[46]

2.2.1 Program Funding

The 2016–2017 HSP received $7.5 million from the California Climate Investment, funds derived from California's carbon cap-and-trade program.[47] There were 114 HSP projects awarded in 2017, 85 Incentive Program projects[48] and 29 Demonstration Projects.[49] The 2018–2019 HSP received $5 million from the California Climate Investment and an additional $10 million from the California Drought, Water, Parks, Climate, Coastal Protection, and Outdoor Access For All Act of 2018 ("2018 California Drought Act").[50] For the second funding cycle, CDFA selected 194 Incentives Projects and 23 Demonstration Projects for award in June 2019, totalling $12.48 million in awards.[51]

[41] *Id.* at 9. Experimental practices allowed only for Type A projects include anaerobic digestate application, microbial inoculation with compost tea, mycorrihizal application, nutrient management with synthetic Nitrogen fertilizer, Nitrification inhibitors, and the use of slow use fertilizers, one-time compost applications, vermicompost application, and whole orchard recycling.

[42] *Id.* at 8–9.

[43] *Id.* at 7.

[44] *Id.* at 4.

[45] *Id.* at 7.

[46] *Id.*

[47] CDFA (2019a).

[48] CDFA (2017b, c).

[49] CDFA (2017d, e).

[50] CDFA (2019a).

[51] CDFA (2019b, c).

2.3 *Discussion*

California is a diverse state with a diversified agricultural sector. Moreover, California is geographically large, spanning more than 160 million square miles.[52] While the potential for sequestering carbon in the state's soils is therefore vast, the challenges of implementing such a solution are also significant. The following sections analyse some of these challenges, including the challenges of (1) implementing a relatively small program to combat a large problem in a state that is geographically large and demographically diverse and (2) ensuring that the program strikes the proper balance between the two goals of the program—climate change mitigation and soil health.

2.3.1 Ensuring That the Healthy Soils Program Equitably Serves a Diverse State

California is extremely diverse in terms of ethnic and cultural diversity. It ranks second in the nation in terms of the diversity of its population, following only Hawai.[53] California's agricultural sector is likewise diverse. In fact, California leads the nation in growing diversity among its farmers and ranchers.[54] And that diversity is increasing. While the overall number of farms decreased from 2007 to 2012, the number of farms operated by Latinx and Asian Americans over this same period in California increased.[55] This trend slowed from 2012 to 2017 for Latinx farmers,[56] but continued for Asian American farmers.[57] California only hosts 2.9% of all farms in the United States, but it has 14.6% of the country's farms principally operated by Latinx farmers and 35.1% of the country's farms principally operated by Asian American farmers.[58]

Even as diversity increases among California farmers and ranchers, racism is embedded in the U.S. food system. Many farm workers, often migrants from Mexico, earn far less than minimum wage.[59] Racial inequity in terms of land access has a long history in the United States. Black, Asian, and Hispanic people were routinely dispossessed of their lands.[60] For decades, government agencies

[52]U.S. Census Bureau, https://factfinder.census.gov/faces/tableservices/jsf/pages/productview.xhtml?src=bkmk. (last visited Jul. 8, 2019).

[53]U.S. Census Bureau (2015).

[54]Pesticide Action Network of North America (2017).

[55]USDA (2012), p. 56.

[56]USDA (2017a), p. 66.

[57]USDA (2017b), p. 704.

[58]CDFA (2017f).

[59]Ayazi and Elsheikh (2015), pp. 9–10.

[60]*Id.*

discriminated against farmers of colour in distributing loans, grants, and disaster relief.[61] In the late nineteeth Century and early twentieth Century, laws such as the Chinese Exclusion Act and the Alien Land Law Act made it illegal or difficult for Asian Americans to own land at all.[62]

Government programs such as the HSP, which must serve all state residents, therefore have a unique challenge in California. How does one create a program that meets the needs and is accessible to a broad range of diverse peoples, especially when considering the historical context of routine discrimination against marginalised groups?

In order to address this issue, California instituted certain protections for historically marginalised communities. The HSP, for example, is funded both by the 2018 California Drought Act (Proposition 68), and the California Climate Investment (SB 535).[63] Both of these funding sources condition funding on prioritising disadvantaged communities. Proposition 68 requires CDFA to award 15% of appropriated funds to "severely disadvantaged communities."[64] Likewise, SB 535 requires that at least 25% funds from the California Climate Investment go towards projects that benefit disadvantaged communities,[65] and that at least 10% of this prioritized funding go to projects located within disadvantaged communities. AB 1550 later amended SB 353, further requiring "at least 25 % of funds [from California Climate Investments] go to projects within and benefitting disadvantaged communities and at least an additional 10 % . . . [go to] low-income households or communities."[66]

In 2017, California went even further and passed the Farmer Equity Act. The Farmer Equity Act requires CDFA "to ensure the inclusion of socially disadvantaged farmers and ranchers . . . in the development, adoption, implementation, and enforcement of food and agriculture laws, regulations, and policies and programs . . . ."[67] The Act defines socially disadvantaged farmers as African Americans, Native Indians, Alaskan Natives, Hispanics, Asian Americans, and Native Hawaiians and Pacific Islanders.[68]

The Farmer Equity Act was passed after the first round of HSP applications had been received.[69] Thus, it did not impact the first round of applications. For the 2018–2019 HSP, however, CDFA made some changes to make the program more

---

[61] *Id.*

[62] Alkon and Agyeman (2011), pp. 68–72.

[63] CDFA (2018b), p. 18.

[64] *Id.* "Severely disadvantaged community" means a community with a median household income less than 80% of the statewide average.

[65] *Id.* Disadvantaged communities are communities disproportionally burdened by multiple sources of pollution and with population characteristics that make them more sensitive to pollution as identified by the California Environmental Protection Agency. *See* CARB (2018d).

[66] CalEPA (2019).

[67] AB-1348 Farmer Equity Act of 2017.

[68] *Id.*

[69] Cal Food & Agr Code § 510 (Deering 2019).

accessible to socially disadvantaged farmers as required by the Farmer Equity Act. Specifically, CDFA prioritized funding to Socially Disadvantaged Farmers and Ranchers.[70] Additionally, CDFA allowed 25% of the funding awarded to be requested in advance and available upfront, before the projects begin.[71] Upfront funding is important for resource-strapped farmers who have no reserves to begin implementation and request reimbursements, which is the standard process for state contracts and grants.[72] Indeed, for farmers with the fewest resources, 25% upfront funding might not be enough. In 2018, the California Farmer Justice Collaborative suggested in public comments that CDFA give 100% of project funding upfront to socially disadvantaged farmers.[73] The Collaborative also suggested that CDFA allow labour costs to be eligible for advanced payments, and noted that the extra bureaucracy required to receive advanced payments might also make the program less accessible to socially disadvantaged farmers.[74] Despite these comments, CDFA chose not to advance more than 25% of funding for 2018–2019 HSP projects.

Although CDFA did not choose to allow 100% advance payments, the agency did make several changes to help socially disadvantaged farmers in response to public comments. Specifically, CDFA hired a Farm Equity Advisor to help the agency incorporate the requirements of the Farmer Equity Act into the HSP.[75] CDFA also increased funding for technical assistance and outreach to Socially Disadvantaged Farmers in 2018, and provided multilingual outreach materials.[76] Additionally, in 2018, AB 2377 (Irwin) was passed, which requires CDFA to develop a technical assistance program.[77] This program will be designed to fund Resource Conservation Districts (RCDs), non-profits and University of California Cooperative Extension in assisting farmers and ranchers to apply for CDFA's climate smart agriculture grant programs, including the HSP.[78] This program will require CDFA to ensure at least 25% of the grant program funds are used to provide technical assistance to socially disadvantaged farmers or ranchers.[79]

---

[70]CDFA (2018a), p. 12.

[71]*Id.*

[72]Smoker (2018), stating "Most farmers face a cashflow challenge every spring and early summer as they pay for months of inputs and labor but have no harvest to earn revenue from. As such, paying tens of thousands of dollars upfront to implement a practice and then waiting 6-12 weeks for reimbursement can put a serious financial strain on farmers during lean times of the year. This is especially true for socially disadvantaged farmers and ranchers, who have to overcome the legacy of racism in agriculture, and on average operate smaller farms, earn less revenue, and have received less government support."

[73]*Id.*

[74]*Id.*

[75]CDFA (2018c).

[76]*Id.*

[77]AB 2377 (2018).

[78]*Id.*

[79]*Id.*

Thus, while the California government is clearly attempting to create state programs that are sensitive to vulnerable populations and historical context, addressing this issue is complex and there is room for improvement. Ongoing efforts through the Farmer Equity Act and AB 2377 will provide information in the future on the success of HSP in reaching socially disadvantaged farmers and ranchers.

2.3.2 Ensuring an Equitable Distribution of Healthy Soil Program Funding Among California Counties

Making the HSP more accessible to ethnically diverse and historically marginalised groups is not the only way to expand the scope of the program. Certain California counties have access to more resources than others. Average income varies, in some cases drastically, from one county to another. For example, the median annual income in Marin county is $104,703 and the poverty rate is 7.9%.[80] In contrast, the median income in Tulare county is $44,871 and the poverty rate is 24%.[81] In fact, four of the five top producing agricultural counties in terms of acres under production--Kern, Fresno, Tulare, and Merced--have poverty rates above 20%.[82] In addition, certain counties receive more federal agricultural loans than others or have more technical infrastructure—such as extension services and RCDs—located within them. These services are often important for growers interested in implementing conservation practices because they provide training, education, and sometimes even access to money. This economic disparity between what is available at county level may influence the implementation of the HSP, particularly the Incentive Program.

Indeed, during the first HSP funding cycle in 2016–2017, the poorest but most agriculturally productive counties were underrepresented in terms of Incentive Program funding. For example, Fresno County, which is located in California's Central Valley, is one of California's most productive agricultural counties. It has more than 4700 farms and over 1.6 million acres under agricultural production.[83] The value of Fresno's agriculture sector is more than $5.7 billion.[84] Fresno county also suffers from some of the worst agriculture-related pollution problems in

[80]U.S. Census Bureau, https://www.census.gov/quickfacts/fact/table/marincountycalifornia,CA/INC110217 (last visited July 5, 2019).

[81]U.S Census Bureau, https://www.census.gov/quickfacts/fact/table/tularecountycalifornia,CA/INC110217 (last visited July 5, 2019).

[82]U.S. Census Bureau, https://www.census.gov/quickfacts/fact/table/fresnocountycalifornia,mercedcountycalifornia,kerncountycalifornia,tularecountycalifornia,CA/IPE120217 (last visited July 5, 2019).

[83]USDA (2017a), p. 262.

[84]*Id.*

California.[85] The poverty rate in Fresno county is sixth highest in the state.[86] Fresno county also has only one RCD for the entire county.[87] Given that Fresno County has so many farms as well as serious pollution problems related to those farms, the county could benefit from more technical and financial assistance.[88] The 2016–2017 HSP funded only two Incentive Program projects in Fresno County, totalling about $96,000.[89] California's other top agricultural counties, including Kern, which received no funding, and Tulare, which received $ 58,368, were likewise proportionately underrepresented in terms of Incentive Program funding during the first funding cycle in 2016–2017.[90] Although Merced, a top agricultural county with a high poverty rate, received the highest amount of funding from the 2016 to 2017 Incentive Program, most of this funding went to the same family farm.[91] In contrast, the 2016–2017 HSP funded seven Incentive Program projects in Marin, fifteen in San Luis Obispo, and eleven in Sonoma.[92] These counties have significantly higher median incomes than Fresno, Tulare, and Kern.[93]

CDFA started to rectify this problem during the second funding cycle. In 2019, CDFA awarded funds to 24 Incentive Projects in Fresno, totalling $962,915.[94] Likewise, Kern, which received nothing during the first cycle, received Incentive Program funding for five projects in 2019, totalling $355,440.[95] Tulare, however, received no funds despite its status as a top agriculturally producing county.[96] Thus, there has been some improvement on this issue. Perhaps the Farmer Equity Act, which directs CDFA to prioritize Socially Disadvantaged Communities and was passed between the two cycles of HSP funding, is already having an effect. Hopefully, the recently passed AB 2377, which requires CDFA to develop a technical assistance program for farmers, will further this progress.

---

85 Marcum (2014).

86 U.S. Census Bureau, https://www.census.gov/quickfacts/fact/table/fresnocountycalifornia,CA/INC110217 (last visited Jul. 8, 2019).

87 Jagannath et al. (2017).

88 Currently, there are no projects implemented by a Resource Conservation District in Fresno County, and hardly any projects located in the San Joaquin Valley as a whole. Instead, most projects are located on the coasts, with the highest density of projects clustered in the wider San Francisco Bay Area, one of the richest regions of the state. *See* RCD Project Mapper (2019).

89 CDFA (2017b, c).

90 USDA (2017a), p. 262.

91 CDFA (2017b, c). See also Appendix A.

92 *Id.*

93 U.S. Census Bureau, https://www.census.gov/quickfacts/fact/table/kerncountycalifornia,tularecountycalifornia,fresnocountycalifornia,sonomacountycalifornia,sanluisobispocountycalifornia,marincountycalifornia/INC110217 (last visited July 8, 2019).

94 CDFA (2018d).

95 *Id.*

96 *Id.*

In terms of Demonstration Projects, Yolo, Fresno, and San Luis Obispo received the highest amount of funding during both funding cycles.[97] Thus, Tulare, Kern, and Merced[98] still received comparatively little funding in the context of Demonstration Projects as well.

For both the Incentive Program and Demonstration Projects, CDFA is limited in deploying grant money by who decides to apply to the program. Farmers with more resources—time, education, and money—are probably more likely to apply. Likewise, for Demonstration Projects, CDFA is limited by which universities, non-profits, RCDs, and other organizations decide to apply—and as stated above, these institutions may be more scarce in counties with fewer resources. Farmers in San Joaquin Valley counties already generally employ fewer conservation practices than other regions, perhaps due to historical and cultural barriers,[99] and thus are potentially less likely to apply without outreach. Because the HSP is an incentives-based program reliant on applicant interest, CDFA has limited discretion in ensuring that all counties (and socioeconomic levels) are proportionately represented. However, new laws such as the Farmer Equity Act and the funding prioritizations discussed above, now mandate that CDFA fund projects in disadvantaged communities, which will hopefully help the HSP to reach a wider group of farmers.

2.3.3 Expanding the Size and Impact of the Program

Another challenge that the HSP faces is the scope of the problem compared to the size of the program. Building healthy soils in California is no small task. There are more than 24 million acres of farmland in California, and over 70,000 farms.[100] Only about 1.07 million of these acres are certified organic,[101] and less than a million acres are under conservation easements.[102] Moreover, only 238,454 acres are farmed with "no-till, practices" compared to over 2.3 million acres where "intensive tillage practices [are] used."[103] Compared to the scope of the problems facing California's soils, the HSP is a tiny initiative. CDFA's Office of Environmental Farming, which administer the HSP, has only 14 staff members, and the HSP is not the only program that the Office administers.[104]

---

[97]CDFA (2017d, e, 2018e).

[98]USDA (2017a), pp. 244–252.

[99]Jagannath et al. (2017) stating "California's San Joaquin Valley is home to the most agriculturally productive farmland in the nation, however historic and cultural barriers to adoption have caused the region to fall behind in the push towards greater environmental stewardship in the state."

[100]USDA (2017b), p. 351.

[101]USDA (2016).

[102]USDA (2017b), p. 644.

[103]*Id.*

[104]CDFA (2019d).

Assuming perfect and continued implementation of the conservation practices funded by the HSP, the program has sequestered an estimated 43,676 metric tonnes of carbon per year.[105] To put that number in context, California's stated goal is to turn the state's working and natural lands into a carbon sink by mid-century.[106] In 2016, California emitted 429.4 million metric tonnes of carbon dioxide equivalent.[107] The agricultural sector was responsible for 8% of these emissions, or 34.4 million metric tonnes.[108] Of the agricultural sector's contribution, crop growing and harvesting accounted for only about seven million metric tonnes.[109] Thus, if California attempted the modest goal of increasing soil carbon content enough to neutralize emissions from crop growing and harvesting, the HSP is still likely not effectuating that goal on its own. Even if the Demonstration Projects truly encourage grower adoption of conservation practices and the effect of the HSP is 100 times greater than the estimated carbon savings from directly funded projects, the HSP would only sequester 6% of emissions from crop growing and harvesting, not enough to neutralize emissions from California's crop-based agriculture. Also, although difficult to determine the carbon sequestration potential of California's agricultural soils, at least one study suggested that California's agricultural soils already sequester an average of 19 g of carbon dioxide per meter squared per year.[110] Other studies suggest that farmlands have the potential to store between 100 and 200 g of carbon dioxide per meter squared per year.[111] Using the 100-gram figure, California farmlands have the potential to store 2.8 million metric tonnes of carbon dioxide per year,[112] again much more carbon than the HSP is currently sequestering. To truly realize the carbon sequestration of California's agricultural soils and to meet California's stated goals, the state needs to expand the scope of the HSP or find additional solutions.

This is especially true because the HSP disproportionately funds farmers already employing conservation practices compared to conventional farmers. Over 50% of funding recipients in both cycles of the HSP already had organic certification or described themselves as using "sustainable" practices.[113] A few of these farmers even describe themselves as farming in a way to benefit the climate.[114] Meanwhile, the U.S. Census of Agriculture demonstrates that a far lower proportion of California

[105] CDFA (2017b, c, d, e).

[106] CARB (2019).

[107] CARB (2018b).

[108] *Id.*

[109] CARB (2018c), p. 13.

[110] Kroodsma and Fields (2006), p. 1975.

[111] Smith et al. (2008), p. 789; Lal (2010), p. 708.

[112] Current CO2 sequestration is 19 g/m^2.y for CA farmland for a total of 0.69 Mton/y. If farmland could sequester 100 g/m^2.y, total sequestration would be 0.69 Mton * 100/19 3.6 Tg/y or an increase of 2.9 Tg/y.

[113] See Appendix.

[114] *Id.*

growers use conservation practices than the HSP funding recipient pool suggests.[115] Moreover, many U.S. farmers do not even believe in climate change much less farm in a way to mitigate it.[116] Because farmers already farming organically or "sustainably" may have healthier soil with higher carbon content,[117] the real climate change mitigation potential lies with farmers with the most degraded soil. But it is unclear how effectively the HSP is reaching these farmers.

Another reason to expand the size of the HSP is because the program is working in opposition to the larger U.S. federal farm subsidy structure. California farms received more than $127 million in agricultural subsidies from the federal government in 2017.[118] Only about 2.79 million of these federal dollars went to conservation programs.[119] The structure of federal agricultural subsidies encourages farmers to rely heavily on annual crops that leave soils bare for parts of the year and to apply excessive amounts of nitrogen fertiliser to maximise crop production.[120] Both of these practices lead to environmental damage—soil erosion and water pollution.[121] As a result, the government spends more money trying to rectify this environmental damage.[122] In 2017, NRCS spent $19 million on air pollution from agricultural sources in the San Joaquin Valley alone.[123] The HSP has deployed a total of $22.5 million over a 3-year period from 2017 to 2019. While this amount is not insignificant, it pales in comparison to the number of federal subsidies going to California farms that promote unsustainable, polluting agricultural practices.

2.3.4 Conflicting Priorities Within the Healthy Soils Program

In addition to the challenges in implementation discussed above, the dual purpose of the HSP can create conflicting priorities. Soil plays an important role in the carbon cycle.[124] Since agriculture is the interface between human civilisation and soils, certain agricultural practices have the potential to store carbon and decrease greenhouse gases in the atmosphere. Moreover, not only can soil store carbon to mitigate climate change, increased carbon content in soil has multiple benefits, including enhanced soil fertility, soil structure stability, water retention, and the capacity to reduce toxic elements.[125] If the same practices sequester carbon and build soil

---

[115]USDA (2017b), p. 644.

[116]Amundson and Biardeau (2019), p. 11653.

[117]Gattinger et al. (2012), p. 18226.

[118]USDA (2017b), p. 322.

[119]*Id.*

[120]Mulik (2016), p. 3.

[121]*Id.*

[122]*Id.*

[123]NRCS (2017).

[124]FAO (2015).

[125]Morgan et al. (2010), p. 6A.

health, then these practices could be a win-win solution for the state. California has been one of the first governments to try and implement this dual solution for soil health and climate change mitigation with the Healthy Soils Program.

However, the implementation of this win-win solution is complicated. Although there is overlap between agricultural practices that increase soil carbon storage and build healthy soils, there are also practices that do one or the other but not both. No-till, for example, may more permanently store carbon in the soil, but also might increase nitrous oxide emissions, a greenhouse gas almost 300 times more potent than carbon dioxide.[126] Thus while no-till may increase soil quality it may not actually serve as a climate mitigation tool.

Indeed, actual sequestration of carbon in soils "would require major changes in cropping systems or significant research."[127] Such major changes include the use of agroforestry and intercropping, which increases the rate of input of organic matter to soils, using perennials in place of annual crops because perennials store more carbon than annuals, and breeding crops to have longer and deeper roots so that they can exude carbon into subsoils for more permanent storage.[128] While agroforestry practices and other practices that increase the use of perennials were mostly absent from the 2016 to 2017 HSP-eligible practices,[129] CDFA added several such practices to the 2018–2019 HSP-eligible practices, including alley cropping, multistory cropping, tree-shrub establishment, and range planting.[130]

Moreover, CDFA acknowledged the need for research on new conservation practices that achieve both goals of the HSP, and designated Type A projects to explore some of those practices. Since Type A projects require more measurements and data collection, Type A projects allow HSP grant recipients to add to the existing scientific literature about which soil conservation practices achieve both goals.[131] Many of these practices were suggested by growers and agricultural organisations during the notice and comment period between the two rounds of HSP funding cycles in 2017.[132]

Thus, while the dual purpose of the HSP poses multiple challenges, CDFA has attempted to address some of these challenges. Still, the complexities of meeting the dual goals of the program remain. Not only must CDFA include conservation practices that achieve both goals, the agency must also make sure that it is giving equal weight to both goals in making funding decisions. Because HSP funding

[126]Powlson et al. (2011), p. S74.

[127]*Id.*

[128]Whitmore (2010), p. 5.

[129]CDFA (2017g).

[130]CDFA (2018a).

[131]CDFA (2018b), p. 9. The experimental practices which are only eligible for Type A funding include anaerobic digestate application, microbial inoculation with compost teas, mycorrhizal application, nutrient management such as replacing synthetic nitrogen fertilizer with manures, the use of nitrogen inhibitors, and the use of slow release fertilizers, one-time compost applications at higher rates than currently approved, vermicompost applications, and whole orchard recycling.

[132]*See* CDFA (2017h).

comes from California Climate Investments, the funding is contingent on demonstrating climate change mitigation, thereby affecting how the HSP is implemented. For example, CDFA requires all projects (both Incentive and Demonstration) to estimate greenhouse gas reductions.[133] In contrast, CDFA does not require Incentive Program projects or Demonstration Program Type B projects to estimate other metrics of soil health.

CDFA does require Type A projects to measure metrics of soil health, but this requirement was added after the non-profit California Climate and Agriculture Network expressed concern in public comments.[134] Previously, CDFA had encouraged, but not required, Type A Projects to measure soil health benefits.[135] In response to comments, CDFA added language on soil health metrics to the data collection requirements in the 2018 Request for Applications.[136] Specifically, applicants who receive HSP funding for Type A projects must now collect data "including but not limited to field measurement of GHG emissions **and soils health indicators.**"[137] Thus, balancing the two goals of the HSP is not always easy. Incomplete science makes implementation tricky due to funding requirements and restrictions. Moreover, the funding source may implicitly prioritise climate change mitigation over soil health.

3 Conclusion

We live in a time where multiple ecological crises are converging at once. Climate change and soil degradation are two of those crises. However, as the Health Soils Initiative shows, sometimes multiple crises have shared solutions. Soils are ecosystems that have manifold benefits to planetary health, and one of those benefits is carbon sequestration. Building healthy soils in California would have tremendous benefits—for the sustainability of agriculture in California, for the carbon health of the planet, for wildlife, for farm workers, for water conservation and purification. While the HSP is a good program, it faces challenges. Implementing an accessible program in a diverse and geographically large state is difficult. Reaching farmers that

[133] CDFA (2018a, b).

[134] Shobe (2018) stating "the statutory language establishing the Healthy Soils Program has made clear . . . the Healthy Soils Program should only incentivize practices if they have *both* a soil health and climate benefit. We maintain that it is a misuse of Healthy Soils Program funding to fund research . . . for proposed practices that CDFA has: (1) Determined have insufficient peer-reviewed research supporting a GHG benefit [or] (2) Not evaluated (e.g. done a literature review) for soil health impacts . . . if CDFA insists on funding such research and demonstration projects on proposed practices, then it must at the very least require those projects to research *both* the soil health impacts and GHG emissions/carbon sequestration impacts of such practices.").

[135] CDFA (2017i), pp. 10–11.

[136] CDFA (2018f).

[137] CDFA (2018b), p. 9.

do not already employ sustainable practices is likewise difficult. Finally, ensuring that the most resource-strapped farmers have access despite bureaucracy and lack of resources remains challenging. But California is trying to rise to these challenges. The Farmer Equity Act is one such solution, and may already have had an impact on the implementation of the Program as demonstrated by the location of projects funded during the second cycle of funding compared to the first cycle. Moreover, California has recognized the need for technical assistance and AB 2377 provides funding for CDFA to invest in this technical assistance. Hopefully AB 2377 will help the HSP reach farmers agriculturally productive but low-income counties such as Tulare. In this vein, CDFA and the California Legislature should consider expanding the Office of Environmental Farming and Innovation, with more staff and funding. As climate change worsens and soil erosion and salinization continue, the Office will only be facing more need, rather than less. California should be prepared for this. Finally, California will still have to decide how it wants to expand the scope of climate smart agricultural practices to conventional, large-scale growers. While incentive programs like the HSP are appropriate for small-scale farmers and early adopters, farms that operate more like factories may require regulation rather than incentive-based approaches. The Healthy Soils Program is an exciting step forward. But it will take more than the Healthy Soils Program to achieve the full range of benefits that California's soils can provide.

Annex: Incentive Project Applicants Already Engaged in Sustainability Efforts

At least 46 of the 85 Incentive Program projects funded by the 2016–2017 Healthy Soils Program were located on farms that were already in engaged with sustainable practices. The remaining Incentive Projects were located on farms that either (1) were not already engaged in sustainable practices or (2) did not advertise sustainability online.

Farm	Website	Language
Acorn Alegria Winery	https://acornwinery.com/sustainable-farming	"We use products and methods that organic growers use, although we are not officially certified organic. We use compost and organic fertilizers. We maintain a cover crop of grass, wildflowers, and weeds between the grapevines to prevent erosion and provide habitat for beneficial insects. . . .we have never had to spray to control insects."
Alexandre Dairy	https://alexandrefamilyfarm.com/ecodairy/organics/	"The organic foods we produce are free of chemicals, herbicides, pesticides, synthetic hormones, and

(continued)

Farm	Website	Language
		antibiotics. These same organic values guide us with regard to soil quality, how to raise animals, pest and weed control, genetically modified organisms, and the use of other input materials. Each of our products is raised on 100% organic land, forage, and feed."
Alexis Robertson/Skyelark Ranch	https://www.skyelarkranch.com/our-ranch	"Part of being a sustainable ranch is reducing our reliance on the use of fossil fuels. We do not use herbicides, chemical fertilizers, or pesticides on our ranch."
B&W Farm LLC	https://www.linkedin.com/company/b&w-quality-growers	"B&W is GlobalGap certified, audited by Primus and certified '100% Organic' for both USDA and EU organic growing standards."
Bordessa Family Dairies	https://www.organicvalley.coop/our-farmers/12578/	"His cows loved the switch to organic pastures, roaming free alongside Jarrid and his family. Happy cows meant a healthier herd and a more profitable operation. "There's no denying [the switch to organic] has been good for us. The results are right there in black and white.""
Campodonico Olive Farm LLC	https://www.campodonico-olives.com/new-page	"our orchard maintains both its COOC Extra Virgin and CCOF Organic certifications,"
Capay Hills Orchard	http://chorganicalmonds.com/	"Farmed organic by the Paddock family"
Chico Flax LLC	https://chicoflax.com/about/chico-flax/	"decided it was time to organize for fiber by creating a locally produced fabric that would represent our place here in Northern California, involve a whole community to create, be sustainable and have the potential to expand beyond our borders."
Crystal Hap LLC	https://www.ccof.org/members/crystal-hap-llc-dba-tarsadia-organic-farms-102	Certified organic
Fat Uncle Farms	https://www.fatunclefarms.com/the-farm.html	"This model blends theory from organic farming, permaculture gardening, and ecosystem management to create a holistic system that is adapted to local rainfall and climate conditions"
Gaytan Family Farm Inc	https://orangehomegrown.org/a-visit-with-gaytan-family-farm-chickens-and-strawberries/	"Gaytan strawberries are grown without pesticides or herbicides,"

(continued)

Farm	Website	Language
Hammond Crossland Vineyard	http://www.sipcertified.org/producers/john-crossland/	"Crossland has sustainably farmed wine grapes throughout Coastal regions of California for the past 40 years."
Joe B. Plummer	http://plummervineyard.net/	"Sustainable farming practices are consistently maintained and the overall operation is being reviewed for certification as a sustainable vineyard."
John P. Wick	https://www.marincarbonproject.org/about/john-wick	"John is co-founder of the Marin Carbon Project and serves as its tireless spokesperson and advisor as it moves from research to implementation."
John Swift	http://swiftsubtropicals.com/	"The goals for Bear Creek Ranch are sustainability, biodiversity, promoting native grasses, and the overall health of our land"
Lucky Dog Farms, LLC	https://www.luckydogfarmstay.com/about.html	"Focusing on wellness, with yoga, massage and fresh, sustainably grown food, we hope to provide our guests with the chance to get grounded."
Matthiasson Family Vineyards	https://www.matthiasson.com/vineyards	"We carefully tailor organic viticultural practices to each vineyard — this is how we harvest healthy and fully ripe fruit with balanced levels of alcohol and acidity. We do our best to protect and enhance the natural environment in the process."
McClelland Dairy	http://mcclellandsdairy.com/our-farm/	"the McClelland cows and chickens are eating organic pasture with no GMO's, pesticides, or commercial fertilizers,"
McGinnis Ranch LLC	https://cuesa.org/seller/mcginnis-ranch	"They are in their third year of transitioning to organic and hope to be certified in Spring 2019."
Michael Broadhurst/ Dragon Spring Farm	https://www.localharvest.org/dragon-spring-farm-M17929	"Dragon Spring Farm is Dedicated to Superior Produce through Sustainable Agriculture."
Michelle Rossow[a]	https://www.ccof.org/members/seth-and-michelle-rossow-farms-87	Certified organic
Michelle Rossow	https://www.ccof.org/members/seth-and-michelle-rossow-farms-88	Certified organic
Nancy E Osborne	https://www.redbluffdailynews.com/2018/09/07/local-farms-receive-funding-for-soil-improvement/	"Nancy and Kelly Osborne have over the last seven years developed a following for their certified organic

(continued)

Farm	Website	Language
		operation through their regular presence at the Chico Certified Farmers Market."
Pacific Gold Agriculture LLC	https://www.pacgoldag.com/	"We are a vertically integrated, sustainability-driven, tree nut orchard platform offering our clients a single portal through which to access orchard investing, orchard management, nut processing, and nut marketing and logistics."
Pie Ranch	https://www.pieranch.org/what-we-do	"Pie Ranch's vision to promote the growth of sustainable organic farming on the San Mateo coast."
Progressive Pastures, LLC	http://www.progressivepastures.com/	"We raise our cattle in a calm, natural environment. Our animals are fed on purely organic certified pastures."
Quaker Oaks Farm	https://www.quakeroaksfarm.org/	"Agroecology can be defined as the balance and respect for all participants in the design of agricultural production systems which are built from traditional knowledge that links environmental, economic, and human health."
Red Gate Ranch LLC	http://redgateranch.us/pasture-raised-meats	"Using regenerative farming practices, we strive to raise our livestock and nurture our garden with a deep reverence for the natural world."
Redwood Empire Vineyard Management, Inc	http://www.revm.net/our-services/farmingviticulture/	Goals are "To continue working collaboratively in the community, developing sustainable, environmentally sound practices; To reduce the impact of our farming operation on the environment each year. . ."
Roberto LeFort	https://certifiedhumane.org/le-forts-organic-crops-creston-california/	"Their organic farm grows vegetables, fruits and herbs and raises free-range chickens, some of which are heritage breeds, for their eggs."
Rock Front Ranch LLC	https://www.rangelandtrust.org/rock-front-ranch/	"In partnership with the landowners, the California Rangeland Trust is seeking to place a conservation easement on the ranch as a strategic buffer to protect the Cuyama Valley and the life it sustains."
Serventi Ranch	http://serventiranch.com/	Organic apples
Seth & Michelle Rossow Farm	https://www.ccof.org/members/seth-and-michelle-rossow-farms-87	Certified organic
Seth Rossow	https://www.ccof.org/members/seth-and-michelle-rossow-farms-87	Certified organic

(continued)

Farm	Website	Language
Sierra Orchards	https://craigmcnamara.org/sierra/	"The orchard is approximately 450 acres in size and produces mostly organic walnuts and olive oil."
Stemple Creek Ranch Inc	https://stemplecreek.com/	"No antibiotics, hormones or growth promotants. Just organic pasture forage, wide open fields, and plenty of sunshine"
Stubbs Vineyard	https://www.stubbsvineyard.com/new-page	"Low-impact and environmentally conscious approach to farming, using organic and sustainable methods. While the vineyard is watered entirely from our rainwater reservoir, the property is powered by an off-the-grid system of photovoltaic panels."
Sun Drenched Organics LLC	https://farmsreach.com/welcome/groups/funding-opportunities/forum/	certified organic; "I am trying to build a farm using sustainable and environmentally conscious practices."
Terra Cultura	https://terracultura.org/	"Terra Cultura's mission is to cultivate resilient communities empowered through education and collaboration in agroecology, sustainable living, and the arts."
The Cloverleaf Farm	https://thecloverleaffarm.com/our-story/	"We have an interesting soil prep method we've been playing around with at the Cloverleaf – no-till no-chemical organic farming!"
The Grove LLC	http://www.growriverside.com/diversification-organic-growing-and-savvy-enable-riverside-family-to-save-farm-and-prosper/	"The Grove, a diversified and certified-organic family farm in Riverside, CA used to grow only citrus fruit and avocados. But in order to survive a changing market, it has diversified to include a wide array of organic produce."
Thomas D. Donati	http://sunfedranch.com/commitment/our-environment/	"The SunFed Ranch commitment to properly managing our grasslands supports the ecological sustainability of our land and watersheds and our climate's health"
Toluma Farms, Inc	http://www.tolumafarms.com/	"We are a Marin County dairy, organic farm and creamery."
Treborce Vineyards	https://www.sustainablewinegrowing.org/swpparticipants.php	A sustainable winegrowing program participant

(continued)

Farm	Website	Language
True Grass Farms	https://truegrassfarms.com/	"Beyond organic, we are part of your bioregional food source."
Wild Farmlands Foundation	https://wildfarmlands.org/	"Wild Farmlands Foundation informs, educates and champions local foodshed development and ecosystem preservation through the strengthening of small farms and ranches."

[a]Michelle Rossow is included twice because the Healthy Soils Program funded more than one Incentive Project on her farm

At least 101 of the 194 recipients funded by the 2018–2019 HSP Incentive Program were located on farms that were already in engaged with sustainable practices. The remaining Incentive Projects were located on farms that either (1) were not already engaged in sustainable practices or (2) did not advertise sustainability online.

Farm	Website	Language
Elkhorn Slough Foundation	https://www.elkhornslough.org/elkhorn-slough-foundation/	"The mission of the Elkhorn Slough Foundation is to conserve and restore Elkhorn Slough and its watershed"
Hiddensprings Ranch	https://www.wildfarmalliance.org/hidden_springs_ranch_case_study	"They are providing wildlife paths, protecting water quality with filter strips of native grasses and wild-flowers, supporting beneficial insects, and restoring the property's oak woodland community sur-rounding the vineyard. Wild Farm Alliance and Community Alliance with Family Farmers (CAFF) are assisting with these improvements with funding through the Wildlife Conservation Board."
Sweet Spot Farm LLC	https://www.thesweetspot.farm/	"Facilitates experiential learning, grows regenerative earth products, stewards diverse plant genetics, restores and regenerates abundant ecosystems, and fosters the healthy lifestyle of all who participate."
The Ecology Center	https://www.theecologycenter.org/	"bring your class to our 28-acre organic farm for a day full of eco-logical experiences."
Tehachapi Grain Project (TGP)	https://www.tehachapigrainproject.org/	"The Tehachapi Heritage Grain Project's aim is to preserve and grow heritage organic grains which are naturally drought tolerant and low in gluten."

(continued)

Farm	Website	Language
Table Bluff Farm LLC	https://www.tableblufffarm.com/	"With Nature's bounty all around us, we at Table Bluff Farm strive to go beyond organic principles to create a fully biodynamic farm focused on preserving and enhancing our natural ecosystem while keeping the low cost/true cost affordability of biodynamically grown, chemical-free, nutrient-dense vegetables, flowers, eggs, and meat."
Stubbs Vineyard LLC	https://www.stubbsvineyard.com/new-page	"a low impact environmentally conscious approach to farming..."
Shinta Kawahara Company	https://www.skberries.com/	"Farmer Rod and Gwen have always been interested in the art and science of growing strawberries so when they fully took over day to day operations of the farm from Kuni and Bea they wanted to start an organic program at the farm. Rod and Gwen view the addition of the organic program as their contribution to the long legacy of our family's farm."
Yael Zaidman, Terra Corazon Farm	https://www.sdsustainable.org/teaching-team	"Yael is currently on the board of SDSLI and is developing the 17 acre permaculture farm at Terra Corazon with her family and friends."
Nagata Bros Farms Inc	https://www.scsglobalservices.com/nagata-bros-blueberries-earn-rigorous-pesticide-residue-free-certification	"Nagata Bros. Farms, based in Oceanside, California, plans to roll out packaged "California Sweet" blueberries featuring the SCS Kingfisher certification mark with the Pesticide Residue Free claim at retail locations, including Albertson's and Stater Bros., later this month."
River Garden Farms Co	https://www.rivergardenfarms.com/	"Today, River Garden Farms is directed by second- and third-generation owners with a long-term vision and strong commitment to the community and the environment. Simply said, we believe in taking care of the land, environment, and community that so greatly take care of us. We actively pursue involvement in environmental and community outreach projects."

(continued)

Farm	Website	Language
Ridge Vineyards, Inc	https://www.ridgewine.com/vineyards/lytton-springs/	"Accordingly, we take our stewardship of the land and the vines into account in both our sustainable farming practices and the winery itself."
John Hanson, Jr.—Willow Creek Ranch	https://www.pointblue.org/science_blog/working-with-ranchers-to-heal-the-land/	"Along with flood irrigating and managing grazing for perennial grass growth, the Hansons are receiving support from NRCS, Bureau of Land management (BLM), and Point Blue to remove light-density juniper on about 4000 acres of adjacent BLM land in priority Greater Sage-Grouse habitat. The Hansons are also working to protect their ranch in perpetuity through a conservation easement."
Pacific Realty Associates L.P. dba: M&T Chico Ranch	https://www.pactrust.com/about/partnerships/mt-chico-ranch/	"The ranch uses modern sustainability practices, including a solar farm that produces power for the operation of the farm and micro-drip irrigation which greatly reduces the amount of water used."
Lone Willow Ranch	https://www.organicheirlooms.com/aboutus.html	"When Lone Willow Ranch was established late 1980's the first organic crop was cotton. The organic cotton started a passion in organics; we then planted fruit trees, row crops for vegetables, herbs, heirloom tomatoes, vegetable transplants, and now raising pastured livestock. We have select heritage breeds of turkeys, pigs, chicken and goats. We now have over 80 acres in production for the pasture livestock. We grow our food for our self as well for our animals. We grow all organic alfalfa, dry deans, wheat and sunflower to supplement the feed for our livestock."
Navarro Vineyards LLC	https://www.navarrowine.com/prerelease/philosophy.php	"Sustainable farming starts in the vineyard. Navarro avoids the use of synthetic insecticides and herbicides. The vineyard rows are banded with flowering cover crops, abuzz with beneficial insects that keep pests in check as well as minimizing erosion and supplying nutrients to the gravelly loam. A flock of Babydoll sheep, with the

(continued)

Farm	Website	Language
		help from some chickens and geese, help keep the cover crop mowed, help fertilize the vines and even sucker the vines' trunks. Navarro Vineyards was one of the first ranches in the US to be certified as a Fish Friendly Farm."
Meadow Farm Community Land Trust	https://www.meadowfarm.org/	"Meadow Farm provides an opportunity for enriching lives through affordable accommodations, permaculture, organic farming, education and other creative endeavors. We co-create via a dynamic and collaborative decision-making model."
Bucio Organic Farms	https://www.localharvest.org/bucio-organic-farm-M69865	Certified organic
Bazzano Azienda LLC	https://www.linkedin.com/in/philipbazzano	"Bazzano Azienda is committed to premium winegrape production and responsible land management by utilizing sustainable farming and environmentally conscious practices."
Deep Roots Ranch	https://www.linkedin.com/in/jean-harrah-814513a8	"raising grass fed beef and lamb, organically fed pork and eggs from pastured chickens."
LaRocca Vineyards	https://ota.com/phil-larocca	"Phil LaRocca of LaRocca Vineyards will receive the 2017 Organic Farmer of the Year Leadership Award. A tireless organic farming advocate and farmer for over 40 years, Phil LaRocca has deep organic convictions that he puts into practice on his farm, in his winery, and in the wider community, particularly in California."
King & King Ranch	https://www.kingandkingranch.com/	"King family farmers have long been devoted to serving as stewards of the land, mindful that growing food serves not only immediate needs but is also a legacy to sustain future generations. In support of our stewardship goal, various unique conservation efforts are underway at the ranch, including large scale mulching, seeding for erosion and weed control, clearing fire breaks, creating wildlife habitat, installing raptor poles and owl boxes, and exploring permaculture

(continued)

Farm	Website	Language
		practices—all contributing to a more sustainable ecosystem."
JSM Organics	https://www.ccof.org/members/jsm-organics-153	Certified organic
Mission Ranches Company, LLC	https://www.joeproduce.com/users/mission-ranches	"Mission Ranches Company, LLC is a Certified Organic/Certified Organic Farm located in King City, CA. Learn more about Mission Ranches Company, LLC and check out their product list for farm fresh, local food. Mission Ranches Company, LLC is an Agrilicious! featured Farm!"
John Muir Land Trust—Family Harvest Farm	https://www.jmlt.org/familyharvestfarm.html	"The mission of Family Harvest Farm is to empower youth and adults to discover and participate in a local food system that encourages healthy living, nurtures the environment, and grows a sustainable community. The goals are to employ transition-age foster youth (ages 18–22) through a job-readiness program, grow organic produce, and host educational workshops for the greater community."
Fat Uncle Farms	https://www.fatunclefarms.com/the-farm.html	"This model blends theory from organic farming, permaculture gardening, and ecosystem management to create a holistic system that is adapted to local rainfall and climate conditions"
Rebekka Siemens	https://www.fatunclefarms.com/the-farm.html	"This model blends theory from organic farming, permaculture gardening, and ecosystem management to create a holistic system that is adapted to local rainfall and climate conditions"
Scott Yenni	https://www.farmtrails.org/farm/yenni-ranch	"All Natural Antibiotic Free Biodynamic GMO Free Grass-fed Hormone Free Humanely Raised Organic Pasture-raised Pesticide Free Rotational Pasture"
Winterport Farm	https://www.facebook.com/pg/WinterportFarm/about/?ref=page_internal	"Grass-fed, small farm beef raised 31 miles from Sacramento with no hormones, no antibiotics, and no grain"
Huerta del Valle	https://www.facebook.com/pg/HDVhuertadelvalle/about/?ref=page_internal	"Our mission is to cultivate an organization of community

(continued)

Farm	Website	Language
		members to grow our own organic crops."
Linda Chu	https://www.ediblemontereybay.com/online-magazine/summer-2017/whats-in-season-demystifying-asian-greens/	"Linda Chu and Chris He of Benito Valley Farm in Hollister grow mature pea greens and a number of other Asian vegetables. Most of the demand for their products comes from conventional produce buyers in Los Angeles and San Francisco, but the farmers hope to begin distributing their crops locally soon. They are especially seeking wholesale and retail outlets for their CCOF-certified produce. "We are working on packaging the vegetables with other ingredients like oyster sauce and garlic with instructions on how to prepare them, hoping that it helps sell the organic in more of a mainstream market," Chu says."
Brian E. Lindberg	https://www.cdfa.ca.gov/oefi/healthysoils/docs/2018-HSPIncentives-SubmittedApplications.pdf	"When we acquired ownership, the land was worn out, and the creek was eroding a new bed every time it ran. We began restoration of the ecosystem and implementation of a permaculture agrarian practice 25 years ago. Ten years ago, we implemented a USDA WHIP project on the land (see Conservation Plan attached)."
Four Elements Farm	https://www.cdfa.ca.gov/oefi/healthysoils/docs/2018-HSPIncentives-SubmittedApplications.pdf	"We seek to farm in a way that has a positive effect on our climate, and the available funds through this incentive program further enable us to do so, by making these practices more feasible for our small, organic farm."
Hummingbird Ranch	https://www.cdfa.ca.gov/oefi/healthysoils/docs/2018-HSPIncentives-SubmittedApplications.pdf	"This project will support implementation of cover crops on 2.1 ac of our organic farm. . . ."
Pamela Posey/Allen Harthorn	https://www.cdfa.ca.gov/oefi/healthysoils/docs/2018-HSPIncentives-SubmittedApplications.pdf	"We purchased the property in December 2017 and have been managing it using only organic practices since that time."
Sunrock Farm	https://www.cdfa.ca.gov/oefi/healthysoils/docs/2018-HSPIncentives-SubmittedApplications.pdf	"as organic food production farmers, we are becoming increasingly aware of the negative impacts that climate change is having on

(continued)

Farm	Website	Language
		our planet. We see these impacts every day, and feel empowered to know that there are numerous things that we can do right here on our farm, to help alleviate this dire situation."
University of California, Riverside	https://www.cdfa.ca.gov/oefi/healthysoils/docs/2018-HSPIncentives-SubmittedApplications.pdf	"Currently, we have two areas for crop production. One part is dedicated to organic farming and the other conventional. In order to become certified organic, we will establish a barrier to prevent drift from spraying that occurs in the conventional farming areas."
Markstein Organic Family Farms	https://www.cdfa.ca.gov/oefi/healthysoils/docs/2018-HSPIncentives-SubmittedApplications.pdf	"Beginning in 2016, I decided to go organic and began the process to certify Markstein Organic Family Farms as USDA and/or CDFA organic."
Las Piletas Ranch, LLC	https://www.cdfa.ca.gov/oefi/healthysoils/docs/2018-HSPIncentives-SelectedProjects.pdf	"Currently there are 2 sections (1280 acres) of relatively level land that is being dry-farmed (organic) for barley (half the land per year, the other half fallowed)."
ACM II California Ranch 4 LLC—Victory Ranch	https://www.ccof.org/sites/default/files/CCOF_Mag_Summer_18%20-%20Web.pdf	Listed as newly organic certified
Michelle Rossow Farms	https://www.ccof.org/members/seth-and-michelle-rossow-farms-88	Certified organic
Seth Rossow Farms	https://www.ccof.org/members/seth-and-michelle-rossow-farms-88	Certified organic
Rincon Farms Inc	https://www.ccof.org/members/rincon-farms-inc-116	Certified organic
Oryza Partnerhsip	https://www.ccof.org/members/oryza-partnership-98	Certified organic
Ryan Christopher Bell (Hummingbird Ridge Farm)	https://www.ccof.org/members/hummingbird-ridge-farm-125	Certified organic
Huerta Family Farms INC	https://www.ccof.org/members/huerta-family-farms-inc-42	Certified organic
Hikari Farms	https://www.ccof.org/members/hikari-farms-llc-dba-nagamine-nursery-2?title=&field_contact_address_locality=&field_contact_address_administrative_area=All&field_contact_address_postal_code=&field_chapter_ref=All&field_products_services=&field_products_services_1=&submit-	Certified organic

(continued)

Farm	Website	Language
	top=Search&reset-top=Reset&page=87	
Sally Negroni	https://www.ccof.org/members/fidel-lazaro-dba-lazaro-organic-farms-0?page=164	Certified organic
Ferreira Dairy	https://www.ccof.org/members/ferreira-dairy-71	Certified organic
Bloom Farms	https://www.ccof.org/members/bertagna-custom-farming-inc-28	Certified organic
Bullseye Orchards 1 LLC	https://www.bullseyefarms.com/about-us	"We strive to understand and enhance our cropping ecosystem. We want to foster a healthy relationship between our crops and the environment in which they grow. It is our hope that our practices will sustain environmental quality and natural resources while satisfying the demand for a healthy, tasty product... A principle that we have focused on in our orchards is soil regeneration. We return hull, shell, and inferior nut meats normally regarded as waste, back the orchard floor. We source and spread local green waste compost, and we plant winter cover crops. Together these practices insulate the soil, prevent erosion, and feed the organisms in the soil that in turn change organic matter and soil minerals into plant food. This improves the biological diversity of our orchards both above ground and below. As a result, we have seen remarkable improvements in our soil and crop health."
BECKSTOFFER VYD XV, LLC	https://www.beckstoffervineyards.com/quality/innovation-sustainability	"We are committed to sustainable farming across our operations and have pioneered new technologies to monitor and analyze vine health and productivity, manage water use and improve vineyard efficiency."
Araceli Farms	https://www.aracelifarms.com/pages/about-us	"We never use chemicals or pesticides in our production process, from the field to our products"
Alturas Ranches, LLC	https://www.alturasranches.com/wild-rice	"Alturas Ranches primary focus is on the health of the land. We have established, converted, and continue to develop vast fields of

(continued)

Farm	Website	Language
		conventionally farmed ground to organic land."
True Grass Farms	https://truegrassfarms.com/	"Beyond organic, we are part of your bioregional food source."
The Abundant Table	https://theabundanttable.org/	"The Abundant Table is a grass-roots, nonprofit organization that seeks to change lives and systems by creating sustainable relationships to the land and local community."
Straus Home Ranch	https://straushomeranch.com/fullscreen/rebels/	"Tomales Bay vacation rental, on our organic ranch in Marshall, CA"
Stemple Creek Ranch	https://stemplecreek.com/	"No antibiotics, hormones or growth promotants. Just organic pasture forage, wide open fields, and plenty of sunshine"
Tolay Springs LLC	https://sonomawinegrape.org/wp-content/uploads/SCW-Annual-Sustainable-Report-2015.pdf	Certified sustainable
DuMOL Estate Vineyard, LLC	https://sonomawinegrape.org/wp-content/uploads/2017-SoCoWinegrowers-Section-LR.pdf	Certified sustainable
SoMar Farms	https://somarfarms.com/about-the-farm	"Our mission is to produce delicious and nutritious food for our community by using regenerative agricultural practices that prioritize the well-being of the planet, animals, and people involved. None of our animals are ever treated with hormones, and while we are not certified organic we do utilize organic practices."
Round River Farm	https://mendolake.localfoodmarketplace.com/Producer/09aabf65-f213-40ca-aff3-0264f14717e7	"In 1998 my wife and I bought the farm from the family that planted the orchard, and soon converted to organic farming methods. The orchard became Certified Organic in 2006."

References

AB 1348. Farmer Equity Act of 2017. Chapter 620. Available at https://leginfo.legislature.ca.gov/faces/billTextClient.xhtml?bill_id=201720180AB1348

AB 2377 (Sept 28, 2018) Cannella Environmental Farming Act of 1995: technical assistance grant program, Available at https://leginfo.legislature.ca.gov/faces/billTextClient.xhtml?bill_id=201720180AB2377

Alkon AH, Agyeman J (2011) Cultivating food justice: race, class, and sustainability. MIT, Boston

Amundson R, Biardeau L (2019) Opinion: soil carbon sequestration is an elusive climate mitigation tool. Proc Natl Acad Sci 115(46):11652–11656. (Nov. 13, 2018)

Ayazi H, Elsheikh E (2015) The U.S. farm bill: corporate power and structural racialization in the United States Food System. Haas Institute for a Fair and Inclusive Society, Berkeley. http://haasinstitute.berkeley.edu/sites/default/files/farmbill_researchbrief_interactive.pdf. Accessed 18 Mar 2019

Cal. Food & Ag. Cod. § 566 (Deering 2019)

CalEPA (2019) California climate investments to benefit disadvantaged communities. https://calepa.ca.gov/EnvJustice/GHGInvest/. Accessed 16 June 2019

CARB (2017) California's 2017 climate change scoping plan. https://www.arb.ca.gov/cc/scopingplan/scoping_plan_2017.pdf. Accessed 16 June 2019

CARB (2018a) Climate change programs. https://www.arb.ca.gov/cc/cc.htm. Accessed 18 Mar 2019

CARB (2018b) California greenhouse gas emission inventory – 2018 Edition. https://ww3.arb.ca.gov/cc/inventory/data/data.htm. Accessed 5 July 2019

CARB (2018c) California greenhouse gas emissions for 2000 to 2016. https://ww3.arb.ca.gov/cc/inventory/pubs/reports/2000_2016/ghg_inventory_trends_00-16.pdf. Accessed 5 July 2019

CARB (2018d) Priority population investments. https://www.arb.ca.gov/cc/capandtrade/auctionproceeds/communityinvestments.htm. Accessed 16 June 2018

CARB (2019) Neutrality--role of natural and working lands. https://ww3.arb.ca.gov/cc/natandworkinglands/nwl-cn-051719.pdf. Accessed 5 July 2019

CDFA (2009) Soil salinization. https://www.cdfa.ca.gov/agvision/docs/Soil_Salinization.pdf. Accessed 16 June 2019

CDFA (2016) Healthy soil action plan. https://www.cdfa.ca.gov/oefi/healthysoils/docs/CA-HealthySoilsActionPlan.pdf. Accessed 16 June 2019

CDFA (2017a) California agricultural production statistics. https://www.cdfa.ca.gov/Statistics/. Accessed 16 June 2019

CDFA (2017b) Selected health soils program incentive projects first solicitation. https://www.cdfa.ca.gov/oefi/healthysoils/docs/2017HSP1stroundIncentivesJuly2018.pdf. Accessed 16 June 2019

CDFA (2017c) Selected healthy soils program incentive projects second solicitation. https://www.cdfa.ca.gov/oefi/healthysoils/docs/2017HSP2IncentivesAwardedProjectsJuly2018.pdf. Accessed 16 June 2019

CDFA (2017d) Selected healthy soils program demonstration projects first solicitation. https://www.cdfa.ca.gov/oefi/healthysoils/docs/2017-HSP-1stRound-DemoProjects-July2018.pdf. Accessed 16 June 2019

CDFA (2017e) Selected healthy soils program demonstration projects second solicitation. https://www.cdfa.ca.gov/oefi/healthysoils/2017SecondSolicitationDocs/2017-HSPDemo2ndSolicitationProjectsAwarded-July2018.pdf. Accessed 16 June 2019

CDFA (2017f) Farms + Data: California's farms are smaller than the US average, but they're big on diversity – and productivity. https://plantingseedsblog.cdfa.ca.gov/wordpress/?p=10952. Accessed 16 June 2019

CDFA (2017g) Incentive project request for grant application. https://www.cdfa.ca.gov/oefi/healthysoils/docs/2017-HSPIncentives_RGA.pdf. Accessed 8 July 2019

CDFA (2017h) Proposed practices for the healthy soils program. https://www.cdfa.ca.gov/oefi/healthysoils/docs/HSP_New_Practices_Proposals_2017-18.pdf. Accessed 16 June 2019

CDFA (2017i) Healthy soils program second solicitation request for grant applications. https://www.cdfa.ca.gov/oefi/healthysoils/2017SecondSolicitationDocs/2017-HSPDemo2ndSolicitation_RGA.pdf. Accessed 8 July 2019

CDFA (2018a) Healthy soils program incentives program: request for grant applications. https://www.cdfa.ca.gov/oefi/healthysoils/docs/2018-HSPIncentives_RGA.pdf. Accessed 16 June 2019

CDFA (2018b) Healthy soils program demonstration program request for grant applications (Dec. 28, 2018), https://www.cdfa.ca.gov/oefi/healthysoils/docs/2018-HSPDemo_RGA.pdf at 4. Accessed 16 June 2019

CDFA (2018c) Comments received on the CDFA healthy soils program comment period: Jul 25, 2018–Sep 12, 2018. https://www.cdfa.ca.gov/oefi/healthysoils/docs/CDFAResponsestoPublicCommentsJuly-Sep2018.pdf. Accessed 16 June 2019

CDFA (2018d) HSP incentive program projects selected for award, https://www.cdfa.ca.gov/oefi/healthysoils/docs/2018-HSPIncentives-SelectedProjects.pdf. Accessed 8 July 2019

CDFA (2018e) HSP demonstration projects selected for award. https://www.cdfa.ca.gov/oefi/healthysoils/docs/2018-HSPDemo-SelectedProjects.pdf. Accessed 8 July 2019

CDFA (2018f) Response to public comments (Dec. 2018). https://www.cdfa.ca.gov/oefi/healthysoils/docs/HSP_PublicCommentResponsesDec-2018.pdf. Accessed 16 June 2019

CDFA (2019a) Healthy soil program. https://www.cdfa.ca.gov/oefi/healthysoils/. Accessed 16 June 2019

CDFA (2019b) Healthy soils program incentives program. https://www.cdfa.ca.gov/oefi/healthysoils/IncentivesProgram.html. Accessed 16 June 2019

CDFA (2019c) HSP demonstration projects. https://www.cdfa.ca.gov/oefi/healthysoils/DemonstrationProjects.html. Accessed 18 Mar 2019

CDFA (2019d) Staff directory. https://directory.cdfa.ca.gov/branch/. Accessed 16 June 2019

FAO (2015) Soils help to combat and adapt to climate change by playing a key role in the carbon cycle. http://www.fao.org/3/a-i4737e.pdf. Accessed 16 June 2019

Gattinger A et al (2012) Enhanced top soil carbon stocks under organic farming. Proc Natl Acad Sci 109(44):18226–11823

Gunasekara A (2013) Environmental farming act science advisory panel: biannual report CDFA. https://www.cdfa.ca.gov/oefi/efasap/docs/Science_Panel_Report.pdf. Accessed 16 June 2019

Jagannath J, Hamilton KD, Aird S (2017) Comments on healthy soils initiative draft framework. In: Public comments on the healthy soil program. CDFA, Sacramento. Available via CDFA. https://www.cdfa.ca.gov/oefi/healthysoils/docs/HealthySoilsComments-Jan19-Mar1_2017.pdf

Kroodsma D, Field CB (2006) Carbon sequestration in California agriculture, 1980–2000. Ecol Appl 16(5):1975–1985

Lal R (2010) Managing soils and ecosystems for mitigating anthropogenic carbon emissions and advancing global food security. BioScience 60:708–721

Marcum D (2014) Fresno ranks No. 1 on California pollution list, Los Angeles Times, April 23. Available at https://www.latimes.com/local/la-me-fresno-pollution-20140424-story.html

Montgomery D (2007) Soil erosion and agricultural sustainability. Proc Natl Acad Sci 104 (33):13268–13272. https://doi.org/10.1073/pnas.0611508104

Morgan JA et al (2010) Carbon sequestration in the agricultural lands of the United States. J Soil Water Conserv 65(1):6A–13A. https://doi.org/10.2489/jswc.65.1.6A

Mulik K (2016) Subsidizing waste: how inefficient U.S. farm policy costs taxpayers, businesses, and farmers billions. https://www.ucsusa.org/sites/default/files/attach/2016/08/Subsidizing-Waste-full-report.pdf. Accessed 16 June 2019

NRCS (2017) $19 Million available to California's farmers to improve air quality. https://www.nrcs.usda.gov/wps/portal/nrcs/detail/ca/newsroom/releases/?cid=nrcseprd1322668. Accessed 16 June 2019

NRCS (2019) More than 80 years helping people help the land: a brief history of NRCS. https://www.nrcs.usda.gov/wps/portal/nrcs/detail/national/about/history/?cid=nrcs143_021392. Accessed 16 June 2019

Olmstead AL, Rhode PW (2017) A history of California agriculture. Regents of the University of California Division of Agriculture and Natural Resources. Available online. https://s.giannini.ucop.edu/uploads/giannini_public/19/41/194166a6-cfde-4013-ae55-3e8df86d44d0/a_history_of_california_agriculture.pdf. Accessed 16 June 2019

Pesticide Action Network of North America (2017) Farmer equity act recognizes racial discrimination, enables state to identify barriers & provide much-needed support to farmers. http://www.panna.org/press-release/farmer-equity-act. Accessed 5 Dec 2020
Powlson DS et al (2011) Soil management in relation to sustainable agriculture and ecosystem services. Food Policy 36:S72–S74
RCD Project Mapper (2019) https://www.rcdprojects.org/Results/ProjectMap. Accessed 16 June 2019
Rees WE (2016) North American soils and world food. Int Yearb Soil Law Policy 1:21
Shobe B (2018) SWEEP and healthy soils program 2nd draft requests for grant application. In: Public comments on the healthy soils program. CDFA, Sacramento. Available via CDFA. https://www.cdfa.ca.gov/oefi/sweep/docs/SWEEP_PublicComments_OctNov2018.pdf. Accessed 16 June 2019
Smith P et al (2008) Greenhouse gas mitigation in agriculture. Philosophical Trans R Soc B 363:789–813
Smoker B (2018) Healthy soils program and state water efficiency and enhancement program 2nd draft requests for grant application. In: Public comments on the healthy soils program. CDFA, Sacramento. Available via CDFA. https://www.cdfa.ca.gov/oefi/sweep/docs/SWEEP_PublicComments_OctNov2018.pdf.available at https://www.cdfa.ca.gov/oefi/sweep/docs/SWEEP_PublicComments_OctNov2018.pdf. Accessed 16 June 2019
State of California. California climate investments. http://www.caclimateinvestments.ca.gov/about-cci. Accessed 11 Apr 2019
Tomich T et al (2015) The California nitrogen assessment executive summary. University of California Division Agriculture and Natural Resources, Davis. Available online. https://asi.ucdavis.edu/sites/g/files/dgvnsk5751/files/inline-files/Executive%20Summary%20Layout_FINAL_reduced.pdf. Accessed 16 June 2019
U.S. Census Bureau (2015) Millennials outnumber baby boomers and are far more diverse, Census Bureau Reports. https://www.census.gov/newsroom/press-releases/2015/cb15-113.html. Accessed 18 Mar 2019
U.S. Census Bureau, https://www.census.gov/quickfacts/fact/table/california,CA/INC110217 (last visited July 5, 2019)
USDA (2012) U.S. census of agriculture 582. https://www.nass.usda.gov/Publications/AgCensus/2012/Full_Report/Volume_1,_Chapter_1_US/usv1.pdf. Accessed 18 Mar 2019
USDA (2016) California: first in the number of certified organic farms and total sales. https://www.nass.usda.gov/Surveys/Guide_to_NASS_Surveys/Organic_Production/2016_State_Publications/CA.pdf. Accessed 16 June 2019
USDA (2017a) U.S. census of agriculture 704. https://www.nass.usda.gov/Publications/AgCensus/2017/Full_Report/Volume_1,_Chapter_1_US/usv1.pdf. Accessed 18 Mar 2019
USDA (2017b) U.S. census of agriculture: California 66. https://www.nass.usda.gov/Publications/AgCensus/2017/Full_Report/Volume_1,_Chapter_2_County_Level/California/st06_2_0002_0002.pdf. Accessed 18 Mar 2019
Water Education Foundation (2019) Salinity in the central valley: a critical problem. http://www.watereducation.org/post/salinity-central-valley-critical-problem. Accessed 16 June 2019
Welle P, Mauter M (2017) High-resolution model for estimating the economic and policy implications of agricultural soil salinization in California. Environ Res Lett 12:9. Available at https://iopscience.iop.org/article/10.1088/1748-9326/aa848e/meta
West G (1890) San Joaquin county biographies. In: Slater JS (ed) A history of The San Joaquin Valley. The Lewis Publishing Company, Chicago, p 526
Whitmore A et al (2010) To evaluate the potential of technologies for increasing carbon storage in soil to mitigate climate change. Department For Environment, Food And Rural Affairs, London. sciencesearch.defra.gov.uk/Document.aspx?Document=SP1605_9703_FRP.pdf. Accessed 16 June 2019
Woodruff v. North Bloomfield Gravel Min. Co., 18 F. 753, 756 (C.D. Cal. 1884)

Soil Protection Governance in Germany

Harald Ginzky

1 Introduction

Soils provide essential functions and services—such as ensuring food security, storage of carbon and biodiversity as well as being a filter and puffer for groundwater—which are of core importance to achieve a sustainable development. However, soil has been a neglected subject for a long time.

Thus, in Germany as in most states, soil protection governance was implemented very late. Water protection provisions in Germany stem from the 50th of the last century.[1] Wildlife regulations are even older.[2] The Federal Soil Protection Act (FSPA),[3] however, came into effect just 1998; followed 1 year later by the Federal Soil Protection Ordinance (FSPO).

This chapter intends to provide an overview of the current German soil protection governance,[4] in particular for experts from outside of Germany. To this end, it will

[1]See the German Federal Water Act in its original version from 1957 (Gesetz zur Ordnung des Wasserhaushalts, BGBl. I S. 1110, S. 1386).

[2]Kloepfer and Neugärtner (2016), p. 90 et seqq.

[3]Gesetz zum Schutz vor schädlichen Bodenveränderungen und zur Sanierung von Altlasten (Bundes-Bodenschutzgesetz - BBodSchG) vom 17. März 1998 (BGBl. I S. 502).

[4]Soils require more than only protection by, for example, limiting the use of fertilizers or by the establishment of protected areas. In fact, approaches, instruments and procedures for "sustainable soil management" are needed to deal with the environmental, economic and social challenges. See Ginzky (2020), p. 38/39. Thus, term "soil protection governance", if used throughout the text, is to

H. Ginzky (✉)
German Environment Agency, Dessau, Germany

Development and Rule of Law Programme [DROP], University of Stellenbosch, Stellenbosch, South Africa
e-mail: harald.ginzky@uba.de

H. Ginzky et al. (eds.), *International Yearbook of Soil Law and Policy 2019*,
International Yearbook of Soil Law and Policy 2019,
https://doi.org/10.1007/978-3-030-52317-6_15

provide some background information on Germany's environmental, economic as well as historical and cultural situation, on the governmental structure in Germany and on the influence of the European Union on Germany's environmental governance. The next section will discuss the main current soil threats and drivers in Germany. The chapter will then turn to outline the various provisions and norms by the Federal State regulating the different soil threats, followed by a section on the soil related regulations of the sixteen "Länder". The analysis will focus on public law. Then, the respective international obligations of Germany are discussed as well as how far Germany has implemented the objective to achieve a "land degradation neutral world", one of the Sustainable Development Goals (SDG). Before concluding, the strengths and weaknesses of the German soil governance will be assessed and some recommendations will be submitted.

2 Background Information

In this section, information on Germany's environmental, economic as well as historical and cultural background will be briefly presented. Moreover, the governmental structure and the influence of the European Union on Germany's environmental policy will be discussed.

2.1 Germany

Germany is located in Central and Western Europe, lying between the Alps in the south and the North Sea and the Baltic Sea in the north. Germany borders with Denmark to the north, with Poland and the Czech Republic to the east, Austria and Switzerland to the south, France to the southwest, and Luxembourg, Belgium and the Netherlands to the west. It covers an area of 357,386 km^2.

Germany's environment is very diverse. High mountains in the south, lakes throughout the country, transboundary rivers like the Rhine and the Danube, hilly regions in central Germany, coastlines to the North and the Baltic sea. Water availability has been still high although Germany has also faced severe droughts in the last years. Having a continental climate, there is no periodic dry season. Due to the continental climate and topographical variety including high mountainous areas and hilly regions down to marsh regions close to the North and Baltic Sea, very

be understood to include all measures, which are potentially needed to achieve "sustainable soil management".

different types of soils can be found in Germany. Soil scientists differentiate at least 12 different soil types based on their texture, compaction and colour.[5]

Germany has a highly diversified industry, which is based on a strong labour force, a large capital stock, a low level of corruption, and a high level of innovation. It has the largest national economy in Europe, which is also the fourth largest by nominal Gross Domestic Product (GDP) in the world. The economic success is mainly built on exports as Germany is the third strongest export country in the world.[6] Main export products are vehicles and vehicle parts, machinery, chemical goods, electronic products, computer/electrical equipment, pharmaceuticals, transport equipment, basic metals, food products, rubber and plastics.[7] Next to stock-rated companies such as Volkswagen, Daimler, E.ON, Allianz and BMW, important economic actors are uncountable specialized—often family-owned—small and medium enterprises.

Although 53% of the total territory are agricultural land[8] and 31% are covered by forests, less than one percent of the GDP was produced by agriculture and forestry in 2018.[9] Only about 2% of employees are working in the agricultural sector (about 940,000 people).[10]

With more than 80 million inhabitants Germany is one of the most densely populated countries in Europe—about 230 persons per square kilometre.[11] About 10 million people are not German nationals as they have immigrated over the years. Although immigrants are mainly Muslim, Christian attitude and perceptions still form the "philosophical" basis of German society.

Over the centuries, Germany has been very rich in culture, such as arts, architecture, music or literature. At the same time, Germany's history is extremely controversial. The two world wars in last century have been at least partly caused by the respective German politics at that time. Moreover, the genocide against the Jewish population during World War II was organized by the German Nazi regime.

---

[5]The number of main soil types, found in Germany, is based on the German classification system. For more information see: https://www.umweltbundesamt.de/en/topics/soil-agriculture/soil-science/soil-types.

[6]Federal Ministry for Economic Affairs and Energy (2018).

[7]Federal Statistical Office (2019a): Exports and imports (special trade) by division of the national Product Classification for Production Statistics 2018.

[8]Federal Statistical Office (2018a): Bodenfläche nach Art der Nutzung, p. 193.

[9]Federal Statistical Office (2019b): Bruttoinlandsprodukt 2018 für Deutschland, p. 10.

[10]Federal Ministry for Nutrition and Agriculture (2019).

[11]Federal Statistical Office (2018b): Statistisches Jahrbuch, p. 26.

2.2 Governmental Structure in Germany

The German Constitution of 1949, the German "Grundgesetz", states that Germany is a "social and democratic federal state" (Article 20 paragraph 1 Grundgesetz).[12] This provision of the German Constitution emphasizes that Germany is a federal state. In fact, the governmental structure could be described as highly decentralized. Hence the sixteen constituent states, which are called "Länder", have both legislative and enforcement competences.

Secondly, Article 20 paragraph 1 Grundgesetz states that Germany is a "democratic ... state". In addition, Article 20 paragraph 2 Grundgesetz emphasizes that "all state authority shall derive from the people." Legislative power rests primarily with the Federal Parliament ("Bundestag"). In cases where a Federal law has implications for the "Länder", the Council of Constituents States ("Bundesrat") must also approve the draft act. This Council of Constituents States consists of representatives of the governments of the Constituents States ("Länder").

The legislative powers are divided between the Federation and the Constituents States ("Länder"). Article 73 Grundgesetz determines the exclusive legislative competences of the Federation, such as foreign affairs, defense, currency, citizenship and others. Article 74 Grundgesetz determines "matters under concurrent legislative powers". In these cases the Federation only has the power to adopt a regulation if federal legislation is needed to ensure "equivalent living conditions" or to maintain "the economic or political unity". These requirements have been interpreted broadly. In practice, thus, the "Länder" only have legislative powers on matters listed in Article 74 Grundgesetz as long as the Federation has not itself approved laws on these matters.[13]

Although the Constitution empowers the Federal Parliament to initiate legislation, in practice, most acts are drafted and submitted by the ministries in charge. Hence, the ministries have the required manpower and expertise to formulate draft acts.[14]

As an important basic rule, the Grundgesetz stipulates the important basic rule that governmental entities are only allowed to interfere with the basic right of people to act and develop their life freely if there is an enabling provision in a formal law. Article 1 paragraph 1 GG states: "Human dignity shall be inviolable. To respect and protect it shall be the duty of all state authority". In addition, Article 2 paragraph 1 and 2 GG states: "Every person shall have the right to free development of his personality ... ". " ... These rights may be interfered with only pursuant to a law." Thus, for all restriction of individual free choice due to necessities of protecting the environment, a legal provision has to be in place.[15]

[12]An English version of the "Grundgesetz" can be found under: http://www.gesetze-im-internet.de/englisch_gg/englisch_gg.html#p0114.

[13]Maunz (1996), Sec. 32 et seqq.

[14]See https://www.bundestag.de/en/parliament/function/legislation/passage-245704.

[15]Kloepfer and Neugärtner (2016), p. 146.

Subsidiary regulations, called ordinance ("Verordnung"), may be released by the Federal Government or specific Ministries. Article 80 Grundgesetz however requires that a formal act, approved by the parliament ("Bundestag") and if needed the Council of the Constituent States ("Bundesrat"), specifically empowers the executive institutions to do so. The purpose of this constitutional provision is to ensure that the legislative entities determine the essential framework conditions and the executive only releases specifications of these "essential framework conditions". Thus, a situation in which the executive in fact takes on essential legislative powers should be prevented.

Federal Government is headed by the Chancellor who is elected by the Federal Parliament—and not by the German population directly. The Chancellor appoints and dismisses the members of his/her cabinet.

Enforcement power rests primarily with the "Länder" (Article 83 Grundgesetz). Even acts by the Federation are normally implemented, enforced and monitored by authorities of the "Länder". The Constituent States ("Länder") have a wide discretion to determine for their territory the competent authority for the implementation of a certain act.

Next to executive and legislative entities, the judiciary has an important role to ensure an effective system of checks and balances. The Constitutional Court is primarily in charge to overlook the legitimacy of actions taken by Federal or the "Länder" institutions and to review legal provisions against the constitutional requirements.

2.3 Germany and the European Union

Germany has been a member of the European Union since 1957. At present stage, the European Union has twenty-eight member states, encompassing about 510 million inhabitants. The European Union counts for one of the strongest economies in the world. When its predecessor, the Coal and Steel Union was founded in 1957, the original purpose was to create an integrated market of the then six member states by establishing the free movements of goods, capital, services and people. From a historical and political perspective, the economic partnership between France and (West-) Germany was supposed to avoid further political tensions between these two countries, which in the past had been one of the core causes of the two world wars in the last century. The first treaty of 1957 was amended and renewed several times. The current treaty is the Lisbon Treaty of 2007 ("Treaty of the Functioning of the European Union"—TFEU).[16] The treaty provisions have been named "EU primary law".

[16]English version under: https://eur-lex.europa.eu/resource.html?uri=cellar:2bf140bf-a3f8-4ab2-b506-fd71826e6da6.0023.02/DOC_2&format=PDF.

Over the years the European Union received more and more legislative powers with regard to topics such as agriculture, fishery, social policy, research, and the protection of the environment. Thus, the European Union is empowered to set legal requirements with regard to inter alia the topics mentioned above which citizens and enterprises have to comply with.[17] Next to these legislative powers, a complex organizational structure has been established. E.g. the EU Parliament is directly elected by the European population.

Furthermore, the European Court of Justice (ECJ) can in general be approached by every European National concerning the legitimacy of acts of EU organs and EU legislation. The governments of the Member States have to abide by the judgements of the ECJ. Thus, the European Union is regarded as a very specific legal structure which was named a "supranational institution".[18]

As stated, the European Union is entitled to deliver legislation on the protection of the environment, pursuant to Article 191 et seqq. TFEU. Article 288 TFEU distinguishes between two categories of legislative instruments. Whereas a *regulation* is binding in "its entity" and "directly applicable in all Member States", a *directive* is only binding with regard to its regulatory goals, leaving the Member States to decide which "form and methods" to be used in order to achieve it. Thus, directives need to be transposed into the legal system of each Member States whereas regulations are immediately binding for all European Nationals.

The European Union has been extremely active with regard to environmental policy. It is no over-statement that the environmental policy within the Member States is widely determined by European environmental legislation. A few pieces of EU environmental legislation which are also relevant for soil protection are mentioned in the following:

- Flora, Fauna, Habitat Directive of 1992[19]
  – concerning nature conservation,
- Water Framework Directive of 2000[20]
  – concerning water protection,
- Environmental Liability Directive of 2008[21]
  – concerning in particular the rehabilitation of contaminated sites,

[17]This concept is called "positive integration". For further reference see European Commission (2017), p. 37 and 100.

[18]European Commission (2017), p. 37.

[19]Council Directive 92/43/EEC of 21 May 1992 on the conservation of natural habitats and of wild fauna and flora, under: https://eur-lex.europa.eu/legal-content/EN/TXT/?uri=CELEX%3A01992L0043-20070101.

[20]Directive 2000/60/EC of the European Parliament and of the Council of 23 October 2000 establishing a framework for Community action in the field of water policy, under: https://eur-lex.europa.eu/legal-content/EN/TXT/?uri=CELEX:32000L0060.

[21]Directive 2004/35/CE of the European Parliament and of the Council of 21 April 2004 on environmental liability with regard to the prevention and remedying of environmental damage, under: https://eur-lex.europa.eu/legal-content/EN/TXT/?uri=CELEX:32004L0035.

- Industrial Emissions Directive of 2010[22]
  – concerning in alia the prevention of the creation of new contamination and
- Environmental Impact Assessment directive of 2014[23]
  – concerning the improvement of the assessment of potential effects on the environment

Most of these pieces of EU environmental legislation were delivered in form of a directive. Thus, they had to be transposed into national law. In other words, the national law has integrated the requirements established by the European Union.

In order to explain the functioning of soil protection governance in Germany, it is therefore sufficient to address the relevant national provisions. For the sake of transparency, if national provisions are based on EU law it will be highlighted in the footnotes.

The European Union has its own soil protection policy. The essential document is the European Commission's 2006 Communication "Thematic Strategy for Soil Protection"[24] which defines four main "pillars" to address soil issues in Europe:

- Putting in place a European framework legislation on soil protection,
- Ensuring that soil issues are taken into account when formulating national and European policies,
- Addressing the existing knowledge gap and
- Awareness raising.[25]

Based on this strategy, the Commission submitted the so-called "Soil Framework Directive".[26] This draft directive intended first to apply a holistic approach addressing all soil threats and second to install a complex planning system. This planning mechanism provided for an assessment of status quo, the determination of appropriate goals and the implementation of the required measures to achieve these

---

[22]Directive 2010/75/EU of the European Parliament and of the Council of 24 November 2010 on industrial emissions (integrated pollution prevention and control), under. https://eur-lex.europa.eu/legal-content/EN/TXT/?uri=CELEX%3A32010L0075.

[23]Directive 2014/52/EU of the European Parliament and of the Council of 16 April 2014 amending Directive 2011/92/EU on the assessment of the effects of certain public and private projects on the environment, under: https://eur-lex.europa.eu/legal-content/EN/TXT/?uri=celex%3A32014L0052.

[24]Communication from the Commission to the Council, the European Parliament, the European Economic and Social Committee and the Committee of the Regions - Thematic Strategy for Soil Protection, under. https://eur-lex.europa.eu/legal-content/EN/TXT/PDF/?uri=CELEX:52006DC0231&from=EN In 2012 the Commission published a report on "the implementation of the Soil Thematic Strategy and ongoing activities". See https://eur-lex.europa.eu/legal-content/EN/TXT/?uri=CELEX:52012DC0046.

[25]Further information see Altvater et al. (2014), p. 25 et seqq.

[26]Proposal for a Directive of the European Parliament and of the Council establishing a framework for the protection of soil and amending Directive 2004/35/EC, under: https://eur-lex.europa.eu/legal-content/EN/TXT/?uri=CELEX%3A32008L0056.

goals.[27] It thus follows the regulatory approach of the Water Framework Directive and the Marine Strategy Framework Directive.[28]

The draft "Soil Framework Directive" did not receive the required political support by EU Member States and was finally withdrawn by the Commission in 2014. The draft was criticized for the following reasons:

First, for not observing the principle of subsidiarity (i.e. issues should only regulated by the EU if they could not better be dealt with by the EU Member states). Second, the holistic approach was expected to lead to immediate and unacceptable requirements for the agricultural sector. Third, the reporting obligations were seen as disproportionately onerous. Fourth, the new directive was criticized for interfering with existing national regimes dealing with soil contamination, in particular with brownfields.[29]

Yet the idea of a legally binding EU instrument for soil protection has not been dropped neither by the EU Commission nor by the EU member states. The 7. Environmental Action Programme (EAP) of 2013[30] states that "the Union and its Member States should also reflect as soon as possible on how soil quality issues could be addressed using a targeted and proportionate risk-based approach within a binding legal framework".[31] Moreover, the EU Commission have set up an expert group on soil protection to discuss appropriate approaches at EU level. However, up to now, the Commission has not submitted a new draft.[32]

3 Main Drivers of Soil Degradation in Germany

The following section will discuss the main threats and drivers of soil degradation in Germany.

[27]Raffelsiefen and Strassburger (2017), p. 404.

[28]Directive 2008/56/EC of the European Parliament and of the Council of 17 June 2008 establishing a framework for community action in the field of marine environmental policy, under: https://eur-lex.europa.eu/legal-content/EN/TXT/?uri=CELEX%3A32008L0056.

[29]See Stankovics et al. (2018) for an in depth-analysis of this criticism.

[30]See: https://www.eea.europa.eu/policy-documents/7th-environmental-action-programme.

[31]See Nr. 25 of EAP.

[32]See conceptual suggestion for a new soil protection directive. The German Environment Agency 2018, under: https://www.umweltbundesamt.de/en/topics/uba-paper-the-need-for-soil-protection-legislation.

3.1 Soil Threats and Drivers: General Observations

Soil protection governance is a cross-cutting issue because almost any human activity uses land and soil.[33] In addition, almost any use of soil will cause degradation—at least to some extent. Thus, the effects on soils need to be considered with regard to basically all human activities.

Soils are vulnerable to different types of threats. The EU Thematic Strategy for Soils of 2006[34] differentiates eight types, including: soil erosion, decline in organic matter, local and diffuse contamination, sealing, compaction, and decline in biodiversity, salinization, floods and landslides.[35] Each of these threats requires a different management regime.

In addition, there are different drivers of soil degradation: Industry, agriculture, urbanization and infrastructure.[36] Some drivers of soil degradation cause mainly certain soil threats. For example, urbanization or the building of new infrastructure will primarily lead to soil sealing, soil compaction and decline of organic matter while industrial uses will primarily result in soil sealing and contamination. Agriculture on the other hand, can be linked to almost all soil threats, even soil contamination by the use of pesticides, manure or fertilizer.

From the perspective of soil governance, it is important to consider which soil driver should be regulated and which soil threat needs to be dealt with. For example, the types of actors which are accountable for the soil degradation may differ significantly. Small-scale farmers are very different in their economic resilience and technical options to industrial operators or to a municipality in charge to plan the housings and infrastructure for its community. In addition, soil threats may be either a point source (industrial site) or diffuse (agriculture).

3.2 Soil Drivers and Threats in Germany

Due to both the industrial and economic development and the population density, nature in Germany is very intensively used for human purposes. There is almost no

---

[33]One exemption is if people act in aquatic settings. For example, fishing or shipping do not necessarily have a direct effect on soils.

[34]EU Commission (2006). Additionally, eutrophication is nowadays discussed as a further threat category.

[35]EU Commission (2012) with further information on the status of degradation with regard to the various threats.

[36]FAO and ITPS (2015), p. 89 et seqq.

There may be other drivers. For example, wars normally cause tremendous negative effects on soil and land.

land left which is not cultivated or modified by human activities or due to human interests.[37]

Industrialization commenced in Germany during the middle of the nineteenth century, thus about 200 years ago. Due to lacking awareness and knowledge concerning the implications for the environment, severe contaminations of soil compartments and the local groundwater bodies on (and below) former industrial sites and landfills have been caused over the years. In Germany, about 425,000 of such brownfields have been detected so far. Due to the partly very toxic contaminants these brownfields causes strong concerns regarding human health and the protection of the environment. The costs for rehabilitation can hardly be estimated. Calculations range up to more than 500 billion Euros.[38]

Another major challenge is to ensure that the myriad of industrial facilities which are currently operated in Germany do not emit contaminants—either via pipelines, the air or directly—which finally end in soil compartments potentially causing soil contamination.

Urbanization and the necessary infrastructural measures such as streets and railways have both quantitative and qualitative implications for soils.[39] The land take in Germany is still unsustainably high with about 58 ha per day although, from 2000 to 2015, the number of daily land take has been reduced by about 50%, which is still too high.[40] Land take also includes green urban areas such as house gardens, parks and graveyards.[41] Thus, only about half of the 58 ha are affected by—partial or complete—soil sealing. Soil sealing is very detrimental as more or less all soil functions are eliminated. In addition, land take for housing or infrastructural measures often affects highly valuable soil compartments such as wetlands or even swampy areas—which is the qualitative aspect of this challenge.

The continuously intensified agriculture is a further important driver of soil degradation in Germany. Contamination by fertilizer or pesticides, erosion through floods, wind or landslides and compaction though heavy machinery are some of the major soil threats caused by agricultural practices in Germany. In addition, reduced carbon storage due to the loss of organic matter and the reduction of biodiversity on agriculturally used soils are further negative developments. Unsustainable agricultural practices such as intense livestock breeding, limited crop rotation and large-scale crop farming, unsustainable use of fertilizer and pesticides induce these negative effects.[42]

Impacts on soils in Germany caused by climate change have not yet been thoroughly scientifically assessed. Nevertheless soils are undoubtedly very

[37] Kloepfer and Neugärtner (2016), p. 1122.

[38] Kloepfer and Neugärtner (2016), p. 1123.

[39] Wunder et al. (2018), p. 47.

[40] German Environment Agency (2018), p. 14.

[41] German Environment Agency (2018), p. 14.

[42] Wunder et al. (2018), p. 47.

important with regard to climate change as they are the second largest biological sequester of carbon.[43]

At the same time, soils may be severely affected by modified weather conditions and further effects caused by climate change, such as droughts, increased water and soil erosion, decline of organic matter and biodiversity.[44]

4 Soil Protection Governance in Germany

In the following section, it will be analysed whether and how soils are protected by the German public law. The protection (or sustainable management) of soils requires—scientifically speaking—the maintenance or enhancement of the following ecological services provided by soils:

- Basis for food production and production of plants for renewable energy;
- sequester for carbon and therefore important in the fight of climate change;
- host for biodiversity;
- vital role in world's biological cycle by *inter alia* storage for nutrients or filter for groundwater from hazardous substances; and
- cultural and biological archive.[45]

From a governance point of view, two main categories of instruments need to be differentiated. One category refers to measures that are supposed to prevent the origin of soil degradation (prevention measures). The other category of measures is applied to rehabilitate or restore already occurred soil degradation (rehabilitation or restoration measures). In addition, planning instruments can be applied in order to manage and reconcile various concerns and interests.

To start with, provisions of the constitution which are particularly relevant for the protection of the environment and of soils are discussed. Then, the Federal Soil Protection Act (FSPA) is presented. The next part elucidates how provisions in other specific acts address the main drivers of soil degradation—such as for example land take. The remaining two parts of this section introduce important procedural regulations and provisions concerning the interface of soil protection and climate change.

[43] Oceans sequester more carbon than soils.

[44] LABO (2010), p. 4.

[45] Ginzky (2020), p. 37.

4.1 Constitutional Law

The German constitution (the Grundgesetz) demarks three framework conditions concerning environmental policy in general, which are obviously also relevant for soil protection policy.

First, Article 20a GG, inserted in 1994, establishes a general responsibility of the governmental institution with regard to environmental policy. It reads as follows: "Mindful also of its responsibility toward future generations, the state shall protect the natural foundations of life and animals by legislation and, in accordance with law and justice, by executive and judicial action, all within the framework of the constitutional order." The steering effect is however limited as this provision is commonly interpreted to be only one objective amongst many others. In addition, a high level of discretion how to implement this "objective" is accepted for the governmental bodies.[46] Furthermore, individuals do not have the right to justify specific claims on this constitutional provision.[47]

Second, the German Constitutional Court has decided that the basic rights do not only require that all governmental interference with the basic freedoms of the citizens need to be legally justified, but the state is also obliged to ensure that these basic rights are protected from negative effects caused by third (inter alia: private) parties. Thus in general, the state is obliged to in general ensure that private actors do not detrimentally affect the environment in a way which would raise concerns concerning for example human health.[48]

Third, the German Constitutional Court (Bundesverfassungsgericht) has interpreted the basic rights in a way that individuals in general have a right to be involved and heard in specific decision taking procedures. This is very fundamental as the Court states that individuals must be involved if they may be significantly affected by decisions concerning the use of the environment.[49]

[46]This high level of discretion has been confirmed in several judgements of the Federal Constitutional Court in recent years. For further reference see Voßkuhle (2013), p. 4.

[47]Kloepfer and Neugärtner, p. 122.

[48]See for example German Constitutional Court Judgements, BVerfGE 49, 89; BVerfGE 53, 30; BVerfGE 56, 54; BVerfGE 79, 174.

[49]Federal Constitutional Court Decision of December 12th 1979 - 1 BvR 385/77 (Mülheim-Kärlich).

4.2 FSPA

Germany is one of the few states which have a specific Soil Protection Act.[50] The Federal Soil Protection Act was adopted in 1998 and therefore 2018 marks its 20th anniversary. One year later, 1999 the Soil Protection Ordinance was put into force.

Concerning the legislative powers, it is generally agreed that the Federation has the competence to enact provisions with regard to soil protection although soil protection is not explicitly mentioned as an exclusive or concurrent legislative competence of the Federation. It is said that this competence derives from a so-called mosaic competence which is based on various topics addressed as concurrent competences—such as land law, waste disposal and economic matters.[51]

Para 1 FSPA stipulates that soil functions have to be sustainably maintained and/or rehabilitated and that to this end both reactive and precautionary measures have to be taken. The overall objective of the law is thus broad enough to encompass all drivers of soil degradation and all soil threats.

The FSPA does not protect soil for itself, but requests that the natural, historical and use functions of soils are maintained, enhanced and rehabilitated. The law specifies these three categories of soil functions:

> . . . 1. natural functions
>
> (a) as a basis for life and a habitat for people, animals, plants and soil organisms,
> (b) as part of natural systems, especially by means of its water and nutrient cycles,
> (c) as a medium for decomposition, balance and restoration as a result of its filtering, buffering and substance-converting properties, and especially groundwater protection,
>
> 2. functions as an archive of natural and cultural history and
> 3. functions useful to man as
>
> (a) a medium that holds deposits of raw materials,
> (b) land for settlement and recreation,
> (c) land for agricultural and forestry use,
> (d) land for other economic and public uses, for transport, and for supply, provision and disposal.

It is a remarkable advancement that the FSPA acknowledges soil as natural resource and the ecological functions which soil could provide for society.

When the FSPA came into force, many other important sectoral topics such as nature conservation, water protection, waste management as well as the environmental effects of industrial sites, mining and forests had already been regulated. In order to avoid overlaps and thus ambiguities, the legislator decided that the FSPA is subsidiary to eleven expressively stated sectoral acts—for example statutes on industrial sites, mining or waste management. The subsidiarity of the FSPA is

---

[50] Gesetz zum Schutz vor schädlichen Bodenveränderungen und zur Sanierung von Altlasten (Bundes-Bodenschutzgesetz - BBodSchG), 17.3.1998.

[51] Kloepfer and Neugärtner (2016), p. 1145 et seqq.

only valid as far as soil protection aspects are specifically addressed by these acts (Para 3 SPA). The scope of application of the FSPA is thus very limited.

However, court decisions have made clear that the specific requirements stipulated by the FSPA and the FSPO have to be considered whenever other sectoral acts are enforced.[52] Thus, the trigger values for certain contaminants which are determined by FSPO have to be complied with when implementing for example mining provisions.

The core operative provisions of the FSPA concern the clean-up of contaminated industrial sites including landfills.[53] The act in conjunction with the ordinance precisely describe the assessment steps in order to decide whether a site is legally to be regarded as contaminated and whether and how it needs to be cleaned up. To this end, the competences of the governmental authorities are explicitly and clearly determined. The assessment distinguishes between a first evaluation and a so-called detailed assessment. Through the first evaluation it should be investigated whether a significant suspicion for a relevant contamination is to be assumed. The necessary investigation measures for this assessment have to be enacted and funded by the competent authority. If such a significant suspicion has been confirmed, the competent authority may oblige either the land owner or the polluter (originator of the contamination) to assess the contamination in detail and examine possible options to remedy the pollution (Para 9 FSPA).

For this assessment, the so-called trigger values are of eminent importance. They are used to document the significant suspicion. Thus, they are also called "suspicion values" ("Prüfwerte"). If they are not exceeded, no significant suspicion is given. If they are exceeded, the amount as well as their location and dispersion need to be further investigated. The quantitative level of trigger values for specific contaminants mainly depends on four criteria: the characteristics of the contaminant (toxicity, bio-accumulation and persistence), the type of soil, the layer in which the contaminant is found, and the use of the area. For example, if a plot is used as a kindergarten the first 10 cm are decisive and the trigger values are very strict given that, children are very sensitive and allegedly "eat" soil from time to time.[54] For agriculturally used areas the relevant layer is down to 60 cm as plants are assumed to root to this depth.

In general, both the landowner and the polluter (e.g. operator of an industrial installation by which the contamination has been caused) can be made responsible for the detailed assessment and the remediation of the site or plot. The legal provisions do not clearly state a ranking between the polluter and the landowner with regard to their legal responsibility. Generally, it is perceived as fair to first consider the polluter.

However, the competent authority has to take into account further criteria, such as the availability of legal evidence regarding the legal responsibility of either the

[52]German Federal Administrative Court Decision (BVerwG) from April 14th, 2005, 7 C 23.03.

[53]§ 4 and 10 of the FSPA—unofficial translation.

[54]Federal Environmental Agency (2013).

landowner or the polluter as well as their respective financial capacity and the technical feasibility to assess and to clean-up the site.

As the contamination may date back to even the nineteenth century, it is often not easy to detect the responsible polluter and its legal successor. This could be a legitimate reason for the competent authority to oblige the landowner instead of the potential polluter.[55]

It has been claimed that the FSPA illegally addresses actions lying in the past (before the law got into force) as the contamination may have occurred decades ago. Courts however decided that this is justifiable if the competent authority can show that the action in the past was not in line with the legal requirements at that time.[56]

There was also a debate whether and to which extent the landowner who has not caused the contamination could be made responsible. The common assumption is that the possession of land also entails a societal obligation to ensure that no harm is caused by this piece of land. "Possession commits" is the short version of this very fundamental concept of the German constitution. The Constitutional Court however decided that this obligation of the landowner is financially limited. First, the landowner must not be obliged to invest for the clean-up of the plot when the costs exceed the value of the plot after remediation. A second limitation is that the landowner must not lose his/her general economic power due to his/her financial commitment for the plot's contamination.[57]

In Germany, security of land ownership is ensured by the so-called "land charge register" which is an official documentation of all land titles. Any sale of land title must be documented in this register in order to be legitimate. In addition, it is legally required that any contract of real estate sale must be notarized by specialized advocates.

The FSPA also entails some provisions concerning the prevention of risks, the so-called precautionary provisions. The landowner is supposed to avoid activities, which could cause the development of soil threats (Para 7 FSPA). So-called "precaution trigger values" for ten contaminants have been determined by the FSPO. If they are exceeded, the additional input per ha per year needs to be determined (Paras 9 and 10 FSPO). The numbers for the additional input per year stated by Para 9 and 10 FSPO must not be exceeded. For several reasons, the precautionary provisions are however quite weak. First, in many cases Para 7 FSPA is not applicable as other sectoral provisions supersede FSPA and FSPO pursuant to Para 3 FSPA. Second, "precaution trigger values" are determined for only ten contaminants, although there are many others. Third, Para 7 Sentence 4 FSPA states that authorities may only enforce the provision against the landowner if specific requirements are determined in the Ordinance. That means, that Para 7 FSPA can only be enforced with regard to the ten "precaution trigger values". Fourth, as specific requirements are only

[55]Kloepfer and Neugärtner (2016), p. 1184.

[56]German Federal Court Decision from September 29th, 2016, I ZR 11/15.

[57]German Constitutional Court Decision from February 16th, 2000, 1 BvR 242/91.

determined for contamination, the precautionary provisions could hardly be enforced for all other forms of soil threats—e.g. compaction, erosion and others.

In addition, the FSPA contains a provision regarding the unsealing of soils (Para 5 FSPA). As a lot of sealed plots exist, this is an important issue in Germany. However, this provision is subsidiary to the German building law and thus has almost no practical effect.[58] Furthermore, Para 8 FSPO deals with erosion caused by water. The norm precisely describes the conditions when a soil degradation by erosion could be assumed and what remedies needs to be implemented.[59]

The FSPA also includes some provisions with regard to good agricultural practices.[60] However, these provisions are not mandatory and very vague. This will be explained in more detail in the section on "agriculture".

4.3 Regulation of Specific Soil Threats and Drivers in Germany

As stated above, the FSPA very much focuses on soil contamination as it declares itself subsidiary to various other sectoral provisions as far as these laws entail soil related requirements (Para 3 FSPA). Thus, as soil degradation is caused in various forms ("threats") and by various drivers, an analysis of the governance structure needs to look at how these drivers and threats are regulated by other sectoral provisions. In Germany, the main drivers—other than the contamination by brownfields—are land take, current industrial installations, mining and waste management as well as agriculture and forests. In this section, the governance of these drivers of soil degradation is explained. In addition, the most relevant procedural provisions are discussed and some remarks on the ongoing debate concerning the interface of soil protection and climate change are presented.

4.3.1 Land Take

As stated above, land take in Germany is still high in quantitative terms. The current number is about 58 ha per day which is caused both by urbanization and by larger infrastructure measures such as the construction of streets, railways and others. From a qualitative perspective, it can be concluded that often very valuable soil is affected by land take.

Negative effects on soils by urbanization are dealt with in the Federal Building Act.[61] The core instrument of the Federal Building Act is the planning instrument for

[58] Kloepfer and Neugärtner (2016), p. 1190.

[59] See Sect. 4.3.4. in this chapter.

[60] § 17 of the FSPA.

[61] Baugesetzbuch (BauGB), 23.6.1960.

urbanization, which rests with the municipalities pursuant to Article 28 of the German Constitution.

The Federal Building Act regulates that the municipalities should establish land-use plans as soon as possible and to the extent that these are required for urban development and regional policy planning (Para 1.3 Federal Building Act). These plans determine the permissible uses (inter alia housing, industry, recreation, graveyards) within certain areas of the municipalities and the admissible form of construction (inter alia: number of stories). The plans are intended to "safeguard sustainable urban development and a socially equitable utilisation of land for the general good of the community, and shall contribute to securing a more human adequate environment and to protecting and developing the basic conditions for natural life" (Para 1.4 Federal Building Act).

With regard to soils, the Federal Building Act requires that soils should be managed economically and carefully. This provision is however not very precise and not at all prescriptive. The responsible municipalities have to weigh all the relevant interests which could result in a decision that more land and soil is required for housing or other uses, consequently leading to more land take.

Furthermore, if a town planning decision causes negative effects on nature, including soils, the act stipulates that these negative effects have to be compensated.[62] This provision is obviously of eminent importance for nature conservation and thus also potentially for sustainable soil management. However, it has also some drawbacks in and of itself.[63] Concerning soils, the provision cannot thoroughly be implemented because the assessment tools generally still focus more on the living nature and neglect soil compartments.[64] Furthermore, a thorough qualitative assessment of soils is still lacking in Germany. This shows that the societal value of soils is still underestimated.

In order to reduce the number of land used for housing, a trading system on land certificates has been investigated and discussed.[65] So far, however, such a system has not been established by law.

For most infrastructural projects, a prior permission is required according to German law. Thus, the effects on soils have to be considered according to the respective laws by the competent authority.[66] For most of these projects, an Environmental Impact Assessment (EIA) needs to be conducted which consists of an Environmental Impact Study. In addition, the EIA requires the participation of relevant authorities and the general public. The effects on soils have to be described, assessed and evaluated within the Environmental Impact Study. The general purpose

---

[62]§ 1a III BauGB.

[63]Bodle (2018), pp. 297–299.

[64]In its Decision of January 31th 1997 (4 NB 27/96), the Federal Constitutional Court determined core principles for the assessment process, thereby focusing on possible impacts on living nature and scenery.

[65]Grimski (2019).

[66]For all infrastructural projects, such as roads, railways, airports, there are specific laws.

of the EIA process is to increase the level of information concerning the potential environmental effects and to ensure that the perspectives and views of the other authorities and the general public are taken into account. However, the implementation concerning soil protection is relatively weak as specific knowledge and data is lacking with regard to the potentially affected soils. In addition, the needs to protect soils could be overruled by other economic and societal interests.

In various plans Germany has formulated objectives to reduce the daily land take to less than 30 ha per day by either 2030[67] or by 2020.[68] A further plan of the German Government requires a limit of 20 ha by 2020.[69] However, these "objectives" only state an ambition and are not to be understood as binding or prescriptive.

4.3.2 Industrial Installations

The clean-up of brownfields which originate from Germany's industrialization over the last two centuries is addressed by the FSPA. This section deals with the currently operated or planned industrial installations, which may affect soil quality, in particular by the release of contaminants.

The German Act on Emissions Control[70] requires an *ex-ante* permission for industrial sites. The permission to build industrial sites may only be granted if detrimental effects on the environment, including soils, are prevented. As for infrastructure projects, an Environmental Impact Assessment (EIA) needs to be conducted for larger industrial installations. An Ordinance, which was adopted for the implementation of this Act, lists the categories of industrial installations for which the conduction of an EIA is mandatory. The effects on soils have to be described, evaluated and assessed within the Environmental Impact Study as a part of the whole EIA process.

In 2010, the Industrial Emissions Directive (IED) of the European Union was revised.[71] New important requirements were established which were then transposed into German law. With regard to the maintenance of soil functions, it is now required that the operator - before production starts - examines and documents the status of soils beneath the industrial site. The operator is obliged to submit a so-called "baseline report" with the application for the operation with the competent authority (Article 22 para. 2 IED). Moreover, after the closure of the site, the operator has to

[67]See Climate Action Plan 2050—https://www.bmu.de/en/topics/climate-energy/climate/national-climate-policy/greenhouse-gas-neutral-germany-2050/.

[68]Germany's National Sustainable Development Strategy, see: https://www.bundesregierung.de/breg-en/issues/suostainability/germany-s-national-sustainable-development-strategy-276504.

[69]Integrated Environment Programme, see: https://www.bmu.de/themen/nachhaltigkeit-internationales/nachhaltige-entwicklung/integriertes-umweltprogramm-2030/.

[70]Gesetz zum Schutz vor schädlichen Umwelteinwirkungen durch Luftverunreinigungen, Geräusche, Erschütterungen und ähnliche Vorgänge (BImSchG), 15.3.1974.

[71]See Sect. 2.3. in this chapter.

rehabilitate the site to the "original" status of the soil and the groundwater, if significant negative effects have been caused (Article 22 para. 3 IED).[72]

The provision is both very important and innovative as it clearly sets an economic incentive for the operators of new industrial installations to avoid negative effects on soils caused. The conceptual approach is probably only applicable in cases when the soil threat is relatively limited in terms of spatial extension. It cannot be applied in case of diffuse sources or very large areas, e.g. agriculturally used plots, as the assessment and documentation of the status of larger parts of soil compartments would simply be too time-, resource- and cost-consuming.

4.3.3 Mining and Waste Management

For mining operations as well as for landfills, an ex-ante permission is required pursuant to the Federal Mining Act [73]and the so-called Federal Recycling Management and Waste Act.[74] Larger projects require an EIA. In any case, soil aspects are to be considered.

Although these acts supersede the FSPA, the "precaution trigger values" and the "suspicion trigger values" as stipulated by the FSPO have to be considered by the competent authority before granting a permission for the respective projects.[75] Although the FSPA is not directly applicable to mining and landfills, this regulatory concept ensures that the needs of soil protection with regard to contamination are considered.

Over the last years, the regulation of the environmentally sound use and recycling of so-called mineral waste has posed a major challenge to the regulator. Mineral waste can derive from construction or demolition work, from metal processing facilities as cinder or from thermal processing as ashes. Mineral waste in Germany amounts to about 240 million tons annually. Thus, it is extremely important that appropriate, meaning environmentally sound, ways of using or recycling mineral waste are determined and enforced. From a regulatory perspective, the requirements concerning the protection of the environment, and in particular groundwater and soil, need to be clearly stipulated in the relevant regulations. To this end, a draft revision of the FSPO was submitted by the competent ministry accompanied by a draft ordinance on mineral waste.[76] The revision of the FSPO is supposed to entail trigger values for more parameters.

---

[72]§ 5 IV BImSchG.

[73]Federal Mining Act, see: https://www.gesetze-im-internet.de/englisch_bbergg/index.html.

[74]Federal Recycling Management and Waste Act, see: http://www.gesetze-im-internet.de/krwg/.

[75]Federal Administrative Court (Bundesverwaltungsgerichts) Judgement 7 C 26.03, April 14th 2005.

[76]See Verordnung zur Einführung einer Ersatzbaustoffverordnung, zur Neufassung der Bundes-Bodenschutz- und Altlastenverordnung und zur Änderung der Deponieverordnung und der Gewerbeabfallverordnung, available at: https://www.bmu.de/gesetz/verordnung-zur-einfuehrung-einer-ersatzbaustoffverordnung-zur-neufassung-der-bundes-bodenschutz-und/.

The draft ordinance on mineral waste determines respective trigger values for eluates from the various types of mineral wastes. These trigger values must be based on the existing trigger values for soil and groundwater in order to ensure an effective protection of soil and groundwater. Thus, this regulatory concept differs from the current concept for mining and landfill in that the draft ordinance on mineral waste will be containing its own trigger value for each mineral waste category, which are however deviated from the trigger values of the Soil Protection Ordinance. However, the draft ordinances are still under discussion, as this transfer from quality standards for soil and groundwater to eluate trigger values is scientifically highly controversial.

4.3.4 Agriculture

In spatial terms, Agriculture is the most extended land/soil use in Germany although the contribution of this sector to GDP is very small (only 1%) and the people employed here account only for about 2% of the employed population.[77] Agriculture in Germany causes almost all soil threats, perhaps with the exception of salinization.[78]

It has to be emphasized that the contamination by fertilizers and nutrients/fertilizers (substance related soil protection) on the one hand and all other soil threats such as erosion, compaction, loss of carbon and organic matter (physical soil protection) on the other hand are regulated very differently.

Specific regulations on pesticides and fertilizers are entailed in the Federal Pesticide Act[79] and the Federal Fertilizer Act[80] in conjunction with relevant ordinances. For pesticides, two different permissions are required. At first, the active substances to be used as pesticide need to be approved at the European level.[81] Secondly, the use conditions for certain pesticide applications, for example for specific categories of wheat or for viniculture, must be permitted by German Authorities according to the Federal Pesticide Act.

For fertilizers, a permission for the use of specific fertilizers is required. In all these permission procedures, the need of an effective protection of the environment has to be assessed and considered.

In theory, these "needs" should be built on the trigger values for the quality of the relevant environmental media that are in particular soil, water and living nature.

---

[77]Federal Ministry for Nutrition and Agriculture (2019).

[78]This may change if severe droughts as in summer 2018, and 2019 will be seen continuously in Germany.

[79]Federal Pesticide Act (Pflanzenschutzgesetz) available at: http://www.gesetze-im-internet.de/pflschg_2012/index.html (German Version).

[80]Federal Fertilizer Act (Düngegesetz), available at: https://www.gesetze-im-internet.de/d_ngg/index.html (German version).

[81]See EU regulation No 1107/2009 concerning the placing of plant protection products on the market, under: https://eur-lex.europa.eu/legal-content/EN/TXT/PDF/?uri=CELEX:32009R1107&from=DE.

Thus, the product related requirements for pesticides and fertilizers need to be in line with the trigger values of inter alia the FSPA and the EU Water Framework Directive.[82] Moreover, both acts determine specific requirements on the use of fertilizers and pesticides under certain circumstances and for certain categories of locations. The general regulatory concept for pesticides and for fertilizers, which is to have a permission regime(s) in place and to determine specific use requirements based on the environmental quality standards for water, soil and nature is in essence a convincing approach. Although in theory this concept is reasonable, in practice the specific use requirements are constantly under political dispute as different views on the required level of environmental protection exist.

For other soil threats which might be caused by agriculture comparably clear and restrictive control mechanisms in public law do not exist. For these other soil threats by agriculture, Para 17 FSPA states so-called principles of good agricultural practice which reads as follows:

1. in general, the soil shall be worked in a manner that is appropriate for the relevant site, taking weather conditions into account,
2. the soil structure shall be conserved or improved,
3. soil compaction shall be avoided as far as possible, especially by taking the relevant soil type and soil humidity into account, and by controlling the pressure exerted on the soil by equipment used for agricultural soil use,
4. soil erosion shall be avoided wherever possible, by means of site-adapted use, especially use that takes slope, water and wind conditions and the soil cover into account,
5. the predominantly natural structural elements of field parcels that are needed for soil conservation, especially hedges, field shrubbery and trees, field boundaries and terracing, shall be preserved,
6. the soil's biological activity shall be conserved or promoted by means of appropriate crop rotation and
7. the soil's humus content, as is typical for the site in question, shall be conserved, especially by means of adequate input of organic substances or of reduction of the intensity with which the soil is worked.

The term "principle" is intentionally chosen to expressively clarify the legal quality of the requirements. The principles of good agricultural practices as mentioned in Para 17 FSPA are in themselves very generic. In addition, there is no mechanism to specify these principles according to the conditions of a specific area or region: The FSPA does not contain a respective enabling clause for the executive to specify the "principles" by ordinances.[83] In addition, competent authorities are not permitted to impose specific requirements—by an administrative order—on farmers regarding the compliance with the principles of good agricultural practice. In practice, the consequence of this non-competence of governmental bodies means

[82]As the current requirements of the German legislation on Fertilizers do not cope with the requirement of the Water Framework Directive, the European Court of Justice has sentenced Germany to amend its legislation accordingly.

[83]Such an enabling clause is constitutionally required to empower the executive branch to deliver subsidiary provisions. See Art. 80 of the German Constitution.

that the "principles of good agricultural practice" cannot be concretized by subsidiary decisions.

Pursuant to Para 17 sub-para. 1 FSPA, the consultative bodies of the "Länder" should assist the farmers concerning the compliance with the principles of good agricultural practices. The consultative bodies can be seen as "extension services". The law thus relies very much on this cooperative model of consultation.[84] However, in practice, the actual involvement of governmental institutions is probably quite weak as these bodies—in some "Länder"—are organized by farmer organizations themselves. In other "Länder", farmers can choose freely to request or not to request the advice of these bodies.

Next to the public law, the funding system of the Common Agricultural Policy of the European Union (EU CAP) is of certain importance concerning environmental protection. Farmers economically very much rely on this funding system as on an average about 50% of farmer's income is based on these financial resources.[85] The programmes of the EU CAP usually run for seven years and are revised accordingly.

The current programme which will terminate at the end of 2020 basically differentiates two pillars. The first pillar concerns the direct payment to farmers for certain agricultural activities which forms the major part of the whole funding system. In order to ensure that the EU wide environmental legal requirements are complied with, the cross-compliance mechanism was introduced in 2003. Farmers could—since then—only receive the direct payments if they comply with the environmental standards of the EU.[86]

In 2009, European governments agreed on the establishment of the so-called "greening" of the EU CAP which required the farmers to abide by some additional environmental standards—set by the EU and specified by the Member States—in order to be eligible for the funding.[87] From the perspective of soil protection, measures to limiting erosion by minimum soil cover, the maintenance of grassland, the crop-rotation and the ban to burn agricultural stubble supportive structure are most important.

The second pillar of the CAP concerns the EU Rural Development Policy in general. Since 2000 this second pillar also contains a chapter concerning the improvement of the environment and the countryside through support for land management as well as the mitigation of climate change (so called

[84] Ekardt (2008), p. 175.

[85] In 2018 the EU supported farmers with a total of 58.82 billion euros, including a total of 41.74 billion euros in income support. See European Commission (2019).

[86] See Council Regulation (EC) No 73/2009 of 19 January 2009 and Commission Regulation (EC) No 1122/2009 of 30 November 2009.

[87] See Regulation (EU) No 1307/2013 of the European Parliament and of the Council of 17th December 2013.

agri-environmental measures). Farmers can get payment for voluntary activities which go beyond what is legally required.[88]

The draft programme for the next period (2021–2027) is currently discussed amongst the EU institutions. The most relevant difference to the previous programme is that the EU will only set generic requirements. It will thus be the mandate of the Member States to determine more specific standards. Thus, the responsibility would be with the German Government to determine appropriate soil related standards.[89]

It could certainly be concluded that the CAP has been supportive in enhancing the protection of the environment in rural areas through its two pillars—conditioned direct payments and the agri-environmental measures. However, the whole system has been criticized for several reasons. Criticism arises because the compliance controls are only carried out in 5% of the funded farms. Moreover, experts demand that direct payments should only be granted for providing public goods; that means for actions which go beyond what is legally required. As this would be a very fundamental change—farmers so long very much depend on the funding by the European Union, the economic consequences for the agricultural sector must be taken into account. From the perspective of soil protection, several aspects have been raised: the requirements of "cross-compliance" and the "greening" are quite vague, they only address some soil threats and are limited to some landscape types.

4.3.5 Forests

Healthy soils are important for the maintenance of forest as soils store water and nutrients and serve as puffer and filter of contaminants. At the same time, forests protect soils against erosion. Healthy soils strengthen the forest's resilience against effects of climate change.

The Federal Forest Act and the forest laws of the "Länder" also have priority over the FSPA (Para 3 FSPA). It requires that forests are managed sustainably due to their economic value and their environmental relevance, in particular with regard to the nature, the climate, the water, air, soil fertility, as well as the recreation of the population. However, the Federal Forest Act is not very specific concerning the instruments to ensure an effective protection of the environmental functions of the forests. The forest laws of the "Länder" are more precise in that regard.

Notably, pursuant to "Länder" provisions which are based on a stipulation of the Federal Forest Act, forest may be cleared or converted to another use form only with an explicit permission of the competent authority ("conversion rule"). Pursuant to

[88]See Council Regulation (EC) No 1698/2005 of 20 September 2005 on support for rural development by the European Agricultural Fund for Rural Development (EAFRD), in particular Art.39.3.

[89]Raffelsiefen (2020).

Para 9 Federal Forest Act the permission shall be denied, if the maintenance of the forest is in the public interest, in particular if the forest is of eminent importance for nature conservation, for forestry use or for the recreation of the population.[90]

In addition, most forest laws of the "Länder" require a compensation in case of a clearing or the transfer to another use form of the forest.[91] The compensation could be either in form of an afforestation or intensified fostering of other forests or in the form of other fostering, protection or management measures.

So far, no specific requirements with regard to the protection of soils in forest are determined by the forest laws of the "Länder". However, the "conversion rule" is important and could potentially be used as a model for soil degradation caused by land use change, too.

4.4 Procedural Provisions

This part will shortly explain some procedural provisions which are of key importance for environment protection governance in general, but also with respect to the protection of soils. First, Environmental Impact Assessments are required for the listed categories of projects pursuant to the Environmental Impact Assessment Act.[92] The purpose of EIA in legal terms is twofold, to provide more detailed information on the environmental effects and to ensure that all relevant stakeholders are involved in the decision making process. All information must be compiled in a so-called Environmental Impact Statement (EIS). The Federal Environmental Impact Assessment Act furthermore demands the participation of all competent authorities and the general public which enables all relevant stakeholders, such as churches, unions, environmental non-governmental organizations and industry associations, to have a word.[93] The responsible authority has to consider the information provided by the "Environmental Impact Statement" and the submissions by the stakeholder, but it is not bound by these "interventions".[94]

In order to ensure transparency and accountability of the work of governmental institutions, access to environmental information is key. The Federal Environmental Information Act uses the very inclusive term "environmental information".[95] There

[90]Altvater (2014), p. 50 et seqq.

[91]See § 15 forest law of Mecklenburg- West-Pomerania. See: http://www.landesrecht-mv.de/jportal/portal/page/bsmvprod.psml?showdoccase=1&doc.id=jlr-WaldGMV2011rahmen&doc.part=X.

[92]Federal Environmental Impact Assessment Act, Annex 1, see: https://www.gesetze-im-internet.de/uvpg/UVPG.pdf.

The act transformed the EU directive on EIA.

[93]Kloepfer and Neugärtner (2016), pp. 342 and 343.

[94]Kloepfer and Neugärtner (2016), p. 346.

[95]Federal Environmental Information Act, see: https://www.gesetze-im-internet.de/uig_2005/UIG.pdf The act transformed the EU directive on access to environmental information.

is hardly any information related to environmental protection which would not fall under this term.[96] Authorities are obliged to grant access to the environmental information available to them given that there are no reasons to deny the access. Paras 8 and 9 Federal Environmental Information Act states that public interests—inter alia international affairs, internal consultation process, penal law procedures—as well as private interests—inter alia commercial confidentiality—could limit or even restrict the access to the information. Interestingly, however, information on emissions must never be regarded as confidential. The basic justification of this provision is that information on effects by emissions on people must always be disclosed.[97] The provision has been interpreted very broadly in the sense that it even encompasses information on test results of chemicals. Based on this interpretation, one could conclude that information on soil degradation always needs to be accessible as degraded soil causes effects on other uses and could thus be regarded as information on emissions. For example contaminated sites may cause contaminant spill-over on plantation or on human beings directly (in the case of kindergarten). This spill-over could be regarded as "emission". However, this has not been decided so far.

An interesting provision is entailed in the so-called Federal Environmental Harm Act[98] which has transposed the Environmental Liability Directive of 2008.[99] Pursuant to Para 10 of this Act citizens and Environmental NGO are allowed to legally request the competent authority to take the appropriate measure to avoid or to clean up soil contamination. The request needs to include the relevant information and evidence. In this case, the authority must at least formally respond to this legal request or take the appropriate measures.

4.5 Soil Protection and Climate Change

Soils and climate change are interrelated. Healthy soils can sequester carbon and thus counteract climate change. At the same time, climate change will most likely have enormous negative effects on soils through inter alia extreme weather events, like droughts or floods and/or intensified uses.

The report of the International Panel on Climate Change in 2018[100] has stressed sustainable agriculture as important and almost indispensable means to deal with climate change by potentially achieving negative emissions. Measures on negative emissions are defined as the extraction of CO_2 off the atmosphere and its

[96]See Kloepfer and Neugärtner (2016), p. 469.

[97]European Court of Justice, Judgment of the Court (Grand Chamber) of 14 February 2012 - Flachglas Torgau GmbH v Federal Republic of Germany (Case C-204/09).

[98]Federal Environmental Harm Act, see: http://www.gesetze-im-internet.de/uschadg/USchadG.pdf.

[99]https://eur-lex.europa.eu/legal-content/EN/TXT/?uri=CELEX:32004L0035.

[100]Special Report, quote.

sequestration into media of the biosphere (soils, water, oceans and vegetation) in order to reduce the atmospheric CO_2 concentrations. The IPCC concluded that measures on negative emissions are—next to maximum emission reduction—undoubtedly necessary in order to halt the world's temperature by 1.5 degree above 1990 level.

Governance basically needs to ensure both that the resilience of soils to climate change is strengthened and that the carbon uptake by soils is increased. Two main studies analysed the regulatory options to achieve the two objectives. Both studies concluded that a further soil function "with respect to climate change, in particular as sequester of carbon" should be added to the soil functions listed in Para 3 of the FSPA.[101] By this addition, it is hoped that administrative decisions in the context of town planning or infrastructure projects will more intensively take into account the carbon sequestering function of potentially affected soil compartments. From the perspective of climate change, it is particularly important that hydromorphic soils are maintained or even enhanced.[102]

In addition, the two studies proposed that FSPA should be amended to include an enabling clause, so that precautionary measures with regard to soil related climate change requirements or measures could be determined and taken.[103]

So far, these proposals have not been materialized as formal amendments to the FSPA and FSPO.

5 Soil Protection Provisions of the "Länder"

In case of concurrent legislative power, the "Länder" are only allowed to enact legal provisions as far as no federal regulation has been adopted. Thus, by the adoption of the FSPA, the available scope for legislative decisions by the "Länder" has been significantly reduced. The constitutional provision in fact means that -by law- the "Länder" can only regulate topics that have not been addressed by the FSPA.[104]

Examples of specific regulations by the "Länder" will be presented. All "Länder" have for example established soil information systems. Moreover, some Länder have established planning instruments with regard to precautionary measures for soil protection. Finally, some "Länder" laws oblige competent authorities to investigate—in the context of town planning or infrastructure projects—whether and to which extent the sealing of soil could be avoided by the use of already sealed areas.[105]

[101]LABO (2011), p. 7 et seqq. and Willand et al. (2014), p 20 et seqq.

[102]LABO (2011), p. 9 et seqq. and Willand et al. (2014), p. 34 et seqq.

[103]LABO (2011), p. 27 et seqq. and Willand et al. (2014), p. 38 et seqq.

[104]Altvater et al. (2014) used the illustrative term "gap filling competence".

[105]See § 4 FSPA of North Rhine-Westphalia, under: https://recht.nrw.de/lmi/owa/br_bes_detail?sg=0&menu=1&bes_id=4898&anw_nr=2&aufgehoben=N&det_id=372782.

As stated above, pursuant to Article 83 et seqq. of the German Constitution, the "Länder" generally are entitled to implement and enforce federal and their own acts. To this end, the "Länder" determine for their territory the competent authority.

First, the "Länder", thus, need to decide in which branch (water, soil, industrial installations) the responsible authority should be located. Second, the appropriate level of administration must be determined. Usually, the administration in the Länder has three levels of competent authorities, starting from a local authority, to a middle authority up to the ministries which usually serve as the highest level.[106] Third, the "Länder" have to codify which other authorities should be involved in a certain decision process and whether this entity should have the right to put a veto or only have the right to be consulted. This determination of responsibilities and competences has to be made by laws and by the "Länder".

To provide an example: Each of the "Länder" has to determine which authority—branch and level—is the leading authority for brownfields caused by past illegal landfills. In addition, it needs to be determined which other bodies should be involved, for example the water and/or the waste management authorities. Finally, whether these bodies have the right to veto or just to provide comments.

By these determinations, the division of labour and the specific competences of the various governmental bodies as well as their responsibilities can be clarified.

These determinations only address procedural aspects, that means how a legal requirement is implemented and enforced. However, they are extremely important in order to achieve an effective soil governance. It needs to be ensured that the competent authority has the required expertise, that the division of labour is effective (no overlaps, no negative competition) and that all required bodies are involved. From the perspective of the citizens, these determinations by law ensure transparency and accountability.

6 International Soil Governance Related Obligations of Germany: Short

In any case and also in the field of soil protection the national policies need to comply with the international obligations. The relevant international provisions on soil protection are entailed in the following three treaties:

- United Nations Framework Convention on Climate Change (UNFCCC) and the 2015 Paris Agreement,[107]

See also § 3 Soil Protection Law of Hesse, under: https://umwelt.hessen.de/sites/default/files/media/hmuelv/haltbodschg_2007.pdf.

[106]Aichberger and Weber (2019).

[107]The Text of UNFCCC could be found under: www.unfccc.int The text of the Paris Agreement could be found under: http://unfccc.int/resource/docs/2015/cop21/eng/l09r01.pdf.

- Convention on Biological Diversity (CBD)[108] and the
- United Nations Convention to combat Desertification in those Countries experiencing serious droughts and/or desertification, in particular in Africa (UNCCD)[109]

The existing international regulations on soil governance can be described as fragmented. None of the treaties provides a comprehensive regime which addresses all soil threats. UNFCCC focuses on climate related issues, CBD on the conservation of biodiversity and UNCCD mainly deals with soil/land degradation as a consequence of desertification processes.[110] Even together, there is no comprehensive regulatory regime in particular as the mandates and the way of cooperation do not allow for a coherent approach.[111] Finally, the treaties entail almost no soil specific obligations, in particular concerning the enforcement instruments. UNCCD which relates expressively to land only requests the State Parties which are affected by land degradation to set up national action programmes.[112]

The Ramsar Convention on Wetlands[113] also deals with soil protection but its regional scope of application is very limited. Moreover, the Convention addresses the sustainable management of wetlands as a whole and does not specifically deal with soil issues.[114]

The German Constitution follows in general a dualistic approach (Article 25 and 59 Grundgesetz). Thus, international obligations need to be transposed into national law in order to become effective internally (Article 59 Grundgesetz). Germany is party to all four international treaties (UNFCCC, CBD, UNCCD and Ramsar Convention). These international treaties have been transposed into national law. However, as they are all not very specific, they do not have an immediate effect on soil governance within Germany.[115]

Finally, Germany is also Party to the Alpine Convention.[116] The Soil Conservation Protocol of 1998 of this Convention contains specific soil related stipulations for the Alpine region. It addresses the drivers of soil degradation and soil threats which are relevant for the region such as traffic, agriculture and tourism. Whereas most provisions are quite vague, some are very precise. As self-executing provisions, they have to be considered by the competent authorities and form an integral part of Germany's soil governance.[117] The Soil Conservation Protocol is a region-specific

[108]The Text of CBD could be found under: www.cbd.int.

[109]The Text of UNCCD could be found under: www.unccd.int.

[110]Boer et al. (2016), p. 58 et seqqu. See also Bodle et al. (2020), p. 40.

[111]Bodle et al. (2020), p. 126 passim.

[112]Bodle et al. (2020), p. 52.

[113]The Text of the Ramsar Convention could be found under: http://www.ramsar.org/.

[114]Bodle et al. (2020), p.73 passim.

[115]Durner (2019), Sec. 147.

[116]See Preamble of the 1995 Alpine Convention, available at: https://www.alpconv.org/.

[117]Bodle et al. (2020), p. 78 passim et seqq.; Markus (2015).

instrument as it responds to the environmental necessities of the Alpine region as a specifically important and fragile ecosystem.

7 Implementation of Land Degradation Neutrality Objective

On September 25th 2015, the UN General Assembly adopted "the 2030 Agenda for Sustainable Development".[118] The Agenda includes seventeen SDGs and 169 accompanying targets. Importantly, the Agenda stipulates commitments for both developing and developed countries.

Several SDG have implications for soil protection. In fact, they need to be considered in their complexity in order to achieve the ambitions of the 2030 Sustainable Development Agenda.[119] SDG 15 and Target 15.3 are the most pertinent ones as SDG 15 requests "to halt and reverse land degradation" and target 15.3 demands that states "... by 2030, combat desertification, and restore degraded land and soil, including land affected by desertification, drought and floods, and strive to achieve a land-degradation neutral world."

The objective of a "land degradation neutral world" (or: "land degradation neutrality"—LDN) has been increasingly perceived as a kind of a leitmotiv of international and national soil governance.[120] Thus, in the following, it is analysed whether and to which extent the LDN objective has been implemented by the German soil protection governance. To begin with, the content of LDN objective and the fundamental steps required to implement it will be exemplified.

7.1 LDN: What Is It?

The UNCCD, which has established itself as the international lead organization for the LDN objective,[121] has unanimously interpreted the term as follows:

> Land degradation neutrality is a state whereby the amount and quality of land resources necessary to support ecosystem functions and services and enhance food security remain stable or increase within specified temporal and spatial scales and ecosystems.[122]

---

[118] https://www.un.org/sustainabledevelopment/development-agenda/.

[119] A thorough analysis of all SDG with regard to soil management is provided by Ehlers in this volume.

[120] Boer et al. (2016), p. 62; Bodle et al. (2020), p. 37 passim. Although the 2030 Agenda, including the SDGs, are not legally binding, they are politically important as the Agenda was endorsed by all States.

[121] Boer et al. (2016), p. 63.

[122] UNCCD 2015.

As soil is an essential element of land, the LDN objective is directly applicable to soils.[123] In addition, the definition underlines that the "amount and quality" has to "remain stable to increase." Or in other words: The amount and quality of soil must not diminish. As land/soil degradation can not be completely avoided, restoration or rehabilitation must be in place in order to achieve neutrality.[124]

UNCCD has developed the so-called "conceptual framework" which includes recommendations on the necessary measures to achieve LDN. UNCCD's "LDN response hierarchy"[125] entails three necessary categories of actions: "avoid", "reduce" and "reverse".[126] "Avoid" intends that degradation is prevented, in other words, that no degradation occurs. The term "reduce" is supposed to require actions ensuring that the negative effects of an activity for soils are minimized to the lowest level possible. "Reverse" should finally include measures to restore or to rehabilitate degradation processes. In the view of UNCCD "restoration" seeks to "re-establish the pre-existing biotic integrity" and "rehabilitation" is intended to "reinstate the ecosystem functionality with the focus on provision of goods and services".[127] Finally, pursuant to the "conceptual framework" "a pro-active focus on planning" needs to be in place by which "anticipated losses" should be counterbalanced by "planned gains".[128] Thus, these four categories of required measures could be used to assess the current law whether and to which extent LDN is already implemented.

7.2 *Implementation of the LDN Objective in Germany*

An effective implementation of a SDG in general and of the LDN objective in particular depends on two core instruments. First a functioning measurement including indicators must be in place in order to assess the overall trend. Second, it needs to be analysed whether appropriate legal actions have been established to achieve the LDN objective. Finally, the "intervention clause" of the Nature Conservation Act should be presented as a paradigmatic approach for the implementation of the LDN objective.

[123]Ginzky (2020), p. 42.

[124]Ehlers (2016), p. 75.

[125]In addition many other matters have to be clarified, defined and arranged, i.a. the following points: A baseline has to be defined, the status of soil/land have to be assessed, indicators for the assessment of the status, of land degradation and of restoration/rehabilitation actions as well as monitoring requirements have to be defined. Further information see Ehlers (2016), Minelli et al. (2016) and Cowie (2018).

[126]Cowie (2018), p. 30.

[127]Cowie (2018), p. 31.

[128]Cowie (2018), p. 31.

7.2.1 Germany's National Sustainable Development Strategy

The first sustainable development strategy was adopted in 2002 and was regularly updated since then. The latest one of 2016 was significantly revised in order to align with the 2030 sustainability agenda. This strategy was partly revised and modified in 2018.[129]

Germany's *National Sustainable Development Strategy* of 2016 includes an indicator on land take ("Built-up area and transport infrastructure expansion"). The government has adopted the objective to reduce the expansion of built up area and infrastructure to less than 30 ha a day by 2030.[130] Moreover, the strategy lists indicators such as nitrogen surplus, organic farming area, and species diversity, which do not specifically and solely address soil protection, but which are relevant.[131]

Most importantly, the German government has requested the ministries in charge to jointly develop a specific soil related indicator.[132] Thus, one could conclude that a comprehensive and coherent measurement system concerning the implementation of the LDN objective has not yet been set up.

7.2.2 Legal Instruments

As stated, the four action categories mentioned in the UNCCD's "LDN response hierarchy"—"avoid", "reduce", "reverse" and "plan"[133]—should be used in order to assess the existing soil protection governance in Germany.

It needs to be highlighted that—after the adoption of the 2030 Sustainable Development Agenda—the German soil protection governance has not been revised and amended with the aim to align the existing provisions with the LDN objective. Thus, the soil governance in Germany remains basically unchanged with regard to the 2030 Sustainable Development Agenda.

Nevertheless, the German soil legislation already contains actions, measures and provisions for the first three categories of the UNCCD response hierarchy—"avoid",

[129]For further information see https://www.bundesregierung.de/breg-en/issues/sustainability/germany-s-national-sustainable-development-strategy-354566.

[130]National sustainable development strategy, p. 212. See under: https://www.bundesregierung.de/resource/blob/992814/730844/3d30c6c2875a9a08d364620ab7916af6/deutsche-nachhaltigkeitsstrategie-neuauflage-2016-download-bpa-data.pdf?download=1.

[131]See https://www.bundesregierung.de/resource/blob/992814/730844/3d30c6c2875a9a08d364620ab7916af6/deutsche-nachhaltigkeitsstrategie-neuauflage-2016-download-bpa-data.pdf?download=1.

[132]See National sustainable development strategy, p. 228.

[133]It could be stated that the categories prevent (1), restore or rehabilitate (2), offset (3) and (4) plan would be more appropriate and clearer. See Bodle (2018) and Ginzky (2020).

However, the UNCCD in fact addresses the same aspects but attributes them differently to the four categories.

"reduce" and "revise". Under the category "avoid" one could group provisions which prohibit certain uses of pesticides or fertilizers which would otherwise cause negative effects on soils.[134] As measures to "reduce" the permission regime for industrial installations or for landfills could be mentioned.[135] "Revise" actions are for example the obligation of the landowner or the polluter to clean up contaminated soil compartments or the commitment of the operator of an industrial site to re-establish the original status of soil and groundwater after the closure of the site.[136] However, although those instruments exist, they do not ensure "neutrality" of degradation and restoration.

Furthermore, on an area level (beyond the project level), there is no planning instrument to manage the achievement of a net balance between ongoing degrading activities and restoration as well as rehabilitation actions.[137] This is the fourth category of the UNCCD response hierarchy.

Thus, although many relevant provisions are in place, the LDN objective is not completely implemented in Germany due to the lack of appropriate legal provisions.

7.2.3 The "Intervention Clause" as a Paradigmatic Approach

In theory, the "intervention clause" of the Federal Nature Conservation Act conceptually entails the first three elements of the UNCCD's response hierarchy. Pursuant to the legal definition in the Federal Nature Conservation Act, "nature" also encompasses soils. Thus, in theory, the intervention clause is also applicable to human actions which cause the degradation of soils.[138]

The intervention rule stipulates a hierarchical, 4 step—approach: (1) Interventions on nature have to be avoided, if possible. At least the negative effects have to be minimized. That relates to the first two elements of the UNCCD response hierarchy "avoid" and "reduce". (2) Unavoidable interventions have to be compensated by measures that establish either (primarily) "similar" or "equivalent" nature functions. (3) If the compensation is not feasible, the authority has to decide whether the interests of nature protection or the project prevails. (4) In the latter case, the project operator has to pay monetary compensation. Step 2 to 4 relates to element three of the UNCCD response hierarchy "reverse". In case of monetary compensation, these resources could be used for future restoration projects.

This clause basically requires that unavoidable interventions in nature are offset by balancing or substitution measures. Thus, if compensation is ensured, neutrality is

[134] See above section.

[135] See above section.

[136] See above section.

[137] A detailed analysis to which extent existing provisions implement the LDN concept concerning erosion, housing and settlements as well as industrial pollution was carried out by Bodle and Stockhaus (2018).

[138] Kloepfer and Neugärtner (2016), p. 1135.

guaranteed. Thus, the intervention clause would be in theory an almost perfect instrument to guarantee "neutrality".

However, in practice, the implementation of the intervention does not achieve its purpose with regard to the protection of soils mainly for two reasons.

First, the intervention clause does not require a compensation of the same nature function, which has been degraded by the project.[139] Guidance documents of the "Länder" entail a measurement system, which attributes numbers to specific nature interventions depending on the level of the negative effects. Rehabilitation must achieve the same number in order to be considered a sufficient compensation in legal terms, irrespective the type of nature function. Second, the existing guidance documents of the "Länder" neglect to a certain extent soil functions as a core element of nature and focus more on the living "nature" such as flora and fauna.[140]

Thus, in practice the intervention clause does not ensure an effective implementation of the LDN objective. Nevertheless, lessons could be learnt from the current implementation of the intervention clause with regard to timing of compensation measures and spatial linkage of degradation and compensation. Concerning timing, it is required that the measures can be compensated "within a reasonable time" which means that the compensatory measures should be carried out at the same time as the intervention. Concerning the spatial proximity, courts have requested that there has to be a "spatial-functional" link between the intervention and the compensatory measures.[141]

Finally, it is stated that agricultural measures which are in line with good agricultural practice are not regarded as an intervention in a legal sense. Generally speaking, this seems to make sense, as the good agricultural practice is supposed to document what is an environmentally sound agriculture. Moreover, otherwise every agricultural action would have to be evaluated by an authority. However, as seen before, the principles of good agricultural practices are so vague that they have almost no steering effect.[142]

As the intervention clause clearly and only addresses the project level, no planning on an area level is entailed.

---

[139]Federal Administrative Court (Bundesverwaltungsgericht) judgment of 10.09.1998 - 4 A 35/97; OVG (Higher Admistrative Court) Koblenz, judgment of 06.06.2000 - 8 C 11556/98.

[140]For example some of these guiding documents focus on the assessment and evaluation of existing biotopes, see i.a. North Rhine-Westphalia (https://www.lanuv.nrw.de/natur/eingriffsregelung/numerische-bewertung-von-biotoptypen/) and Saxony-Anhalt (https://lvwa.sachsen-anhalt.de/das-lvwa/landwirtschaft-umwelt/naturschutz-landschaftspflege-bildung-fuer-nachhaltige-entwicklung/eingriffsregelung/).

[141]Schmidt et al. (2014), p. 434; BVerwG (Federal Administrative Court) judgment of 10.09.1998 - 4 A 35/97.

[142]See Sect. 4.3.4.

8 Strength and Weaknesses of German Soil Governance: Lessons Learnt

This section will provide an assessment of the strengths and weaknesses of German soil protection governance. Based on this assessment, some recommendations in the sense of lessons learnt will be presented.

It has to be emphasized that to establish an effective soil governance regime which complies with the aspirations of sustainability is for several reasons very demanding. First, almost every human activity affects soils and almost inevitably causes—to a certain extent—the degradation of soils. Agriculture, forestry, viniculture, streets, even parks or graveyards, all these uses have—mostly detrimental—effects on soils. Thus, soil is a cross-cutting issue which needs to be considered for various sectoral regulations. Second, soil threats are very different with regard to the challenges they pose. This fact becomes even more complex as soil in itself significantly differs too. Third, the ecological services and functions of soils and in particular fertile soils have been neglected for a long time. Thus, the importance of soils for the society has been increasingly recognized only within the last twenty years and—as a consequence—specific soil protection governance came late.

From a regulatory perspective, that means that the then new soil provisions needed to be aligned with the already existing environmental regulations.

To begin with, it has to be highlighted as a remarkable advancement that the FSPA acknowledges soil as natural resource and the ecological functions which soil can provide for society.

In addition, the regulatory concept of the FSPA to declare itself subsidiary to other already existing sectoral provisions—as far as these sectoral provision deal with soil aspects—seems to be reasonable. Thereby the existing implementation tools could be used—at least in theory—for establishing an effective soil governance. In addition, German jurisprudence has clarified that the specific scientific requirements of the FSPA and FSPO, in particular the trigger values for contaminants, must be considered with the implementation of the other sectoral provisions.

This combination of the subsidiarity of the legal provisions of the FSPO, the enforcement of soil aspects via the other sectoral acts and the obligation to consider the scientific requirements of the FSPA/FSPO seems to be reasonable to avoid overlaps and unnecessary and over-complex procedures.

The analysis has shown that the German soil governance is particularly strong with regard to soil contamination. The FSPA clearly stipulates scientific requirements which are legally binding. In addition, the FSPA clearly defines the responsibilities of competent authorities and the private "culpable"—either the landowner or the polluter.

Moreover, the scientific requirements, the trigger values, need to be considered for the implementation of other sectoral norms, as in the case of mining law, or are taken as a basis to from which product related specifications can be derived, as in the case of pesticides or fertilizers.

Thus, as a first lesson learnt it could be concluded that an effective soil protection governance needs clear specifications of scientific requirements in a legally binding manner. In addition, as a second lesson learnt, clarity of the responsibilities of the competent authority and the private actors clearly supports the effectiveness of soil protection governance.

Compared to soil contamination, no similar clear scientific requirements have been determined for other soil threats. The principles of good agricultural practices are very generic and there is no mechanism to specify them—neither by an ordinance delivered by the ministries in charge nor by administrative orders of the competent authorities. Thus, in order to have an effective regime for the other soil threats as well, it seems to be recommendable to enable the competent authorities of the "Länder" to concretize these principles site-specifically—by an amendment to the FSPA. This is the third lesson learnt.

More generally, it seems that the steering effects of non-binding objectives such as the principles of good agricultural practices, the land take reduction objectives (20 or 30 ha by 2020/2030) or the LDN objective is low. Given the importance to maintain soil quality, it needs to be seriously considered whether certain objectives should be imposed as mandatory requirements—which is lessons learnt number four.

The regulatory approach with regard to currently operated industrial installations, to request a documentation of the status of soils at the beginning of the operation combined with the duty to re-establish the original status after the closure of the site, is certainly very sensible. It factually enacts an economic incentive for the operators of the industrial site to avoid negative effects on soils and groundwater. That means that such an approach is generally recommendable.

However, this concept only works for point sources as the documentation of the status-quo and the assessment of effects after the closure for spatially extended areas would not be feasible. Thus, this concept is probably not appropriate for very diffuse and wide-spread soil contamination caused for example by the use of pesticides or fertilizers (lessons learnt number 5).

As climate change could severely affect soil quality and soils as carbon sequester are extremely important for climate change mitigation and adaptation, it is to be recommended that this ecological service of soils is explicitly addressed in Germany's soil protection governance, probably at best in the FSPA. Such a statement could trigger the development of relevant scientific requirements and should motivate the competent authorities to consider this aspect when implementing the provisions (lessons learnt number 6).

As outlined, the "Länder" precisely determine by law the competences of the authorities, the division of labour, the responsibilities and the involvement of other bodies. Thereby, the effectiveness of the work of authorities and—from the perspective of citizens—transparency and accountability are ensured.

It should be stressed that from the perspective of effective soil protection governance the specification of the competences of the authorities and the relevant procedures, including the involvement of other authorities and the general public, is similarly important as the substantial regulations (lessons learnt number 7).

With regard to agriculture, it seems to be important to at least strengthen the extension services by giving them a clear mandate and sufficient financial resources and manpower (lessons learnt number 8).

Procedural regulations such as free access to environmental data and the necessity of an EIA are also of eminent importance. In this chapter it was not feasible to substantiate this assumption by a sociological assessment. However, it seems obvious that the probability of an effective soil protection governance increases if the general public or civil society organizations could control the acts of governmental entities. In this context the provision of the German Act on Environmental Harm, which empowers civil society organizations to formally request a justification from the competent authority as to why it has dealt with a contaminated site in a specific way is remarkable (lessons learnt number 9).

The implementation of the LDN objective is so far weak in Germany. Although the "intervention clause" of the Federal Nature Conversation Act in theory entails almost perfectly the first three elements of the UNCCD "LDN response hierarchy", in practice the steering effect in order to achieve the LDN objective is low. It seems to be logical to consider a "pure" soil related intervention clause as the recognition of soils within the "intervention clause"—based on the Federal Nature Conservation Act—has been debated in Germany for more than a decade—with limited success. It could be argued that such a topic-specific "intervention clause" is not without precedent in German law, as the "conversion clause" of the forest acts of the "Länder" only refers to forestry (lessons learnt number 10).

With regard to the LDN objective two additional instruments seem to be necessary. First, it would make sense to establish a permission regime for all substantial land use changes. By this permission procedure, competent authorities would get legal means to ensure that the effects on soils are minimized.

Moreover, such a permission regime should be combined with a compensation requirement by which the just mentioned soil related intervention clause could be operationalized (lessons learnt number 11). Finally, it seems to be instrumental to impose the LDN objective as a mandatory requirement. Moreover, the net balance of ongoing degradation and restoration/rehabilitation measures needs to be ensured by planning instruments on an area level (lessons learnt number 12).

9 Outlook

As shown, the analysis of legal instruments for soil protection requires a very broad approach as soil protection is a cross-cutting issue and is addressed in various sectoral legislation.

The above analysis has clearly disclosed that clear scientific requirements are most important for an effective soil governance. In Germany, they primarily exist for soil contamination. For the other soil threats, in particular with regard to agricultural uses, these specific requirements must be worked out and adopted as binding standards. Moreover, effective instruments need to be established for the implementation of the LDN objective.

The ambition of the recommendations in this chapter is certainly not just to comply with the internationally adopted Sustainable Development Goals. The overarching reason is that sustainable management of soils is required to be able to cope with fundamental challenges to humankind in the close or intermediate future—climate change, extreme whether events, hunger, poverty, migration. Considering these challenges to keep the status quo of land quality and degradation—which neutrality is of essence—can only be the first step. Thus, we need to see how the quality of soil compartment can be improved in the long run.

References

Aichberger T, Weber K (2019) Verwaltungsbehörden (Aufbau). Creifelds, Rechtswörterbuch, 23. Edition 2019

Altvater S, Dooley E, Roberts E (2014) Legal instruments to implement the objective of "land degradation neutral world" in international law. Umweltbundesamt, Texte 09/2015. Available at: https://www.umweltbundesamt.de/sites/default/files/medien/378/publikationen/texte_19_2015_legal_instruments_to_implement_the_objactive_land_degradation_neutral_world.pdf

Bodle R (2018) Implementing land degradation neutrality at national level: legal instruments in Germany. Int Yearb Soil Law Policy 2:287–308

Bodle R, Stockhaus H (2018) Geeignete Rechtsinstrumente für die nationale Umsetzung der bodenbezogenen sustainable development goals, insbesondere des Ziels einer „land degradation neutral world", Berlin

Bodle R, Stockhaus H, Oberthuer S, Wolff F (2020) Improving international soil governance – analysis and recommendations, Berlin

Boer B, Ginzky H, Heuser I (2016) International soil protection law: history, concepts and latest developments. Int Yearb Soil Law Policy 1:49–72

Cowie AL et al (2018) Land in balance: the scientific conceptual framework for land degradation neutrality. Environ Sci Policy 79:25–35

Durner W (2019) Umweltverfassungsrecht, 1. Umweltvölkerrecht (Sec. 147–149). In: Beckmann M, Durner W et al (eds) Landmann Rohmer Umweltrecht. Band 1. 89 Ergänzungslieferung, Stand 1. Februar 2019, C.H. Beck

Ehlers K (2016) Chances and challenges in using the sustainable development goals as a new instrument for global action against soil degradation. Int Yearb Soil Law Policy 1:73–84

Ekardt F, Heym A, Seidel J (2008) Die Privilegierung der Landwirtschaft im Umweltrecht. Zeitschrift für Umweltrecht (ZUR) 2008 Heft 4, 169

EU Commission (2006) Communication: Thematic Strategy for Soil Protection - COM(2006)231 final. Available at: https://ec.europa.eu/environment/archives/soil/pdf/SEC_2006_620.pdf

EU Commission (2012) Report: The implementation of the Soil Thematic Strategy and ongoing activities, COM (2012) 46 final. Available at: https://eur-lex.europa.eu/legal-content/EN/TXT/?uri=CELEX:52012DC0046

EU Commission (2017) The ABC of EU law. Available at: https://publications.europa.eu/en/publication-detail/-/publication/5d4f8cde-de25-11e7-a506-01aa75ed71a1

European Commission (2019) The common agricultural policy at a glance. Available at: https://ec.europa.eu/info/food-farming-fisheries/key-policies/common-agricultural-policy/cap-glance_en#howitspaidfor

Federal Environmental Agency (2013) Umwelt und Kindergesundheit. Available at: https://www.umweltbundesamt.de/sites/default/files/medien/377/publikationen/umwelt_und_kindergesundheit.pdf
Federal Ministry for Economic Affairs and Energy (2018) Facts about German foreign trade. Available at: https://www.bmwi.de/Redaktion/EN/Publikationen/facts-about-german-foreign-trade.html
Federal Ministry for Nutrition and Agriculture (2019) Arbeitskräfte in der Landwirtschaft (Tabellen zur Landwirtschaft). Available at: https://www.bmel-statistik.de/landwirtschaft/tabellen-zur-landwirtschaft/
Federal Statistical Office (2018a) Bodenfläche nach Art der tatsächlichen Nutzung - Fachserie 3 Reihe 5.1 – 2017
Federal Statistical Office (2018b) Statistisches Jahrbuch 2018 - Kapitel 2: Bevölkerung, Familien, Lebensformen
Federal Statistical Office (2019a) Exports and imports (special trade) by division of the national Product Classification for Production Statistics 2018. Available at: https://www.destatis.de/EN/Themes/Economy/Foreign-Trade/Tables/imports-exports.html
Federal Statistical Office (2019b) Bruttoinlandsprodukt 2018 für Deutschland. Begleitmaterial zur Pressekonferenz am 15. Januar 2019 in Berlin
Food and Agriculture Organization of the United Nations, Intergovernmental Technical Panel on Soils (2015) Status of the World's Soil Resources (SWSR) – Main Report, Chapter 5 – Drivers of Global Soil Change
German Environment Agency (2018) The need for soil legislation at European level
Ginzky H (2020) Good governance of "sustainable management of soils" at national and international level: How to do it? In: Yahyah H, Ginzky H, Kibugi R, Kasimbazi E, Ruppel O (eds) Legal instruments for sustainable soil management in Africa. Springer, Berlin, pp 35–54
Grimski D (2019) Tradable land planning certificates to reduce land take: results of a simulation game with communities in Germany. In: Ginzky H, Dooley E, Heuser I, Kasimbazi E, Markus T, Qin T (eds) International yearbook of soil law and policy 2018. International yearbook of soil law and policy, 2018. Springer, Cham.
Kloepfer M, Neugärtner RD (2016) Umweltrecht, 4. Auflage, C.H. Beck
LABO (Advisory body on soils of federal and Länder ministries of environment, department soil protection) (2010) Positionspapier – Klimawandel. https://www.labo-deutschland.de/documents/LABO_Positionspapier_Boden_und_Klimawandel_090610_aa8_bf5.pdf
LABO (Advisory body on soils of federal and Länder ministries of environment, department soil protection) (2011) Möglichkeiten der rechtlichen Verankerung des Klimaschutzes im Bodenschutzrecht. https://www.labo-deutschland.de/documents/BORA-Stellungnahme_zu_Klimawandel-Bodenschutzrecht_Veroeffentlichung_Nov_2011_7a0.pdf
Markus T (2015) Verbindlicher internationaler Bodenschutz im Rahmen der Alpenkonvention. Zeitschrift für Umweltrecht:214–221
Maunz T (1996) Art. 74 Grundgesetz. In: Maunz T, Dürig G (Founders), Herdegen M, Klein H et al (eds) Grundgesetz Kommentar, 87. Ergänzungslieferung März 2019
Minelli S, Erlewein A, Castillo V (2016) Land degradation neutrality and the UNCCD: from political vision to measurable targets. Int Yearb Soil Law Policy 1:85–104
Raffelsiefen (2020) The EU Common Agricultural Policy and its contribution to the protection and sustainable use of soils: new approaches for the period 2021–2027. Int Yearb Soil Law Policy 2:?? - ??
Raffelsiefen M, Strassburger T (2017) The protection of soil: does the European Union live up to its own ambitions? In: Ginzky H, Heuser I, Qin T, Ruppel O, Wegerdt P (eds) International yearbook of soil law and policy 2016. Springer, Cham, pp 389–410
Schmidt R et al (2014) Umweltrecht, 9th edn. Berlin
Stankovics P et al (2018) Identifying gaps between the legislative tools of soil protection in the EU Member States for a common European soil protection legislation. Sustainability 10.8 (2018):2886

Voßkuhle A (2013) Umweltschutz und Grundgesetz. Neue Zeitschrift für Verwaltungsrecht 2013. Heft 1:1–8

Willand A, Burkmeister D, Höke S, Kaufmann-Boll C (2014): Erarbeitung fachlicher, rechtlicher und organisatorischer Grundlagen zur Anpassung an den Klimawandel aus Sicht des Bodenschutzes: Teilvorhaben 1: Erarbeitung der fachlichen und rechtlichen Grundlagen zur Integration von Klimaschutzaspekten ins Bodenschutzrecht

Wunder S, Kaphengst T, Freilih-Larsen A, MacFarland K, Albrecht S (2018) Land Degradation Neutrality, Berlin

The EU Common Agricultural Policy and Its Contribution to the Protection and Sustainable Use of Soils: New Approaches for the Period 2021–2027

Markus Raffelsiefen

1 Introduction

Soil is under increasing pressure in the European Union,[1] with threats varying between Member States. The Soil Thematic Strategy of the European Commission[2] mentioned soil erosion, soil organic matter decline, decline in soil biodiversity, salinisation, soil compaction and floods/landslides as well as soil contamination and soil sealing as the main threats to soil.

Discussions on whether these threats should be addressed by a comprehensive, dedicated legal framework at the Union level are going on despite the official withdrawal of the Commission proposal for a Soil Framework Directive in 2014.[3] No outcome is in sight in the short run. While a certain number of soil-related provisions exist in European Union law,[4] other environmental media, namely water and air, are subject to a much more comprehensive and coherent set of rules (Industrial Emissions Directive, Water Framework Directive, etc.).

One field in which legislative activity at the Union level regularly stirs up hopes for a better level of environmental protection is the Common Agricultural Policy (CAP), being reviewed every seven years in connection with the adoption of the Multiannual Financial Framework of the European Union.

[1]European Environment Agency (2015); Joint Research Centre (2012).

[2]COM(2006)231 final, p. 2.

[3]Raffelsiefen and Strassburger (2017), pp. 403, 405.

[4]Raffelsiefen and Strassburger (2017), pp. 391ss.

M. Raffelsiefen (✉)
Federal Ministry for the Environment, Nature Conservation and Nuclear Safety, Division 'Soil Conservation Legislation and Remedial Soil Conservation; Mining Law', Bonn, Germany
e-mail: markus.raffelsiefen@bmu.bund.de

H. Ginzky et al. (eds.), *International Yearbook of Soil Law and Policy 2019*,
International Yearbook of Soil Law and Policy 2019,
https://doi.org/10.1007/978-3-030-52317-6_16

There is a somewhat special relationship between soil protection and the relevant players in agricultural policy. Soil is the basis for agricultural production. Thus, caring for soil quality serves the very own interest of farmers, landowners and agricultural authorities alike. Still, short-term economic interests often overrule the concern for a sustainable long-term management of agricultural soils.[5] Therefore, binding rules as well as voluntary incentives for soil protection is necessary in the CAP framework.

The CAP framework already contains a certain number of environmental provisions, including some soil protection-related rules, which could be upgraded in the current review process—or even lose their edge, if things go wrong. The purpose of this chapter is to analyse the Commission proposal and the current state of discussions in the European legislative bodies around the question: how much progress for European soils can be expected from this CAP review?

To answer this question, this chapter will briefly look at the history of the CAP and of soil protection rules at the Union level. It will then take a general look at the Commission proposal for the new CAP regulations—which at first glance seem to aim at a paradigm shift by leaving more responsibility to Member States. A detailed look at the soil protection provisions in the current CAP framework, including national implementation using the example of Germany and England. The soil protection provisions in the Commission proposal will show that this proposal is indeed much less revolutionary, but still could make some positive impact through national implementation. However, a look at the current state of negotiations, particularly in the Council, reveals that the final outcome of the legislative process will probably be even less ambitious than the Commission proposal.

There is a close link between soil protection and mitigation of climate change, as soils are able to store large amounts of carbon, depending on how they are used. The role of agricultural soils with regard to climate is currently emphasised by the '4 per 1000' initiative[6] launched by the French government. The analysis of the soil protection provisions in the Commission proposal will also take a look on whether they are instrumental in this regard.

Meanwhile, soils also suffer threats from various sources outside of the influence of the farming sector. Among the most important ones are soil sealing and—in particular in the Southern Member States—climate-related problems.[7] These threats cannot be addressed through the CAP regulations, as good as they may be. Apart from the piecemeal solutions in the existing environmental acquis of the European Union, a new attempt at a dedicated European legal framework for soil protection could be an option. These ideas will also be discussed in the last section.

[5]See for the change in agriculture since World War II: Brodt et al. (2011).

[6]https://www.4p1000.org/.

[7]To address water shortage problems, the Commission has recently issued a proposal for a Water Reuse Regulation (COM (2018) 337 final).

2 The Development of the CAP and Its Environmental Components

2.1 History of the CAP and Its Environmental Components[8]

2.1.1 Beginnings of the CAP

The Common Agricultural Policy itself goes back to the creation of the European Communities (the precursors of today's European Union). The *Treaty of Rome* establishing the European Economic Community,[9] signed in 1957, on the one hand, established a Common Market, including for agricultural products (Article 38 (1) subpara. 2 of the Treaty on the Functioning of the European Union—TFEU[10]). On the other hand, Member States wanted to maintain a high level of state intervention in agriculture. This intervention had to be transferred to the Community level.[11] Therefore, the Treaty of Rome also provided for the establishment of a Common Agricultural Policy (CAP) (Article 38 (4) TFEU) which was realised in 1962.[12]

This policy initially focused on maximising agricultural production and food security by providing price guarantees to farmers and thus promoted a short-term maximisation of agricultural land use.[13] The objectives of the CAP, laid down in Article 39 TFEU, did not deal with the environmental aspects of food production. The first major reform of the CAP in 1992 ('MacSharry reform'[14]) aimed at limiting rising production. It replaced the price guarantee system by 'direct payments' depending on the size of arable land cultivated by a farmer. It also created 'set-aside' payments to withdraw land from production, payments to limit stocking levels, and introduced measures to encourage retirement and afforestation. While these measures were for the most part not directly aimed at the protection of the environment, they helped to relieve some of the pressure linked to the intensification of land use with its negative effects such as soil compaction and pollution.[15]

---

[8]Section 2.1 of this chapter is based on Raffelsiefen and Strassburger (2017).

[9]Now the Treaty on the Functioning of the European Union (TFEU).

[10]For the sake of convenience, all references to the European treaties cite the provisions of the current version of the treaties after the entry into force of the Lisbon Treaty.

[11]European Parliament (2016).

[12]Tangermann and von Cramon-Taubadel (2013), p. 17; Zobbe (2001), pp. 4–5.

[13]Delayen (2007), p. 1.

[14]The basic document of this reform is Commission Communication COM (91) 100 final: 'The development and future of the CAP – Reflections paper of the Commission'.

[15]Lee (2006), p. 73.

2.1.2 Introduction of Environmental Components into the CAP

After nearly forty years of existence of the CAP, two major steps were taken one shortly after the other which introduced environmental aspects into the CAP: First, agri-environment measures were introduced as part of the so-called second pillar of the CAP. Second, cross-compliance was introduced in the process of the so-called decoupling of direct payments. The introduction of these rules into the CAP is covered by the legal basis for the adoption of secondary law for the implementation of the CAP (Article 43 TFEU) which offers the basis for environment and soil-related legislation, insofar as the provisions concerned are related to agricultural practices.[16]

2.1.2.1 Agri-Environment Measures

First, the Agenda 2000[17] adopted by the European Council in 1999 introduced the well-known 'two pillars' structure of today's CAP. The second pillar of the CAP, the EU Rural Development Policy, contains a second axis which concerns the improvement of the environment and the countryside through support for land management as well as helping to fight climate change ('agri-environment measures'). Such projects could for example concern preserving water quality, sustainable land management, planting trees to prevent erosion and floods. For 2007–2013 EU expenditure on agri-environment measures amounted to nearly EUR 20 billion or 22% of the expenditure for rural development.[18]

Agri-environment measures provide payments to farmers who subscribe, on a voluntary basis, to environmental commitments related to the preservation of the environment and maintaining the countryside.[19] Farmers commit themselves, for a minimum period of at least five years, to adopt environmentally friendly farming techniques that go beyond existing legal obligations. In return, farmers receive payments that provide compensation for additional costs and income foregone resulting from applying those environmentally friendly farming practices in line with the stipulations of agri-environment contracts. Agri-environment measures may be designed at the national, regional, or local level through rural development programming (RDP) to be adapted to particular farming systems and specific environmental conditions (e.g. intensive or extensive farming, local landscape features). Depending on the political will and the environmental situation as assessed before the programme period, Member States may incentivise agricultural management practices with direct or indirect positive effects on soil. It is only fair to mention

[16]Epiney (2018), margin number 26.

[17]Bulletin of the EU, Supplement 5/97, p. 26.

[18]European Commission (2015a).

[19]European Commission (2015a).

that soil-related measures have been part of the Rural Development Policy, but at a decent level. Soil has never been a top priority.

2.1.2.2 Cross-Compliance

Since 2003, all farmers receiving direct payments are subject to compulsory cross-compliance.[20] The current legal basis is Regulation (EU) 1306/2013 and Commission Implementing Regulation 809/2014.

Cross-compliance includes two elements, of which the second one has the largest potential to contribute directly to soil protection:

Statutory Management Requirements (SMR) are 13 legislative standards of the acquis in the field of the environment, food safety, animal and plant health and animal welfare listed in Annexe II of Regulation (EU)1306/2013. Due to the lack of dedicated soil legislation at the European level, there are only a few standards that have some relevance for soil protection: the sewage sludge directive 86/278/EEC, the Nitrates Directive 91/676/EEC and the Plant Protection Products Regulation (EC) No. 1107/2009.

Good agricultural and environmental condition (GAEC) is a requirement first introduced in 2009 on top of the respect of the SMR. These standards go beyond requirements that are already part of existing EU legislation and can play a positive role for soil protection. As of today, three GAEC requirements relevant to soil and soil carbon are listed in Annexe II of Regulation (EU) 1306/2013 (see Sect. 3).

Member States shall define minimum requirements on the basis of these standards. Minimum GAEC requirements should be defined by taking into account the specific characteristics of the areas concerned, including soil and climatic condition, existing farming systems, land use, crop rotation, farming practices and farm structures. The implementation of the GAEC framework results in a high variety of minimum requirements throughout Europe and sometimes even within the Member States when GAEC standards are defined at regional level. This flexibility is, in principle, desirable to account for the heterogeneity of agricultural conditions across Europe. Farmers who decide to ignore the restriction of cross-compliance will be penalised by the loss of a certain percentage of their payments. However, the criticism is made that Member States in the past have interpreted these standards in a lax way and also that they have not really been enforced sufficiently.

It is to be stressed here that GAEC reflects minimum standards only, not ensuring *best* practices. As such, its aspiration level may be below the standards of national farming advisory services. As such, national services are sometimes perceived by farmers as too demanding when compared to GAEC.

[20]European Commission (2015b).

2.1.3 The 2013 'Greening' of the First Pillar of the CAP

A long-awaited reform—agreed upon in 2013 for the CAP framework 2014–2020—introduced a new environmental aspect into the first pillar by creating 'Greening payments' now making up for 30 percent of direct payments.[21] Greening is supposed to reward farmers for preserving natural resources and providing public goods thus enabling the CAP to be more effective in delivering its environmental and climate objectives.[22] Each holding receives a payment per hectare for respecting certain agricultural practices beneficial for the climate and the environment. An independent evaluation in 2017 found all greening measures to have at least some relevance for addressing specific environmental and climate priorities in Member States.[23]

Functionally, the greening measures can be compared to the GAEC standards, while the requirements of the greening measures are set out in much more detail in the text of Regulation (EU) No 1307/2013.

The three basic measures in Article 43 (1) of Regulation (EU) No 1307/2013 are maintaining permanent grassland, crop diversification and maintaining an 'ecological focus area' of at least 5% of the arable area of the holding. One GAEC standard under the previous CAP rules—before 2013—had been the protection of wetlands and carbon-rich soil. Following the 2013 reform it has been moved into the basic text to become part of the permanent grassland eligibility condition.

With the CAP framework 2021–2027, the greening system will be obsolete again, as the current greening standards will be transferred into GAEC standards (see Sect. 3). The system will thus be harmonised.

2.2 *The Commission Proposal for the New CAP 2021–2027*

On June 1, 2018, the European Commission has submitted its legislative proposal[24] for a set of regulations laying down the legislative framework for the CAP in the period 2021–2027. The budgetary framework for the CAP is set by the Multiannual Financial Framework (MFF) for 2021–2027, for which the Commission has presented its proposal on May 2, 2018.[25] Both legislative proposals are—as of April 2019—still being discussed in the Council and the European Parliament.

The budgetary framework has an important impact on the environmental orientation of the CAP, as it notably defines the distribution of funds between the first pillar (direct payments and eco-schemes) and the second pillar (agri-environment-climate commitments). The Commission proposes a substantial reduction in 2nd

[21] European Commission (2015c).

[22] Alliance Environnement, Thünen-Institut (2017), p. 1.

[23] Alliance Environnement, Thünen-Institut (2017), p. 5.

[24] COM (2018) 392 final.

[25] COM (2018) 322 final.

pillar resources, but Member States will have flexibility to redistribute funds between the 1st and the 2nd pillar, so it is not clear yet if there will indeed be a real reduction in 2nd pillar funding.[26]

The most important part of the legislative framework is the CAP Strategic Plan Regulation, which also contains the soil-related aspects of the CAP.[27] Equally important to understand the new CAP as a whole is the so-called Horizontal Regulation with rules on financing, management and monitoring.[28]

According to the Commission's own announcement,[29] the central approach of the reformed framework is to leave more responsibility to Member States for defining national or regional requirements of the CAP, including environmental requirements. The EU should only set out fundamental general and specific objectives of the CAP (Title II of the Strategic Plan Regulation), such as a viable farm income and also efficient management of natural resources such as water, soil and air. These objectives are to be pursued by Member States which have to specify interventions, via their national strategic plans, based on the types of interventions set out in the regulation in accordance with certain common requirements (Art. 8 of the Strategic Plan Regulation). Member States are even required to aim at making, through their Strategic Plans, a greater overall contribution to the environmental- and climate-related objectives in comparison to the period 2014–2020 (Art. 92).

Among the eligible types of interventions are direct payments and interventions for rural development (Art. 64), including agri-environment-climate commitments. Thus, the substance of the current 1st and 2nd pillars of the CAP is preserved. Among the common requirements is a system called 'conditionality', under which an administrative penalty shall be imposed on beneficiaries receiving direct payments who do not comply with the statutory management requirements (SMR) and the standards for good agricultural and environmental condition of land (GAEC), as listed in Annexe III of the regulation (Art. 11). Conditionality is therefore essentially a new name for the current cross-compliance system. Greening as a system on its own is abolished and its requirements are transferred into the cross-compliance system.

A new environmental element, voluntary schemes for the climate and the environment ('eco-schemes'), is created in the 1st pillar (Art. 28). The eco-schemes are somewhat comparable to agri-environment measures, but they will be part of the

[26]Matthews (2018).

[27]Proposal for a Regulation of the European Parliament and of the Council establishing rules on support for strategic plans to be drawn up by Member States under the Common agricultural policy (CAP Strategic Plans). Financed by the European Agricultural Guarantee Fund (EAGF) and by the European Agricultural Fund for Rural Development (EAFRD) and repealing Regulation (EU) No 1305/2013 of the European Parliament and of the Council and Regulation (EU) No 1307/2013 of the European Parliament and of the Council, 2018/0216 (COD).

[28]Proposal for a Regulation of the European Parliament and of the Council on the financing, management and monitoring of the common agricultural policy and repealing Regulation (EU) No 1306/2013, COM (2018) 393 final.

[29]The Future of Food and Farming, COM (2017) 713 final.

direct payments regime and will be calculated per year and per hectare. It will be up to Member States how much funding they reserve for these eco-schemes.

It will also be the task of Member States to set up efficient management and control systems in order to ensure compliance with the Union legislation governing Union interventions, i.e. the allocation of European Union funds to farmers in the 1st and 2nd pillars (Art. 57 para. 2 of the Horizontal Regulation). This includes the adoption of a regulatory framework for controls, penalties and the recovery of undue payments (Art. 57 para. 1). The controls are supposed to include systematic checks which shall also target the areas where the risk of errors is the highest (Art. 58 para. 1). The control system for conditionality (Art. 84 para. 1) shall include dedicated on-the-spot checks which may be replaced by an equivalent control system which is already in place for the requirement in question (Art. 84 para. 3 lit. a and b). Thus, Member States can task environmental authorities with the application of CAP penalties in the framework of their routine control of environmental standards, which is already done under the present CAP rules. The penalty for non-compliance with conditionality rules shall reach from a 3% reduction of direct payments in the case of first-time non-compliance due to negligence (Art. 86 para. 2) up to a total exclusion from payments in the case of severe intentional non-compliance (Art. 86 para. 4).

While the basics of the environmental requirements addressing farmers are mostly going to stay the same, two other aspects of the new CAP could possibly lead to a more efficient implementation of these requirements. First, the Strategic Plans—in which Member States are supposed to specify, inter alia, the conditionality rules and the 2nd pillar commitments—have to be submitted to the Commission (Art. 106 para. 1 of the Strategic Plan Regulation). The Commission shall approve the Strategic Plan only if it is satisfied that the Plan is compatible with the requirements set out in the Strategic Plan Regulation (Art. 106 para. 4).

Second, Member States shall submit annual performance reports to the Commission on the implementation of the CAP Strategic Plan by reference to, inter alia, output and result indicators showing progress in the pursuit of the CAP objectives (Art. 121). If the output is not corresponding to the objective of the expenditures, the Commission shall adopt implementing acts prior to 15 October of the year following the budget year in question determining the amounts to be reduced from Union financing. (Art. 52 of the Horizontal Regulation). Thus, Member States could face reductions in the amount of Union financing they receive if their implementation of the CAP fails to meet the objectives set out in the Strategic Plan. Meanwhile, the proposed output indicators in Annexe I of the Strategic Plan Regulation that cover environmental aspects (O. 13 and O. 14) only refer to the 'number of ha covered by environment/climate commitments going beyond mandatory requirements', i.e. 2nd pillar measures. As soil protection in the CAP relies mainly on conditionality (see Sect. 3 of this chapter), which applies to all areas that benefit from direct payments, the annual performance clearance will probably be less relevant for soil protection. Likewise, all other output indicators refer to sheer quantities, i.e. the number of beneficiaries or the area covered by a certain measure.

Both the required pre-approval of national Strategic Plans and the annual performance clearance seem to be powerful instruments, at first glance, for a strict control of the pursuit of the objectives of the CAP by the Commission. The effectiveness of the annual performance clearance is nonetheless severely limited as it relies on quantitative outputs only. It would be much more effective—and more difficult to handle—to have indicators that cover qualitative aspects such as the evolution of the state of soils during a budget year. However, it appears to be extremely unlikely that qualitative output indicators will be introduced as a result of the discussions in the Parliament and the Council. On the contrary, it remains to be seen how much of the proposed control system will survive in the final text, after discussions with Member States in the Council. It also remains to be seen how effectively the Commission will use its powers with regard to performance on environmental objectives.

3 A Detailed Look on Soil Protection Standards Within the CAP Today and Tomorrow

As eco-schemes and 2nd pillar measures depend completely on national implementation and are not predefined in the CAP Regulations, this chapter will concentrate on the analysis of the soil-related GAEC standards, including those standards that are currently part of the Greening system and will be transferred into GAEC standards. Soil-related eco-schemes and 2nd pillar measures could, for example, support the acquisition of low ground pressure vehicles and machinery or the implementation of soil-conserving tillage methods going beyond GAEC standards.[30]

Among the current GAEC and greening standards and the upcoming GAEC standards, five standards aim directly at soil protection, while other standards might make an indirect contribution (such as protection of ground water against pollution or retention of landscape features). These five standards are

- minimum soil cover (currently a GAEC standard according to Annexe II of Regulation (EU) 1306/2013.
- Limiting soil erosion (currently a GAEC standard)
- Maintenance of soil organic matter and ban on burning arable stubble (currently a GAEC standard)
- Crop diversification (currently a greening measure according to Article 44 of Regulation (EU) No 1307/2013; this will become a GAEC standard for crop rotation).
- Maintaining permanent grassland including wetland and peatland (currently a greening measure according to Article 45 of Regulation (EU) No 1307/2013; this will be transferred into two GAEC standards).

[30] LABO (2018), p. 9.

In the following, these five requirements will be analysed in detail with regard to the current wording of the standard, with regard to national implementation using the example of the implementation in Germany and in England (*not* United Kingdom), and with regard to the new version of these requirements in the Commission proposal of the Strategic Plan Regulation.

In the Commission proposal, three GAEC standards are explicitly designated as mainly addressing the issue of climate change by preserving the carbon stock in soils (maintenance of permanent grassland, protection of wetlands and peatland, and ban on burning arable stubble for maintenance of soil organic matter). Implementation of these standards will also contribute to the preservation of soil quality in general. Likewise, implementation of the other soil-related standards (limiting soil erosion, minimum soil cover, crop rotation) will serve to protect soil quality in general as well as carbon stocks in particular.

3.1 Minimum Soil Cover

The basic idea of the current GAEC standard 4 ('minimum soil cover') is to protect soils from degradation, in particular by erosion, by maintaining a minimum soil cover throughout the year. The standard leaves the areas concerned by this requirement as well as the type of cover to the implementation by Member States.

The German implementation has a very reduced scope of application. It applies only to areas that have been declared, by the farmer, as certain types of ecological focus areas according to article 46 of Regulation (EU) No 1307/2013 (i.e. as part of the 'greening' obligations).[31] The rules also apply to arable land lying fallow but not declared as an ecological focus area.[32] Thus, the minimum soil cover standard in Germany does not cover most of the arable land that is actually in use.

On the contrary, the English implementation of this GAEC standard in principle requires all areas to have a minimum soil cover—unless one of the various agronomic reasons recognized by the Rural Payments Agency applies or an individual derogation has been granted.[33] Minimum soil cover can be provided in different ways, for example by vegetative cover by all types of crop, grass and herbaceous forage; cover crops; overwintered stubble from combinable crops.

[31] § 5 Abs. 1 Agrarzahlungen-Verpflichtungenverordnung (AgrarZahlVerpflV).

[32] § 5 Abs. 4 AgrarZahlVerpflV.

[33] Rural Payments Agency (2016a).

3.1.1 What's New in the Commission Proposal?

In the Commission proposal, this requirement is transferred into GAEC 7: 'No bare soil in most sensitive period(s)', listed in Annexe III of the proposed Strategic Plan Regulation among the soil-centred conditionality rules.

The implications of the new wording 'no bare soil' do not seem to be different from those of 'soil cover', as the wording equally allows for soil cover by vegetation as well as by other means, e.g. overwintered stubble or mulch. The term 'in most sensitive period(s)' appears to be a cutback on the scope of the requirement, but it probably will not have negative implications in practice, because bare soil is mostly an issue during winter.

The practical effect of the new GAEC standard, as of the old one, thus fully depends on the ambition of the national implementation, in terms of areas covered and of eligible types of soil cover. It is to be hoped that the German implementation of the new standard will be more ambitious than the current one.[34]

3.2 Limiting Soil Erosion

The general idea of the current GAEC standard 5 ('minimum land management reflecting site-specific conditions to limit soil erosion') is to require farmers to have appropriate land management in place in order to prevent site-specific soil erosion risks.

Soil erosion can be caused by a variety of factors, including soil cropping practices and cropping structures, leading to soil compaction; livestock management causing overgrazing and poaching; wind; vehicles, trailers and machinery.[35]

Appropriate measures to reduce erosion risks include[36] deep cultivation to remove compaction; establishing crops early in the autumn during dry conditions that ensure good soil structure and good crop cover over the winter; creating banks and diversion ditches within the field to intercept and slow down runoff; choosing relatively flat fields for growing row crops; moving stock regularly and loosening the soil as soon as conditions allow; using low ground pressure vehicles and machinery and use established tracks to avoid vegetation damage; choosing tyres that allow lower pressures.

The implementation of this GAEC standard in England is based on a rather general approach: Farmers lose some of their scheme payments if they have not taken all reasonable steps to prevent erosion over a single area of 1 or more hectares. Minor erosion of less than 1 hectare resulting in minimal soil loss will not be

[34]For ideas see LABO (2018).

[35]Rural Payments Agency (2016b).

[36]Rural Payments Agency (2016b).

penalised.[37] The UK system is thus based on penalising soil erosion that has already occurred and has been traced by farming authorities.

The German implementation, on the other hand, is based on precautionary requirements: First, soil erosion-related requirements only apply to certain areas with a high erosion risk due to either water or wind. These areas are delimited by regulations of the *Länder* attributing certain erosion classes to each area of arable land.[38]

An area being labelled as being under high erosion risk is subject to limitations on ploughing: On areas with high water erosion risk, ploughing is prohibited between December 1st and February 15th.[39] On areas with very high water erosion risk, ploughing is inadmissible also during the rest of the year if crops are not established directly afterwards.[40] On areas with high wind erosion risk, ploughing is only admissible when crops are established before March 1st or directly after the ploughing.[41] All rules are subject to a variety of exceptions applying, inter alia, to specific cropping arrangements.

3.2.1 What's New in the Commission Proposal?

The soil erosion requirement is transferred into GAEC 6: 'Tillage management reducing the risk of soil degradation, including slope consideration,' likewise listed in Annexe III of the proposed Strategic Plan Regulation among the soil-centred conditionality rules.

The wording of the new GAEC standard is quite different from the old one, but it is unclear if a different meaning is intended. 'Tillage management' is a narrower term than 'land management'. In Germany, the old GAEC standard has only been implemented as a ban on ploughing, which is obviously a part of tillage management. On the other hand, interpretations such as the English one that oblige the farmer to prevent soil erosion in general, which might also require measures that are not strictly part of tillage, might at first sight be difficult to reconcile with the new wording. On the other hand, the table in Annexe III gives the old wording of the GAEC, 'minimum land management reflecting site-specific conditions to limit soil erosion', as the main objective of the standard, which hints at the interpretation that things might be implemented in the same ways as before. The latter interpretation seems to be the better one, because it appears to be more reasonable to leave it to farmers which steps they are taking to prevent soil erosion, and to penalise erosion that has happened if farmers have failed to do so.

[37] Rural Payments Agency (2016b).

[38] § 6 Abs. 1 AgrarZahlVerpflV.

[39] § 6 Abs. 2 AgrarZahlVerpflV.

[40] § 6 Abs. 3 AgrarZahlVerpflV.

[41] § 6 Abs. 4 AgrarZahlVerpflV.

The term 'soil degradation' is wider than 'soil erosion', while it is not clear whether it is intended to include degradations such as soil compaction by machinery.

3.3 Maintenance of Soil Organic Matter and Ban on Burning Arable Stubble

The current GAEC standard 6 ('maintenance of soil organic matter through appropriate practices including a ban on burning arable stubble, except for plant health reasons') could, in theory, prompt a wide variety of practices that serve to maintain soil organic matter.

However, in the national implementation it has been difficult to find appropriate measures for this standard. Thus, many national implementations seem to be limited to the very minimum of this standard, the ban on burning arable stubble.

For example, the German implementation only includes the ban on burning arable stubble albeit without the exceptions laid down in GAEC 6.[42]

The English implementation accurately repeats the GAEC minimum in prohibiting burning arable stubble, except for plant health reasons. Where a notice has been served under the Plant Health (England) Order 2005 and the burning of stubble is thus allowed, a multitude of safety requirements apply to the burning process.[43]

As a positive example, Ireland, in the former CAP period, had once followed a broader approach.[44] If the organic matter percentage on land that is in continuous tillage was found to be below a threshold value of 3.4%, farmers were obliged to seek the advice of the national farming advisory service (Cross Compliance—Farm Advisory System (CC—FAS)) on any remedial action that may be considered necessary. The CC—FAS report would set out in detail any remedial actions as may be required. Where necessary, these remedial actions had to commence in the year following the organic matter sampling. Examples of such practices were incorporating straw, applying farmland manures (while respecting limits to nutrient supply under the nitrates regulations), crop rotation, reduced tillage and returning the land to permanent pasture.[45] But Ireland has since discontinued this requirement and has also limited its implementation to the ban on burning arable stubble.

[42]§ 7 Abs. 4 AgrarZahlVerpflV.

[43]Rural Payments Agency (2016c).

[44]Department of Agriculture, Food and the Marine (2009).

[45]Department of Agriculture, Food and the Marine (2009).

3.3.1 What's New in the Commission Proposal?

The current soil organic matter requirement is transferred into GAEC 3: 'Ban on burning arable stubble, except for plant health reasons', listed in Annex III of the proposed Regulation among the climate-centered conditionality rules.

As Member States have struggled to define other appropriate measures for the maintenance of soil organic matter, the Commission proposal consequently reduces this GAEC standard to the ban on burning arable stubble. This is a bit disappointing especially because soil organic matter is directly climate-related. However, the new crop rotation GAEC standard has the potential to contribute to maintaining and improving soil organic matter (see below).

3.4 Crop Diversification vs. Crop Rotation

The current crop diversification greening measure (Art. 44 of Regulation (EU) No 1307/2013) requires farmers to have a certain diversity of crops on their arable land. Crop diversification does not mean, however, that farmers have to rotate the crops which they grow on a certain parcel of land over time. In detail, the measure has the following principal requirements: Where the arable land of the farmer covers between 10 and 30 hectares, there shall be at least two different crops on that arable land. Where the arable land of the farmer covers more than 30 hectares, there shall be at least three different crops. In each case, the main crop shall not cover more than 75% of the arable land.

Crop diversification without crop rotation does not entail any relevant benefits for soil quality. While the crop diversification measure has been conceived to improve, inter alia, soil resilience, the positive effects of this greening measure are consequently believed to be very little.[46] Although the measure has a certain number of exceptions, it also puts additional burden on farmers who already have a crop rotation system in place which is not compatible with the strict framework of this greening measure.

3.4.1 What's New in the Commission Proposal?

This greening standard is transferred into a new GAEC standard, GAEC 8: 'Crop rotation,' listed in Annexe III of the proposed Strategic Plan Regulation among the soil-centred conditionality rules.

As crop rotation is now a GAEC standard, the Commission proposal does not give any detailed definition of the requirements under this standard. This is now completely up to Member States to define in their Strategic Plans.

---

[46]Cf., for example: Matthews (2015).

The term 'crop rotation' and the main objective of the standard, 'preserve soil potential', raise hope that the standard can be implemented in a more efficient and practical way than the old crop diversification standard. The well-known benefits of crop rotation such as preservation of soil fertility, better nutrient management with less reliance on fertilisers, better pest and weed control without an excessive use of plant protection products should be pursued as effectively as possible. Crop rotation has also been found to increase soil carbon stocks especially if it includes a perennial crop or a cover crop.[47]

Still, farmers have adopted mono-cropping systems for a reason. In addition to the short-term economic benefits of being able to concentrate on one crop, mono-cropping is part of the process of separation between farmers specialised in crop farming and those specialised in stock farming, as crop rotation is normally associated with stock farming. A generalised return to crop rotation would thus severely disrupt the current trend of industrial livestock production. It is very unlikely that such a radical change can be induced by the rewording of a GAEC standard.

The challenge for national strategic plans is, if the version of the Commission proposal is upheld, will be to define requirements that, on the one hand, effectively exclude the negative effects of mono-cropping on as many areas as possible and, on the other hand, leave enough freedom to farmers to choose the crop rotation system that is adequate to the land they are cultivating.

3.5 Maintaining Permanent Grassland

With regard to the maintenance of permanent grassland, individual farmers are subject to a ban on ploughing and converting of certain environmentally valuable grasslands (Art. 45 para. 1 Regulation (EU) No 1307/2013). Member States must designate such environmentally sensitive permanent grasslands in Natura 2000 areas, i.e. in areas covered by the Habitats Directive[48] or the Birds Directive,[49] including in peat and wetlands situated in these areas. They may also designate environmentally sensitive permanent grasslands outside such areas in which farmers are subject to this requirement.

A second obligation under this greening standard is directed at Member States: They have to ensure that the ratio of permanent grassland to the total agricultural area does not fall by more than 5% compared to a reference year (Art. 45 para. 2 Regulation (EU) No 1307/2013). If the ratio has fallen below the 5% threshold, Member States are obliged to take action by imposing obligations on farmers to reconvert land into permanent grassland (Art. 45 para. 3 Regulation (EU) No 1307/2013).

---

[47] King and Blesh (2018), p. 10.

[48] Directive 92/43/EEC.

[49] Directive 2009/147/EC.

The effect of this standard is considered to be limited. This is partly due to the reluctance of many Member States to designate relevant grassland areas, and also to the lack of a pre-authorisation system for grassland conversion in the majority of Member States.[50]

3.5.1 What's New in the Commission Proposal?

This greening standard is transferred into two of the climate-centred GAEC standards, GAEC 1: 'Maintenance of permanent grassland based on a ratio of permanent grassland in relation to agricultural area', and GAEC 2: 'Appropriate protection of wetlands and peatland'.

The new GAEC 1 basically continues the obligation to maintain a ratio of permanent grassland to the agricultural area which is directed at Member States under the current greening rules. As a GAEC standard, it now directly addresses farmers. It remains to be seen how Member States are going to implement this requirement. One way would be to oblige farmers directly to preserve the ratio on their own arable land, which would be a very strict requirement. The other option is to introduce a general pre-authorisation system for grassland conversion in which farmers would be obliged to show, if they want to convert grassland, that appropriate compensation is provided by the reconversion or remediation of grassland on other sites within the same region.[51]

To implement GAEC 2, Member States will have to define appropriate requirements to protect wetlands and peatland in order to protect carbon-rich soils. This should include restrictions on the agricultural use of such areas and specifically a ban on the drainage of moorlands.[52]

4 State of Play

At the time of the finalisation of this chapter (April 2019), discussions in the Council and the European Parliament are still ongoing with no position agreed upon by the institutions. The position of the European Parliament will only be decided upon by the new Parliament elected in May 2019, which is not bound by the Committee opinions issued before the elections.

With regard to the position to be taken by the Council, compromise texts by the Romanian Presidency for the Strategic Plans Regulation[53] and the Horizontal

---

[50] Alliance Environnement, Thünen-Institut (2017).

[51] Such a system is proposed in LABO (2018), p. 7.

[52] LABO (2018), p. 7.

[53] Romanian Presidency (2019a).

Regulation[54] have been made public in February and March 2019. While compromise texts by the Austrian Presidency for the Annexes of the Strategic Plan Regulation exist since December 2018.[55] They reflect the conclusions which the presidency has drawn based on the views expressed by Member States in the Council Working Groups, but it remains to be seen which version the Commission, the Council and the European Parliament will agree upon in the trilogue.

The Romanian Presidency texts do not contain substantial changes in the provisions discussed in this chapter.

With regard to the GAEC standards, the following changes in the wording have been proposed in the Austrian Presidency text concerning Annexe III of the Strategic Plan Regulation:

> *GAEC 1: Maintenance of permanent grassland based on a ratio of permanent grassland in relation to agricultural area* ***at a level defined by Member States***
>
> *GAEC 2:* ~~*Appropriate*~~ ***Minimum*** *protection of wetland and peatland*
>
> *GAEC 6:* ***In the case of a significant risk of soil degradation****, tillage management reducing* ***that*** ~~*the*~~ *risk* ~~*of soil degradation*~~*, including slope consideration*
>
> *GAEC 7:* ~~*No*~~ ***Avoiding*** *bare soil in period(s)* ***that are*** *most sensitive* ~~*period(s)*~~
>
> *GAEC 8: Crop rotation* ***or other practices with an equivalent effect***

As a common observation, all changes indicate a trend towards more flexibility and less strict soil protection requirements. Nonetheless, what 'appropriate' protection needs to be defined at the national level in all of the GAEC standards, anyways, so the changes to GAEC 1, 2, 6 and 7 might be rather seen as cosmetic. The change in GAEC 8 is more substantial, as it allows for other practices than crop rotation to be applied to pursue the objective of preserving the soil potential. These measures might in reality be much less effective. As the new wording of GAEC 8 was the most obvious improvement in the Commission proposal in terms of soil protection and protection of the environment in general, its rewording would be disappointing, albeit not surprising.

5 Conclusions and Outlook

The new CAP regulations need to be ready for the 2021–2027 financial period. As the current presidency compromise texts suggest, the final version of the new CAP regulations will probably be even a bit less ambitious in terms of environmental aspects than the Commission proposal.

---

[54] Romanian Presidency (2019b).

[55] Austrian Presidency (2018).

The new CAP could still be seen as progress in terms of soil protection if effective requirements are introduced in the national Strategic Plans and if the Commission effectively uses its preapproval powers. The challenge in designing the Strategic Plans will be to find appropriate rules for implementing the GAEC standards, for the eco-schemes and for 2nd pillar measures.

As the analysis of the implementation of the current GAEC standards in Sect. 3 has shown, it has been difficult to find appropriate implementing rules—and, if they are found, they tend to be complicated and bureaucratic. A way forward could be to favour results-based approaches, where possible. One example is the English implementation of the erosion GAEC standard that penalises soil erosion when it has happened. This could be transferred, for example, to the objective of preserving soil organic matter: Farmers could be obliged to preserve soil organic matter by the way they consider appropriate (e.g. crop rotation). If a substantial decline in soil organic matter is measured, they could be penalised by a reduction of direct payments. Implementing rules that are reasonably simple to understand and to apply will also favour a more efficient control system as required by Art. 57 para. 2 of the Horizontal Regulation.

Aside from to the implementation of the future CAP, the idea to create a dedicated legal framework for soil protection at the European Union level is still on the table. The 7th Environment Action Programme of the European Union[56] states that 'the Union and its Member States should also reflect as soon as possible on how soil quality issues could be addressed using a targeted and proportionate risk-based approach within a binding legal framework'. Against this background, the Commission has, in 2015, established a Soil Expert Group composed of experts mandated by Member States to allow for a structured and formal dialogue. The work of this expert group has so far been focused on the stocktake of existing soil policy instruments in the EU.[57] The German Federal Agency for the Environment, for example, has developed fresh ideas for what could be done in terms of European soil protection legislation[58] if enough political momentum is present.

A look on the state of environmental legislation in Europe and its Member States shows that currently, progress is hard to be achieved, especially if there is a conflict with agricultural interests. Things are moving when certain topics get a strong presence in the media (nitrates; glyphosate; decline in insect populations).

Increased focus on the value and protection of soils could notably emerge from the debate on climate change and climate adaptation, which could also foster an ambitious national implementation of the new GAEC standards, especially the permanent grassland/wetland and peatland and the crop rotation standard. But up until then, within the CAP and beyond, it's about small steps. And preferably, small steps forward.

[56]European Commission (2015d).

[57]Raffelsiefen and Strassburger (2017), p. 407.

[58]German Environment Agency (2018).

References

Alliance Environnement, Thünen-Institut (2017) Evaluation of the CAP Greening Measures. https://ec.europa.eu/agriculture/sites/agriculture/files/leaflet_en.pdf. Accessed 6 May 2019

Austrian Presidency (2018) Presidency drafting suggestions (Annexes II and III) for the Strategic Plan Regulation. https://data.consilium.europa.eu/doc/document/ST-15058-2018-ADD-1/en/pdf. Accessed 6 May 2019

Brodt S, Six J, Feenstra G, Ingels C, Campbell D (2011) Sustainable agriculture. Nat Educ Knowl 3 (10):1. https://www.nature.com/scitable/knowledge/library/sustainable-agriculture-23562787. Accessed 6 May 2019

Delayen C (2007) The Common Agricultural Policy: a brief introduction. http://www.iatp.org/files/451_2_100145_0.pdf. Accessed 6 May 2019

Department of Agriculture, Food and the Marine (2009) Maintenance of soil organic matter. https://www.agriculture.gov.ie/media/migration/farmingschemesandpayments/crosscompliance/soilorganicmatter/Soil%20organic%20matter%20guidelines.pdf. Accessed 6 May 2019

Epiney (2018) TFEU Article 192. In: Landmann, Rohmer (2018) Umweltrecht, 88. El September 2018

European Commission (2015a) Agri-environment measures. http://ec.europa.eu/agriculture/envir/measures/index_en.htm. Accessed 6 May 2019

European Commission (2015b) Cross-compliance. In: Direct Support. http://ec.europa.eu/agriculture/direct-support/cross-compliance/index_en.htm. Accessed 6 May 2019

European Commission (2015c) Greening. In: Direct Support. http://ec.europa.eu/agriculture/direct-support/greening/index_en.htm. Accessed 6 May 2019

European Commission (2015d) General Union Environment Action Programme to 2020. Available via http://ec.europa.eu/environment/action-programme/. Accessed 6 May 2019

European Environment Agency (2015) Soil. In: State of the Environment Report 2015. https://www.eea.europa.eu/soer-2015/europe/soil. Accessed 6 May 2019

European Parliament (2016) Rationale behind the CAP. In: The common agricultural policy (CAP) and the Treaty. http://www.europarl.europa.eu/factsheets/en/sheet/103/the-common-agricultural-policy-cap-and-the-treaty. Accessed 6 May 2019

German Environment Agency (2018) The need for soil protection legislation at EU level. https://www.umweltbundesamt.de/sites/default/files/medien/1410/publikationen/181101_uba_hg_bodenschutz_engl_bf_low.pdf. Accessed 6 May 2019

Joint Research Centre (2012) The State of Soil in Europe – a contribution of the JRC to the European Environment Agency's Environment State and Outlook Report – SOER 2010. http://esdac.jrc.ec.europa.eu/ESDB_Archive/eusoils_docs/other/EUR25186.pdf. Accessed 6 May 2019

King AE, Blesh J (2018) Crop rotations for increased soil carbon: perenniality as a guiding principle. https://esajournals.onlinelibrary.wiley.com/doi/pdf/10.1002/eap.1648. Accessed 23 June 2019

LABO – Bund/Länder-Arbeitsgemeinschaft Bodenschutz (2018) Bodenschutz in der Gemeinsamen Agrarpolitik (GAP) nach 2020. https://www.labo-deutschland.de/documents/Eckpunkte_des_Bodenschutzes_fuer_die_GAP_nach_2020.pdf. Accessed 6 May 2019

Lee YL (2006) Nachhaltiger Bodenschutz international, europäisch und national. Universitätsverlag der TU Berlin, Berlin

Matthews A (2015) Scrap the crop diversification greening requirement and find a sensible replacement. http://capreform.eu/scrap-the-crop-diversification-greening-requirement-and-find-a-sensible-replacement/. Accessed 6 May 2019

Matthews A (2018) CAP spending in the next MFF. http://capreform.eu/cap-spending-in-the-next-mff/. Accessed 6 May 2019

Raffelsiefen M, Strassburger T (2017) The protection of soil: does the European Union live up to its own ambitions? In: Ginzky H, Heuser I, Qin T, Ruppel O, Wegerdt P (eds) International yearbook of soil law and policy, vol 2016. Springer, Cham, pp 389–410

Romanian Presidency (2019a) Presidency drafting suggestions for the Strategic Plan Regulation. https://data.consilium.europa.eu/doc/document/ST-7485-2019-INIT/en/pdf. Accessed 6 May 2019

Romanian Presidency (2019b) Presidency drafting suggestions for the Horizontal Regulation. https://data.consilium.europa.eu/doc/document/ST-6981-2019-INIT/en/pdf. Accessed 6 May 2019

Rural Payments Agency (2016a) GAEC 4: providing minimum soil cover. In: Guide to cross compliance in England: 2016. https://www.gov.uk/guidance/guide-to-cross-compliance-in-england-2016/gaec-4-providing-minimum-soil-cover. Accessed 6 May 2019

Rural Payments Agency (2016b) GAEC 5: minimising soil erosion. In: Guide to cross compliance in England: 2016. https://www.gov.uk/guidance/guide-to-cross-compliance-in-england-2016/gaec-5-minimising-soil-erosion. Accessed 6 May 2019

Rural Payments Agency (2016c) GAEC 6: maintaining the level of organic matter in soil. In: Guide to cross compliance in England: 2016. https://www.gov.uk/guidance/guide-to-cross-compliance-in-england-2016/gaec-6-maintaining-the-level-of-organic-matter-in-soil. Accessed 6 May 2019

Tangermann S, von Cramon-Taubadel S (2013) Agricultural policy in the European Union – an overview – https://www.uni-goettingen.de/de/document/download/468756dd26772ba40606fb7034c7995d.pdf/Diskussionsbeitrag-1302.pdf. Accessed 6 May 2019

Zobbe H (2001) The economic and historical foundation of the Common Agricultural Policy in Europe. http://ageconsearch.umn.edu/bitstream/24212/1/ew010012.pdf. Accessed 6 May 2019

Part IV
Cross-Cutting Issues

Perceptions of Soil in Catholic Theology

Markus Vogt

In September 2016, the German bishops published a socio-ethical expert advice on the protection of soil.[1] In it, soil protection is considered as a central field of probation for contemporary responsibility for creation, because of the indispensable affordances that fertile farmland effects not only for human nutrition, but for the whole network of living processes. From the analysis—how much these functions are endangered worldwide by a creeping process of degradation—results the postulation of a paradigm shift towards a nature-compatible agriculture and area planning, a shift that touches upon some aspects of agricultural policy, development cooperation and consumption habits.[2] The way churches handle their considerable land ownerships has come under scrutiny, too. The following theses take up the Catholic text "*Der bedrohte Boden*" (The Endangered Soil) and sketches some guidelines of Christian environmental ethics for the example of soil conservation. In times of anthropogenic climate change, population growth and increasing international land speculation, soil protection has a special importance for sustainable development not only in Germany but all over the world.

Our present times are often described as undergoing a "great transformation": a time of accelerated changes in economy, work and life.[3] In particular, the ecological distortions that have already occurred or are to be expected, together with the associated social exclusions, challenge our societies to search for new ethical

[1]Die Deutschen Bischöfe – Kommission für gesellschaftliche und soziale Fragen [DBK] (2016), Vogt (2018a), pp. 265–275.

[2]Keestra et al. (2016).

[3]Fundamental for the topos "Great Transformation" in Germany is: Wissenschaftlicher Beirat der Bundesregierung Globale Umweltveränderungen [WBGU] (2011).

M. Vogt (✉)
LMU Munich, Faculty of Catholic Theology, Munich, Germany
e-mail: m.vogt@lmu.de

H. Ginzky et al. (eds.), *International Yearbook of Soil Law and Policy 2019*,
International Yearbook of Soil Law and Policy 2019,
https://doi.org/10.1007/978-3-030-52317-6_17

standards for global development as well as for production and consumption patterns. This debate affects the foundations of our self-concept and thus, exhibits a substantial religious and cultural dimension.

In the shadow of the "Defiant Earth"[4] a radical change in our models of development is postulated. In the face of such high demands for social change, it seems to be important to neither overestimate nor underestimate the role of the churches: they are no "moral agencies"[5] undergirding socio-ecological imperatives with theological pleas. Instead, their specific competence lies in a "moral beyond moralizing" assumption.[6] In the given context, the ecclesiastical starting point of reflection and advice has to be the message of liberation together with the praise of creation and the awareness of its/her beauty. This brings the longing for lost integrity, justice and fullness of life to the forefront. What we need is not a rearmament with moral outrage, but a re-centering on the grand narratives of guilt and responsibility, anthropology and the perception of nature as well as culture and technology, in order to allow us to cope with the challenges of today's life. It is foremost this narrative and anthropological integration that constitutes the surplus value of Christian theology of creation compared with a secular ethics of nature.

1 It's a Matter of Power

In today's global society, the unequal access to resources is closely related to a massive imbalance of power. This signifies a systematic challenge to include ecological aspects into the theory of justice insofar as it implies a considerable part of humanity to be cut off from their means of existence and bereft of their liberty. "The power of man over nature turns out to be a power wielded by some men over others with nature as their instrument."[7] Ecological power can be defined as the faculty to privatize the benefits of environmental consumption and to externalize the costs by shifting them to the general public comprising of present or future generations. From the systematic ethical perspective, what is essentially relevant here are the power and system questions combined with their underlying cultural patterns. It is precisely this extension of the ethical discourse beyond the mere discourse of giving reasons and beyond mere "orientational knowledge" in relation to the goals of action, which I see as the core of the arousing new language in the encyclical *Laudato si'*: in liberation theological diction, "power" is one of the most frequent lexemes in the text (67 uses). The encyclical does not limit itself to desirables in an idealistic fashion, but instead addresses questions of system and power.

---

[4]Hamilton (2017).

[5]Joas (2016); Bischof and Sautermeister (2017).

[6]Sautermeister (2017), p. 299.

[7]Wuppertal Institut (2006), p. 40.

The struggle for soil, therefore, is and has often been a conflict of power. The worldwide headword for this today are "land grabbing" and "green grabbing".[8] The massive seizure of fertile areas by countries like China, Saudi-Arabia and others especially in Africa has reached a level that can be dubbed ecological aggression with neo-colonial features. However, one should not overlook that the influx of capital into agriculture can also be conducive to a more productive cultivation. To this end, enhanced transparency is needed in combination with functional governments and administrations, which are capable of acting and not corroded by corruption. In addition, smallholders have to enjoy special protection. Even though the International Assessment of Agricultural Knowledge, Science and Technology for Development (IAASTD) report on global agriculture (2008) distinctly highlighted and empirically backed up the grand socio-economic and ecological potentials of smallholder structures,[9] these are nonetheless globally and constantly driven back by centralized interests of power.

Power always comes in the additional form of the power to define key concepts, which are coining the scientific and socio-political discourses in cognitive and normative respects. The radical "soil oblivion" in current economic guiding theories, in this regard, is an essential reason for soil protection to be unable to get beyond more or less non-effective pleas. While "land" has long been one of the central dimensions of reference for economic theories, since the nineteenth century it is typically only labor and capital that are taken into account as significant factors in theory building.[10] Soil is not addressed as a physical coordinate but as a dependent variable of price formation. Consequently, soil protection only comes into view as a financial problem, which is why the privileging of land as area for settling or traffic in comparison to its use as agricultural or simply close-to-nature area can hardly be opposed in a systematic way. Extensive and multifunctional usages are unable to stand their ground against utilization that is quantified and absorbed in monetary value.

2 Soil as Cross-Sectional Topic

A necessary, but by no means sufficient, foundation for soil protection is to raise awareness. We lack the awareness that the dirt under our feet does indeed contain value. We are neither accustomed nor institutionally prepared to recognize soil in its absolutely basic functions for all living things.[11] Politically, soil protection doesn't

---

[8]Brot für die Welt et al. (2013).

[9]International Assessment of Agricultural Knowledge, Science and Technology for Development [IAASTD] (2009).

[10]For a criticism of this kind of modeling economy and the search for alternatives cf. Rogal (2008), pp. 95–118.

[11]Cf. the concise summary in Netzwerk Boden (2015); Umweltbundesamt (2014).

stand a chance, because it is hardly appreciated in its connectedness to the protection of climate, biodiversity, water and food security. Which average citizen would know that far more CO_2 is bound in soil than in the atmosphere? If we were to make progress in climate protection, we would have to bind more CO_2 in the ground (e.g. by means of deep rooting plants and moor protection). Life arises from soil. Without its protection, we will not succeed in conserving the diversity of species. The manifold eco-systemic services of soil are decisive for uncovering its ethical significance.

Already in 1998, the ecological working group of the German Bishops' Conference formulated a striking passage on soil protection in their text "*Handeln für die Zukunft der Schöpfung*" (Acting for the Future of Creation) with reference to global nutrition.[12] The discrepancy between the shrinking availability of fertile soil on the one hand and humanity growing in numbers as well as demands on the other, did already at that time upset the authors to considerable extent. Although the situation has by no means relaxed since then, food security is still not seen as consequent enough in its connection to soil protection. This is all the more the case, if we speak of "food sovereignty", meaning the right to grow one's own nourishment instead of just being supplied, which is socio-culturally decisive and has led to a paradigm change in development policies concerning the interpretation of the human right to food.[13]

Furthermore, soil protection requires the building of awareness of the many different actors that are to be considered. What is at stake here is the land use regulation on a local and transregional scale: long-ranging and effective courses have to be set to determine potential ways of land use, which in Germany and elsewhere, still substantially bear the mark of utopian dreams of progress from the 1960s and which have not been transformed systematically in the face to the challenges of sustainability.[14] Townships play a key role in this issue. As a result of insufficient participation in tax revenue in combination with false expectations regarding economic development, however, they are quite often forced to sell their land and property. In addition, every single consumer is co-responsible for the ways our society deals with soil: Our dietary habits lay claim to considerable amounts of land, in our own country as well as elsewhere, due to imports of foods and animal feedstuffs. Increasing mobility, too, is a booster for the growing sealing of valuable soils. To this day, 66 hectares of soil are buried under asphalt, concrete or other settlement areas in Germany each day. In the German sustainability strategy, the original aim to reduce the daily consumption of soil to less than 30 ha before 2020 has been postponed by 10 years to 2030. A path to at least come closer to this target is, however, not within sight.

[12]Cf. Die Deutschen Bischöfe (1998), p. 21f.

[13]For the concept of food sovereignty cf. the IAASTD report (footnote 8), pp. 13 and 221; For the socio-ethical aspects cf. Vogt and Hagemann (2010).

[14]Cf. Lendi and Hübler (2004).

3 Human as Double-Being: "Earthbound" and Spiritual

Etymologically, there is a close relation between soil and culture: the Latin *cultura* is derived from *colere*, to grow, till, cultivate, and further, to inhabit, foster, worship. Agriculture was seen as the archetype of culture in general: to embrace and adopt one's environment, to get to feel at home in it, while still being dependent on the forces of nature that we cannot control, but instead have to maintain. The religious understanding of *cult* is also derived from this origin. Thus, we arrive at dimensions of soil that exceed today's mere instrumental-rational relation by far. The peasant relationship to soil denotes the consciousness that it is the earth itself, which nourishes us and on which we subsist. Thus, soil becomes the symbol for the ecological roots of human existence. At the same time, a cultural dimension resonates, since it is always in part human labor, which opens up soil as nourishing living spaces. In peasant cultures, this synthesis of nature and culture is often closely interconnected with religious contents, such as harvest festivals or blessings of soil, animals and herbage.[15]

The Bible dubs human *adam*, "earthling" (cf. *adamah*, soil): he belongs to earth and soil, he is made from it. The biblical anthropology contains a theology of "earthiness". The earth is not just our possession, but rather we are part of her, we belong to her. The Latin language is also aware of this connection between human and soil: *homo*, human, is probably derived from *humus*, soil. Only those who do not forget their closeness to earth, their roots and limits, will stay humane. Humor, also derived from *humus*, can help with this task and could often do some good to environmental ethics.

Humans are double-beings, combining earth and spirit. What emerges is a concept of "grounded spirituality", tying sense and sensuality, heaven and earth, but nevertheless staying down to earth, in spite of intellectual and spiritual flights. For Christian spirituality, this close to earth consciousness as creature in the middle of creations constitutes a prerequisite as well as a benefit of its attachment to God.

In his encyclical *Laudato si'*, Pope Francis strongly develops such an anthropological dimension. He combines this approach with Latin American traditions of close relations between humans and "Mother Earth" (*Pachamama*). The encyclical specifies the consequences of such a Christian earthbound anthropology, especially in the context of smallholder agriculture as well as for a return to simple, close to nature ways of life and nutrition, which are paraphrased as "everyday ecology".

To this end, the Bible contains numerous directives, which at least in parts are still noteworthy today, e.g. the concept of Sabbath as a period of rest for human and cattle. As sabbatical year (waiving crop cultivation every 7th year), this had immediate consequences also for soil protection. In such directives, some scientists see the reason for the complete lack of evidence of even a single greater famine in ancient Israel, despite its precarious ecologic circumstances in the mostly barren landscape

---

[15]Cf. Vogt et al. (2004).

of Palestine on the edge of the desert.[16] As jubilee, the sabbath protection gained an enormous socio-political importance in this peasant culture.[17]

4 Property Entails Responsibility

Since, pursuant to the Bible, God himself is the owner of his creation and therefore of its soil, the land was considered inalienable during Israel's early periods: "The land shall not be sold in perpetuity, for the land *is* mine. For you *are* strangers and sojourners with me." (Lev 25,23; cf. Ex, 9,29; *Laudato si'*, 67) The earth as a whole together with everything in it and on it is subject to God's legal claim to power (cf. Ps 89,12; Dtn 10,14). Pursuant to the Bible, the soil and all the land are regarded as a loan that may be worked on, inhabited and managed, but has to be passed on to the next generation in full integrity (cf. *Laudato si'*, 116). As seen by Thorsten Philipp, this differentiated concept of ownership is the most important contribution to environmental ethics from the side of Catholic social doctrine.[18] It has been clearly expressed by Thomas Aquinas, who gave private property a pragmatic foundation instead of one in natural law. In his view, private property helps to motivate responsibilities and to demarcate usage rights, while at the same time it remains subject to limitations as a *bonum commune*.

In contrast to the concept of ownership by John Locke, who asserts an absolute right acquired by labor, the first social encyclical *Rerum novarum* (1891) resorts to Aquinas to search for a compromise between liberalism and communism. In this tradition, Catholic social teaching never considered the right to property to be absolute and indefeasible, but subject and committed to the principle of common welfare. This applies especially to the goods of creation that primarily belong to the whole human family for common use. From here, the Compendium of the Social Doctrine deduces its principle that a stable climate is a public good and has to be internationally protected by corresponding statutory duties.[19] In secular international environmental law since the 1960s, this concept of an overall purpose of the goods of creation finds a widely analogous form under the heading "nature as a common heritage of mankind".

---

[16]Hüttermann and Hüttermann (2002).

[17]The biblical instances hereto are numerous, cf. exemplarily Ex 20,8-11; 23,12; Lev 19,3.30 and Dtn 5,12-15. For soil protection, especially Lev 25,4-7 (sabbatical year as fallow) and Lev 25,8-31 (year of Jubilee) are relevant. Characteristic for the Old Testament Sabbath commandment is the unity of social and ecological aspects. Pope Francis takes up the Sabbath commandment as an eco-social guiding maxime, cf. Francis (2015) no. 71 and 237.

[18]Philipp (2009), pp. 112–119.

[19]This would have to be developed further in dialogue with the economic differentiations of collective/public and club goods as well as the problems of commons. Cf. Ostrom (1990), Ostrom (2010).

Following these stipulations, the concept of ownership should not be conceptualized as unqualified with regard to soil; the owner is not free to arbitrarily do whatever comes to his or her mind. Instead, use of soil is limited insofar as it must not violate the public interest in its stand and functions.[20] The *ius abutendi*, the right to misuse, is to be excluded in regard to soil, because of its sensitive importance for affected ecosystems. According to the polluter pays principle, those who harm the soil have to pay for the harm done. This applies also to the pollution of groundwater that occurs during soil cultivation, which is considered one of the most explosive environmental problems worldwide. Since the established term "obligation to the common good" (*Gemeinwohlpflichtigkeit*) is easily ignored, the episcopal expert advice on soil protection sharpens it by the introduction of "obligation to ecology" (*Ökologiepflichtigkeit*).[21]

5 The Concept of Resource Justice

The normative questions of soil protection can only be reflected adequately if they are integrated in more general categories of environmental ethics. Thus, it can be assigned to one of the three fields of *environmental justice*, *ecojustice* or *resource justice*. Each of these conceptions has different theoretical models as its background that each lead to different evaluations. For instance, the concept of environmental justice, which originated in the 1980s in the context of the American civil-rights movement, is oriented socio-politically, socio-spatially and anthropocentrically, while ecojustice is most often linked to eco-centric patterns of argument. Distributive problems of ecologically unequal access to environmental goods are expressed best by the term resource justice.[22] In virtue of this "moral grammar" that focuses on the relevant areas of conflict in dealing with soil, the concept of resource justice seems to be suited best as a normative theoretical framework for soil protection.

In social philosophy, this kind of contemplation within the context of a broader theory of justice can be annexed to the *capability approach*. It views resources in the light of the *functionings* that they facilitate and unfolds a concept of development on this basis, which does not begin with the mere supply of goods, but instead with its concept of *development as freedom*.[23] With this normative approach, soil protection can be grounded on the basis of human rights: Where humans and social groups depend on access to fertile soil for their development of basic functionings, they have the right of assurance that this access will not be denied.

[20]Cf. Hansjürgens et al. (2018), pp. 762–767.

[21]Cf. Die Deutschen Bischöfe (2016), pp. 39–41.

[22]For an empirically and conceptually differentiated introduction cf. Wuppertal Institut (2006), pp. 125–155; German Minister of Development Gerd Müller also uses "ressource justice" as a guiding concept, Müller (2017), pp. 56–67.

[23]Cf. Sen (1999).

After the questions of resource justice have long been discussed primarily with regard to the future, the last few years witnessed a growing consciousness of countless people, who already today suffer from the costs of ecological externalization. This is prominently expressed by Stephan Lessenich in his title "*Neben uns die Sintflut*" (The Deluge next to us).[24] Analytically concise, he demonstrates how implausible it is to disregard societal environmental problems as merely relevant for the distant future. In variation of a much-quoted key phrase of Pope Francis, it could be put as follows: Our contemporary treatment of soil already today kills thousandfold every day.[25] To put it positively: soil reforms could save thousands of human lives every day.

The decisive ethical-political challenge is marked by the overcoming of short-sighted and fragmented perspectives, in order to activate the moral, political and economic resources of solidary action for providing protection of soil and environment. To this end, it is necessary to strengthen global control institutions for the implementation of resource justice. Soil protection therefore requires a fine-tuned interplay of regional, national, international as well as transnational regimes.

The expert advice "*Der bedrohte Boden*" (The Endangered Soil) published by the German Bishops sums up its theological and ethical reflections in ten guiding principles for sustainable soil protection.[26] These are practice-oriented priorities, which aim at politics, agriculture and society as well as the Church itself. Even though there remain many open questions in relation to their concrete interpretation and implementation, they nevertheless offer a strong normative compass. Following these guiding principles, I summarize the theses of this paper in the style of ten priorities, rules or "commandments" of soil protection.

6 Ten Commandments of Soil Protection

1. Stop soil loss and degradation: The ongoing and worldwide loss in fertile soils in terms of quantity as well as quality, due to erosion, steppe formation, desertification, floods, pollution and the recession of land surface brought about by rising sea-levels, is not compatible with the ethical principle of sustainability. Considering the growing world population, it is an urgent imperative of justice to stop this creeping process as soon as possible and to compensate for unrecoverable losses.
2. Reduce land consumption: The persistent trend to extensively seal areas in favor of settlement and traffic contradicts public welfare in the medium and long-term perspective. In Germany and comparable countries, the net new sealing has to

[24] Lessenich (2016), pp. 9–36; 171–199.

[25] Here I refer to Pope Francis' phrase "Such an economy kills" in the Apostolic Exhortation Evangelii gaudium, Francis (2013), no. 53.

[26] Die Deutschen Bischöfe (2016), pp. 46–50.

stop, to which renaturalization and land recycling can contribute as well. Reduction of land consumption must feature as an indispensable component of sustainability strategies in all countries.

3. Render intensive agriculture more compatible with soil conservation and expand sustainable agriculture: Intensive and extensive forms of agriculture are to be enabled site-specifically within the framework of a well-balanced overall concept under strict stipulations for the protection of soil, water and biodiversity. Transfer payments to agriculture (e.g. as part of the EU agricultural support) have to be tied to compliance with soil friendly methods of cultivation. Innovations in agricultural engineering have to be stronger oriented towards the ends of soil protection. New concepts of agriculture without using humus, which are experienced in the context of bio-economy, can help to compensate the loss of soil.[27]
4. Limit nutrient inputs and minimize pollutant inputs: The input of nutrients into soil as well as of pesticides (including glyphosate) are to be significantly reduced. Necessary measures have to be scientifically determined. Compliance has to be controlled independently and transparently.
5. Assert social responsibility and the polluter pays principle with regard to soil: The burdens of soil damages must not be passed on to third parties. The costs have to be internalized as comprehensively as possible through adequate legal frameworks; polluters have to be charged with the liabilities.
6. Establish food security and sovereignty: Within the conflict of competitive uses of fields between food, fodder and energy production, the human right to food enjoys systematic primacy. The access of smallholders to soil in the global south has to be promoted as a key strategy in the fight against poverty as well as on the level of international environment, development and agricultural politics.
7. Tie land consumption to social and ecological norms: Foreign direct investments for the acquisition and use of soil in poorer countries (so-called *land grabbing* and *green grabbing*) have to be strictly bound to the rules of socially and environmentally compatible cultivation methods and to be designed transparently. Hereto, the international community should formulate and enforce property and participation rights for the local population.
8. Accommodate the significance of soil in climate and biodiversity protection: Because of the considerable, and to this day widely underestimated, potential of soil to store carbon (and conversely, to release it), a repositioning of soil in the discussions and negotiations on global climate protection is needed. The role of soil as carbon sink has to be systematically furthered. In addition, soil protection should be incorporated as a key category in national and international biodiversity strategies.
9. Change consumption habits: Consumers carry an essential co-responsibility for soil protection. Therefore, they have to be accordingly informed and motivated as well as to systematically organize themselves within civil society. Areas for

[27]Vogt (2018b).

prioritized action are to avoid food waste (in Germany currently 18 million tons per year, equivalent to one third of the total amount), to reduce meat consumption and to procure fair trade, organic and regional foodstuffs.

10. Make sustainable use of church areas: Churches have a decisive function as role models in soil protection. This arises from the fact that second to the state, they are the largest land owners in Germany, holding about 5–7% of all the farmland and being able to influence its cultivation via lease contracts. In some African countries, churches also own plenty of land.[28] Here, the criteria of social obligations to ecology for soil cultivation should be consequently demanded. Since in the interplay of the different legal entities, the parishes or monasteries (which own a considerable portion of soil) are autonomous, ecclesiastical soil responsibility has to be promoted primarily via the raising of awareness as well as support in formulating and implementing framework specifications. On the basis of credible practice, the churches could and should take public action as advocates of soil and the responsibility for creation.

References

Bischof F-X, Sautermeister J (eds) (2017) Christliche Weltverantwortung – Kirche als moralische Instanz in der Gesellschaft. Themenheft der Münchener Theologischen Zeitschrift, MThZ 68 (4/2017)

Brot für die Welt, FIAN Deutschland, Misereor, Oxfam Deutschland, Welthungerhilfe (2013) Land Grabbing. Transparenz alleine reicht nicht! Positionspapier zur G8 Landtransparenzinitiative, auf: http://www.fian.de/fileadmin/user_upload/dokumente/shop/Land_Grabbing/Positionspapier_G8_Landtranzparenzinitiative_4_2013.pdf

Deutsche Kommission Justitia und Pax (Ed.) (2010) Food security and energy supply between self-interest and global justice (justice and peace 121). Bonn

Die Deutschen Bischöfe – Kommission für gesellschaftliche und soziale Fragen [DBK] (1998) Handeln für die Zukunft der Schöpfung, hg. v. Sekretariat der Deutschen Bischofskonferenz. Bonn, esp. no. 21f

Die Deutschen Bischöfe – Kommission für gesellschaftliche und soziale Fragen [DBK] (2016) Der bedrohte Boden. Ein Expertentext aus sozialethischer Perspektive zum Schutz des Bodens, hg. v. Sekretariat der Deutschen Bischofskonferenz. Bonn

Francis Pope (2013) Apostolic exhortation Evangelii gaudium, Vatican City

Francis Pope (2015) Apostolic exhortation Laudato si', Vatican City

Hamilton C (2017) Defiant Earth. The fate of humans in the anthropocene. Cambridge

Hansjürgens B, Lienkamp A, Möckel S (2018) Art. Boden. In: Staatslexikon, Bd. I, Sp. 762–767

Hüttermann AP, Hüttermann AH (2002) Am Anfang war die Ökologie. Naturverständnis im Alten Testament. München

International Assessment of Agricultural Knowledge, Science and Technology for Development [IAASTD] (2009) Global Report. Washington. https://www.weltagrarbericht.de/fileadmin/files/weltagrarbericht/IAASTDBerichte/GlobalReport.pdf

Joas H (2016) Kirche als Moralagentur? München

Keestra SD et al (2016) The significance of soils and soil science towards realization of the United Nations Sustainable Development Goals. SOIL 2:111–128

[28]Deutsche Kommission Justitia und Pax (Ed.) (2010), pp. 19–27; Vogt and Hagemann (2010).

Lendi M, Hübler K-H (eds) (2004) Ethik in der Raumplanung. Zugänge und Reflexionen. Hannover
Lessenich S (2016) Neben uns die Sintflut: Die Externalisierungsgesellschaft und ihr Preis. Berlin
Müller G (2017) Unfair! Für ein gerechte Globalisierung. Hamburg
Netzwerk Boden GIZ (2015) Boden. Grund zum Leben. Pressedossier Boden & Ernährung. Bonn https://www.grund-zum-leben.de/fileadmin/user_upload/dateien_-_grund_zum_leben/Dokumente/boden_grundzumleben_pressedossier_01_ernaehrung.pdf
Ostrom E (1990) Governing the Commons. The evolution of institutions for collective action. Cambridge, UK
Ostrom E (2010) The challenge of common pool resources. Environ Sci Policy Sustain Dev 50 (4):8–21
Philipp T (2009) Grünzonen einer Lerngemeinschaft. Umweltschutz als Handlungs-, Wirkungs- und Erfahrungsort der Kirche. München
Rogal H (2008) Ökologische Ökonomie. Eine Einführung, Wiesbaden, esp. 95–118
Sautermeister J (2017) "Kirche als Moralagentur?" (H. Joas). Theologisch-ethische Überlegungen zur moralischen und politischen Relevanz von Kirche in der Gesellschaft. MThZ 68(4):292–305
Sen A (1999) Development as freedom. New York
Umweltbundesamt [UBA] (2014) Stark unter Druck. Lebensgrundlage Boden. Daten und Fakten aus globaler Sicht (UBA Fact Sheet 12/2014). Dessau
Vogt M (2018a) Die Zehn Gebote des Bodenschutzes. Schöpfungsethische Leitlinien zum Umgang mit Land. Stimmen der Zeit 4:265–275
Vogt M (2018b) Bedingungen ethisch verantwortbarer Bioökonomie. Forum Wirtschaftsethik 26. Jg. (special edition), pp 31–51
Vogt M, Hagemann H (2010) Zwischen Ernährungssouveränität, Exportorientierung und Energiegewinnung. Sozialethische Analysen zu Landwirtschaft und Ernährungssituation in Afrika. Amosinternational 4:19–27
Vogt M, van Saan-Klein B, Dirscherl C (eds) (2004) "... es soll nicht aufhören Saat und Ernte" (Gen 8,22). Ein Praxisbuch zum Mehr-Wert nachhaltiger Landwirtschaft. München
Wissenschaftlicher Beirat der Bundesregierung Globale Umweltveränderungen [WBGU] (2011) Welt im Wandel: Gesellschaftsvertag für eine Große Transformation. Berlin
Wuppertal Institut (2006) Fair Future. Begrenzte Ressourcen und globale Gerechtigkeit. München

The UN-Habitat Urban-Rural Linkages Guiding Principles: Assessment of the Adoptability to Topical Land Management Challenges in Germany, Kenya and Tanzania

Stephan Bartke, Thomas Forster, Grace Githiri, Almut Jering, Jackson Kago, Sina Schlimmer, and Remy Sietchiping

We are grateful for the support and recommendations of the editor and two reviewers. The work presented here has been partly funded by the German Federal Ministry of Education and Research (BMBF) through the project "Stadt-Land-Plus – Wissenschaftliches Querschnittsvorhaben" (FKZ033L200), and by the hosting institutions of the authors.

S. Bartke (✉)
German Environment Agency, Dessau, Germany

Helmholtz Centre for Environmental Research – UFZ, Leipzig, Germany
e-mail: stephan.bartke@uba.de

T. Forster
New School University, Practice2Policy LLC, New York, NY, USA

G. Githiri · R. Sietchiping
UN Habitat, Nairobi, Kenya
e-mail: grace.githiri@un.org; remy.sietchiping@un.org

A. Jering
German Environment Agency, Dessau, Germany
e-mail: almut.jering@uba.de

J. Kago
Kenyatta University, Department of Spatial Planning and Urban Management, Nairobi, Kenya
e-mail: jackson.kago@ku.ac.ke

S. Schlimmer
Institut Français des Relations Internationales, Paris, France

UN Habitat/Global Land Tool Network, Nairobi, Kenya
e-mail: schlimmer@ifri.org

H. Ginzky et al. (eds.), *International Yearbook of Soil Law and Policy 2019*,
International Yearbook of Soil Law and Policy 2019,
https://doi.org/10.1007/978-3-030-52317-6_18

1 Introduction

Urban and rural areas depend on each other—especially in urbanizing regions with competing interests in scarce environmental resources, such as soil and land. The development of cities, urban surroundings and rural areas is functionally and closely intertwined: food, building materials, energy, water and waste are typical material flows next to the pervading networks of housing markets, culture or tourism. The networks and material flows are all linked to contending land-use demands. Urban and rural stakeholders can cooperate or do compete in the exploitation of soil functions and services. However, often decision-makers are not aware of or do not fully encompass the meaning and impacts of urban-rural linkages for un-/sustainable land management. This unawareness is not least due to the complex nature of the interdependencies and urban-rural linkages.

Since the global population is growing faster in urban than in rural areas, countries with growing cities and shrinking regions are facing major logistical, financial and ecological challenges. In prospering regions, interest in the **use of soil and land resources** often intensifies, whereas shrinking regions face increasing inequalities in living conditions and exploitation of apparently abundantly available land. Governments are called to facilitate the best possible living conditions for people, while taking also care of the efficient and sustainable use of the natural basis—the soil—through integrated sustainable measures and investment.

It seems to be obvious: **Urban and rural areas must not be separated but need to be understood as a coherent system** as the interdependencies are fundamental for sustainable development, and thus, for the future viability of regions. The **UN-Habitat "Urban-Rural Linkages: Guiding Principles" (URL-GP)** were elaborated in a stakeholder engaging process in the intention to provide a framework for action to advance integrated territorial development.[1]

This contribution was triggered by a workshop on international sustainable soil policies in Berlin in October 2018, when two of the contributing authors met and discussed the consultation phase of the URL-GP. The idea emerged to **investigate specific topical land-use related URL debates in Germany and Africa** and investigate (1) if the GP at their meta level can provide a global framework even though the specificities of the cases are numerous and heterogeneous, and (2) in how far the GP are soil and land-use specific.

Next, **this contribution** introduces the URL-GP and links them to the global agenda. Subsequently, we shed light on four specific cases in which rural urban linkages are currently discussed and debated in Germany, Kenya and Tanzania (mainland)—these cases represent topical fields of research in these countries. The examples are selected to be linked to sustainable land-use questions. Notwithstanding, the examples go beyond a narrow soil focus because of the complex nature of the URL. Consequently, we assess commonalities and specificities/differences of the cases and the role of the URL-GP to support the assessment and for providing a

[1] UN HABITAT (2019).

framework for action. Here, we discuss how the URL-GP are linked to the soil specific topics.

Before starting, we point out some **limitations of our analysis**: These limitations mainly relate to the specificities of the case studies and to the disciplinary background of authors. Although being an international team, bringing in diverse backgrounds from economics and sustainability science, agricultural science, urban planning, political science and geography, we cannot yet present a full picture of the complex URL and soil relationships. Nevertheless, we do focus on illustrative examples, which **aim at deepening the understanding of global and regional challenges for sustainable rural urban linkages.**

2 The UN-Habitat Urban-Rural Linkages: Guiding Principles

2.1 Overview of the URL-GP

In both the 2030 Agenda for Sustainable Development (with its Sustainable Development Goals (SDGs)) and the New Urban Agenda (NUA), United Nations Member States agreed to **normative policies supporting integrated urban and territorial planning and development,** effectively calling for new, inclusive approaches and enhanced synergies between urban and rural communities and spaces. The reciprocal and repetitive flows of people, goods, information, financial and environmental services between specific rural, peri-urban and urban locations (*defining urban-rural linkages*) are interdependent and constitute the reality of socio-spatial arrangements, creating places with a distinct socially-constructed identity, yet interwoven functional territories. The complexity and uniquely local character of urban-rural linkages are critical to meeting the challenges of sustainable development with positive and lasting benefits. This same complexity and uniqueness has been called to demand a level of shared comprehension of overarching principles that are universal for strengthening urban-rural relations. As will be noted here, many of the principles and actions are relevant to soil and land management.

The goal of the **URL-GP** is to inform pragmatic strategies and propose a framework for action to build an enabling environment for more inclusive and functional urban-rural linkages. Convened by UN-Habitat in January 2018, development of the *Guiding Principles* and accompanying *Framework for Action* incorporated inputs from actors and sources, including case studies and experiences, policy decisions and reports, expert meetings of UN agencies, governments at all levels, international development partners, city networks, academic and civil society experts. The final draft version was published in February 2019 and can be found at https://urbanrurallinkages.wordpress.com/.

The ten *Guiding Principles* and eleven fields in the *Framework for Action* address the major flows across the urban-rural continuum, and together, address

Fig. 1 The Urban-Rural Linkages: Guiding Principles. Source: UN HABITAT (2019), p. 4

both functional and spatial complexity of the issues that face integrated territorial development. The URL-GP (see Fig. 1 and Annex for detailed description) are: locally grounded interventions, integrated governance, functional and spatial system based approaches, financially inclusive, balanced partnerships, human-rights based, do no harm and provide social protection, environmentally sensitive, participatory engagement, and data driven and evidenced based.

2.2 Global Agenda and the URL-GP

Up to twelve paragraphs in the **New Urban Agenda** refer to urban-rural linkages as commitments by member states (see Fig. 2). These commitments exemplify the role that the guiding principles will play as tool for member states to strengthen the

Fig. 2 Paragraphs of the New Urban Agenda linked to urban-rural linkages. Source: UN HABITAT

urban-rural linkages and integrated territorial development. The National Urban policies were selected as tools/indicators for monitoring the implementation of the NUA and the SDGs. As countries develop and review their policies, strategies and plans referring to the guiding principles, aspects related to sustainable soil/land management should be addressed. The NUA explicitly acknowledges the international guidelines for urban and territorial planning for territorial development. The guiding principles exemplify what actions, plans and policies should be incorporated for integrated territorial development.

Most **Sustainable Development Goals** (SDGs) are interlinked and complementary, and the achievement of a particular goal hinges on achieving most of the others. Implementation of the guiding principles of urban-rural linkages complements that of the SDGs. Some of the indicators of the SDGs are directly linked with the framework of action for the URL-GP; this linkage squarely positions the policies mainstreamed with regard to urban-rural linkages as tools for implementing the SDGs including issues related to land and land use. The illustration in Fig. 3 gives an indication of how the URL-GP *Framework of Action* is linked to the SDGs. The links are particularly obvious for SDGs 2, 11 and 15 for the context of sustainable soil use and land management.

2.3 URL-GP Consultation and Implementation Steps

The URL-GP are founded on several events, including the expert group meetings (EGMs) and workshops, which have been held towards **development and advancement of the URL-GP work**. The *Monteria Communique* is the outcome of an EGM titled 'The role of intermediate cities in strengthening urban-rural linkages towards the New Urban Agenda', held in Monteria in 2016. The EGM recognized the need to apply territorial approaches that view urban, peri-urban and rural areas as 'part of the same system'. It was also proposed that governance structures for the implementation of urban–rural linkages should be 'multi-sectoral and multiscale'.[2] This claim was followed by a working session in World Urban Forum (WUF) 2018 presenting a first draft of the URL-GP developed and adding more contributing partners.

In June 2018, there was an EGM in La Paz titled; 'Developing and implementing the National Urban Policy in Bolivia from a territorial perspective'. One outcome included recommendations to mainstream urban-rural linkages in national urban policy in Bolivia. A regional awareness workshop was held in Nairobi in September 2018, where six African countries participated and discussed the significance of **mainstreaming urban-rural linkages** in national urban policies. Other partner events took place in November and December 2018, where the URL-GP were recognized and included in resolutions including the Committee on Food Security (CFS), CBD COP14, the Global Land Forum, and UNFCCC COP24. In 2019, UN

[2]Kago et al. (2019).

Fig. 3 Sustainable Development Goals linked to urban-rural linkages. Source: UN HABITAT

Food and Agricultural Organization launched the FAO Framework for the Urban Food Agenda in a high level political forum, where linking food systems and urban-rural linkages was recognized. During the UN Environment Assembly in March 2019, sustainable food systems dialogue, the URL-GP were recognized as a tool for understanding the territorial dimension of food security.

A **set of URL-GP tools** are being co-developed with partners linked to the *Framework of Action* for policy proposals, planning frameworks and other implementation efforts. Tools include an assessment toolkit to define URL priorities in specific geographic contexts accompanied by a toolkit for developing policy recommendations. The toolkits support national and subnational efforts to assess the challenges, opportunities, capacity, stakeholders and policy gaps for urban-rural linkages, and provide templates for making recommendations. A guide for policymakers also exists, titled "Mainstreaming URL in National Urban Policies". Other tools, including explanatory notes and thematic URL guides, among others, will be developed as Fig. 4 illustrates for the implementation of the URL-GP.

UN-Habitat is currently **implementing projects** in five African countries towards mainstreaming urban-rural linkages in national urban policies. These countries include: Cameroon, Guinea Conakry, Zanzibar, Niger state and Mozambique. The respective toolkits that have been developed are also being implemented in workshops and interviews among other context specific approaches.

Moreover, UN-Habitat is currently co-designing tools with partners for the **implementation of the URL-GP**, where some partners plan to adapt the assessment toolkit and thematic guides in ongoing and new projects. Noteworthy are those projects and tools that are relevant to soil and land management: As of April 2019, these projects include:

1. FAO support for integrating food into national urban policies at country and subnational levels, where soil and land management are one important dimension for planning and policy.
2. Network of Regional Governments for Sustainable Development (nrg4SD), whose regional governments are linking protection of biodiversity with food systems and climate change.
3. United National Standing Committee on Nutrition (UNSCN), for which a thematic approach to the URL through a focus on nutrition includes the link of good nutrition to effective management of soil and land for sustainable diets.
4. CIRAD and partners' plans to use the URL-GP in the context of promoting territorial development in countries of the African Sahel.

3 Case Studies

In the following, we introduce four topical case studies to illustrate the role of urban-rural linkages in different contexts. The case studies were selected in order to test the URL-GP in various geographical and contextual backgrounds. As will become obvious, the URL materialize in very specific and heterogeneous settings.

3.1 Governing Land Markets on the Rural Fringes of Dar es Salaam: Informal Land Deals and Speculative Strategies

Rampant land markets in peri-urban areas commonly result from current urbanization dynamics, pushing urban dwellers and speculators towards the cities' rural fringes. In fact, land in peri-urban zones is the most affected by urbanization trends,

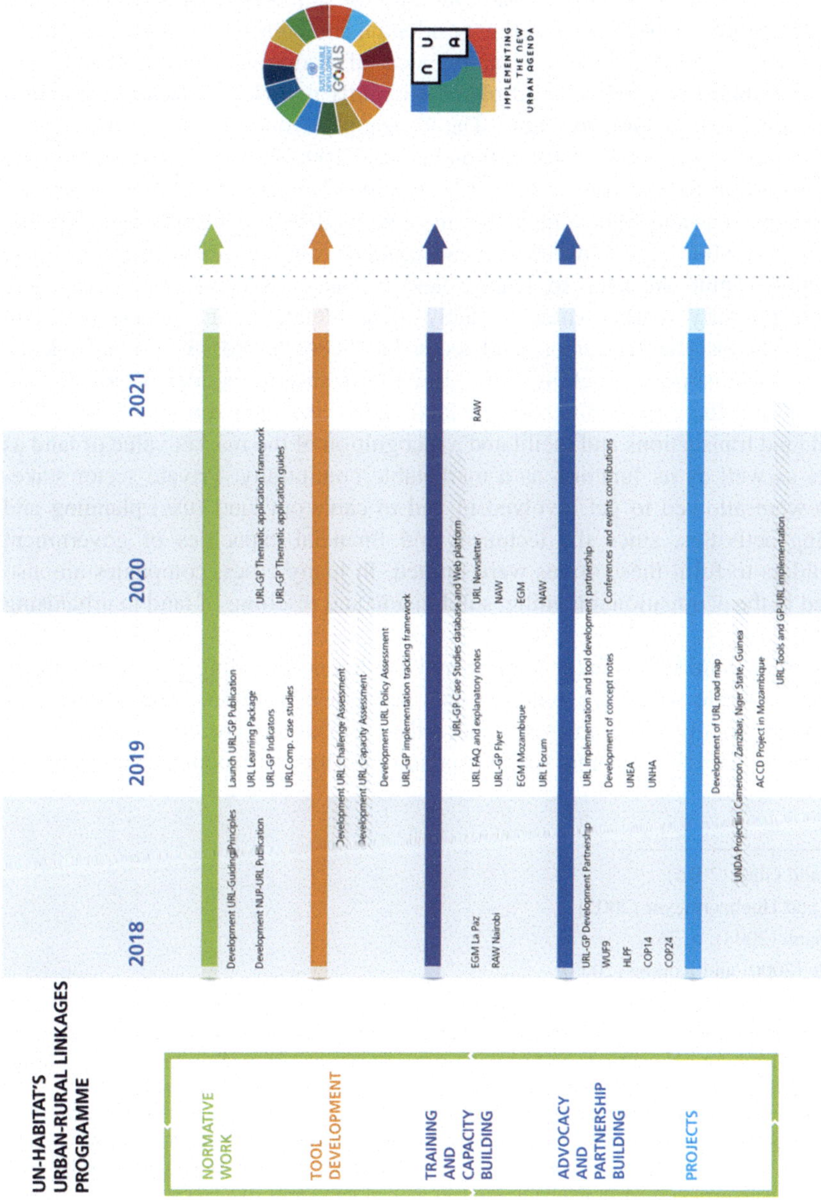

Fig. 4 Overview of tools to support implementation of the urban-rural linkages. Source: UN HABITAT

not only in terms of land transactions, but also through land use transformation and land commodification.[3] The term "peri-urban" is a broad concept that refers to geographically and socially heterogeneous environments.[4] Peri-urban zones "completely surround city centers and fade into the adjacent rural areas [. . .]: they include the entire suburban zone and its residential and industrial suburbs, as well as satellite towns, exurbia and the surrounding areas of predominately rural character".[5]

Dar es Salaam is a prime example of a rapidly growing African city with a dynamic peri-urban land market.[6] The Tanzanian economic capital has experienced a population increase of 1.9 million between 2002 and 2012, almost doubling from 2.5 million to 4.4 million in 2012.[7] Rapid urban growth in Dar es Salaam affects the land use and ownership in the surrounding districts of Bagamoyo, Kibaha, Kisarawe and Mkuranga.[8] Middle-income residents and members of the political and economic elite are keen to acquire land in Dar es Salaam's outskirts where economic property market value is likely to accelerate in the course of urban growth.[9] Whereas the Tanzanian land sector has been shaped since independence (1961) by socialist policy-making and a strong resistance to the idea of private land ownership and the commodification of land rights, the land reform of the 1990s enabled land transactions and facilitated a recognition of the market value of land as an asset as well as its function as a marketable commodity. Private sector stakeholders were allowed to get involved in and to carry out land (use) planning and surveying activities, since the technical and financial capacities of government stakeholders to fulfil these duties were limited. In many cases, companies are also involved in the acquisition, planning, subdivision and reselling of land in urbanising settings.

Most of the land sellers in the villages and hamlets surrounding Dar es Salaam are indigenous people who have acquired land through inheritance. **The reasons for selling land** are diverse: numerous land sellers aim to acquire "fast cash" for the payment of school fees or other social expenditures, for constructing or improving a

[3]Locke and Giles (2016).

[4]Mbiba and Huchzermeyer (2002).

[5]Wehrmann (2008), p. 76.

[6]Kironde (2000) and Kombe (2005).

[7]Andreason (2013).

[8]This study is part of a post-doctoral research program in political science. A ten-week field work (February–March and November 2018) has been conducted in Dar es Salaam as well as in the districts of Bagamoyo and Kisarawe. In both districts, different wards and villages have been selected in order to reflect the diversity of land uses and degrees of urbanization in peri-urban settings. Bagamoyo consists of eleven wards (which are subdivided into hamlets), of which nine have been transformed into townships and two still undergo the process. The research method was based on a qualitative approach involving semi-structured interviews, documentary research and a press review. In Tanzania, around 60 interviews were conducted with land experts; central, district, ward and village government stakeholders, land buyers, land sellers, brokers/middlemen and civil society stakeholder.

[9]Tomitho and Chiombola (2015).

home, for establishing a small shop or even to buy a cheaper piece of land situated in more remote areas. Land selling has helped some villagers to become financially capable of diversifying their economic activities and of partly shifting from agriculture to business and shop-keeping activities. Land subdivision and -selling can also be encouraged by local authorities in peri-urban villages, which are administratively transformed into townships and where the ownership of farms extending three acres is prohibited.

Peri-urban land is mostly bought by urban dwellers from Dar es Salaam seeking land for housing, investment, speculation or for weekend-farming. Businessmen, public servants, normal workers, and also members of the East-African diaspora are among the most recurrent profiles of land buyers. Urban and middle-class land buyers do not necessarily intend to develop the land. The expected growth of land values in peri-urban areas, increased by investment and infrastructure development attracts land speculators and individuals looking for lucrative and relatively easy and riskless saving strategy.

The **rapid growth of peri-urban land markets** implies a certain number of challenges regarding land governance: despite existing regulations framing land selling, peri-urban land markets are characterized by informal negotiations, procedures and stakeholders.[10] For instance, the involvement of dubious land agents and middlemen lead to opaque procedures and artificially rising land values. The lack of selling records in local government offices further complicates the follow-up and formalization of land deals. The regulation of the peri-urban land market is further restricted by poor and sporadic planning, which is linked to the limited technical and financial resources of local governments. Rather, pockets of planned zones were observed in areas where farmland had been acquired by land purchasing companies for subdivision and plot selling. The quality and accuracy of the planning process by local government land agents is also subject to increasing pressure by private stakeholders, such as land purchase companies driven by the interest of economic profit.

Finally, the poorly regulated business of land subdivision and selling also involves land use **transformation from agricultural to residential and commercial land use**. In fact, land sales trigger fundamental agrarian changes, breaking cultural ties and subsistence activities.[11] The transformation of villages into townships and of farmland into residential and commercial land affects the status, value and the use of land as well as the natural, social and economic environment of indigenous communities. These changes come along with a confrontation of different conceptions of land rights (collective and customary vs. individual and more formalized land rights). Unequal access to knowledge between sellers and buyers, e.g. regarding the potential appreciation of peri-urban land reflects the power imbalance at the expense of the indigenous. Poor villagers and local community members selling land in a short-term perspective for finding "fast cash" risk losing

[10]Nelson (2007).

[11]Becker (2013), p. 115.

access to their livelihoods. Consequently, locals are pushed towards the remoter parts of the rural fringes of metropolitan cities. These risks are yet to be addressed by both urban and land policies.

The above-mentioned challenges of land governance in peri-urban settings are more or less directly **addressed by several of the URL-GP**, but principles 2, 3, 6 and 10 are of a particular importance: the management of land rights within the urban-rural nexus is characterized by hybrid forms of land administration, involving traditional, municipal, rural etc. authorities. Tenure security can only be achieved by joint forces of these stakeholders that favor a horizontally integrated governance approach (GP 2). An integrative territorial approach of policy and planning (GP 3) taking into consideration the diverse needs in terms of land use of the various socio-economic groups inhabiting peri-urban areas would be a cornerstone of the regulation of urban sprawl towards the cities' outskirts. The guidelines also stress the need to embed human rights—such as the access to housing, and thus, also land—into all policy initiatives (GP 6). The URL-GP thereby foster existing initiatives and international frameworks promoting the recognition of land rights and the strengthening of tenure security. Yet, improving land tenure security of various social groups co-habiting in peri-urban areas involves the collection of knowledge and empirical evidence on the various forms of existing land claims and land uses existing along the continuum of land rights (GP 10). Policy-initiatives and measures to regulate land in peri-urban zones, which has yet to be investigated by further research, must be based on the collection and analysis of valid and accurate data.

3.2 Land Use and Soil Degradation in the Urban Fringe in Kenya

Nairobi City has continued to expand to the peri-urban regions due to demand of land for housing and commercial facilities at the expense of these rich agricultural zones. The effect of this expansion has not only led to the loss of fertile land but also to the disruption of livelihoods in the zones. The City is surrounded by fertile agricultural regions; that is the Kiambu highlands on the Western and north-western peri-urban interphase that have fertile volcanic loam soils conducive for dairy, tea and coffee farming. The expansion of the city of Nairobi as a result of rapid urbanization has encroached on the rich agricultural lands that feed the city. Livelihoods of smallholder farmers in Kiambu are being disrupted and the area is transforming to fast-growing gated neighborhoods and informal settlements with a mixture of residential and commercial functions. 'Tatu City', a Kenyan flagship project for Kenya's Vision 2030 is a private real estate development in Kiambu, 24 km from the City of Nairobi. The project is a mixed-use development project on a 5000 acre piece of land previously under coffee. Within the same locality, Migaa Golf Estate, which is located 21 km from the City, is an upcoming gated residential neighborhood on a 774 acre piece of land also previously under coffee. Tilisi

Logistics Park, on the other hand, is set to be constructed in Limuru, 27 km from Nairobi on a 400 acres piece of land previously under tea. 'Northland City', located 20 km from the City, is a mixed-use development on 11,576 acres of land potential for ranching that will be converted to residential, educational commercial, industrial, recreational and ranching use.

Additionally, the **water catchment zones** of the rivers flowing through Nairobi city also face pollution and encroachment. An example is Ondiri Swamp, the source of Nairobi River that is threatened by increased human activities stemming from the growing peri-urban Kikuyu satellite town located just one kilometer away. The underground water source of Kikuyu springs provides water to Nairobi and the adjacent Kikuyu town and is also home to a range of bird species. The 74-acres peat bog faces environmental challenges stemming from solid waste dumping, harvesting of macrophyates, extensive deforestation, over extraction of water, overgrazing, and encroachment for residential and commercial use. The existing land use practices are not conducive for its existence and the growing town seems to surround the wetland.[12] The authors note that success in conservation efforts calls for a participatory approach that is based on informed stakeholders. This note is in line with URL-GP 9 that calls of participatory engagement and also GP 1 that calls for locally grounded interventions. On the southern part of the city, the urban foot print has expanded into the savannah grasslands conducive for cattle ranching and home to the indigenous Masaai community, leading to disruption of their livelihoods and affecting their pastoral lifestyle. The urban expansion has additionally interfered with the wildlife migration corridor disrupting the sustenance of wild animals in the Nairobi National Park. An article in Reuters illustrates how rapid urbanization is taking toll on Maasai communal land.[13] The URL-GP note the significance of cultural diversity and the need to reduce inequality, with GP 7 emphasizing the need to strengthen URL to overcome conflict, provide social protection and improve diversity. Further, GP 6 notes that interventions should be human rights-based.

The **expansion of cities' along the urban-rural fringe** should be managed more effectively to ensure a balance between various land uses to prevent urban sprawl achieve sustainable urbanization. Land use planning along the urban fringe should encompass a territorial approach that "emphasizes the functional and spatial interconnection of urban, peri-urban and rural areas, forming an urban-rural continuum".[14] This ties with the URL-GP that advocate for integrated governance horizontally, vertically and sectoral. The URL-GP, like the New Urban Agenda, outline the need to create an integrated and balanced urban development that is inclusive and beneficial to both urban and rural dwellers and ensures the protection of ecologically sensitive areas. The URL-GP 8 calls for environmentally sensitivity in line with Rio+20 outcome document "The Future: we want" and advocates for the need to "Invest in the protection of ecosystem functions and services from peri-urban

[12]Macharia et al. (2010).

[13]Kavilu (2016).

[14]UN-Habitat (2019).

and rural areas near cities to improve resilience and to provide basic services in disaster preparedness and restoration".[15]

The **land use planning of the peri-urban region poses challenges**[16] where the main concerns lie in the push by developers to make profitable use of the land adjacent to the city. Can farmers be convinced not to dispose of their farms to developers? In the case of the land adjacent to the City of Nairobi, the decline in coffee prices (leading to unprofitable agricultural activities) has aggravated the conversion of land in Kiambu to housing and industrial use. No visible efforts for diversification of these large-scale coffee plantations to other agricultural ventures exist. Innovative models like Payments for Environmental Services (PES)[17] to give incentives to farmers and landowners to sustain agricultural activities amidst competing land use could be explored in line with GP 4 on financial inclusivity. Peri-urban land falls between urban and rural jurisdictions with differing capacity to deal with urbanization and sometimes different legislation on land use and capacity to enforce development. Thus, developers can take advantage of the situation and expand into these areas. Further legislation that protects peri-urban land is weak, focusing more on the environmentally sensitive wetland areas and river riparian reserves, but not the agricultural-rich land.

The **Kenyan Government** recognizes the challenge of loss of vegetation cover, and eventually, soil degradation as a result of competing land uses like real estate development, infrastructure, and other land uses. Through the National Land Use Policy, it seeks to ensure that there is sustainable urbanization to "protect agricultural land from indiscriminate extension of urban boundaries and other encroachments and re-planning of peri-urban areas for agricultural and pastoral communities".[18] Kiambu County is in the process of developing an integrated spatial framework to guide spatial development and regulate land use in the county.[19] This government initiative is a late intervention that comes after large-scale real estate developments have already taken place. Due to a lack of this County Spatial Plan, there currently exist no specific zoning guidelines of various land uses in the county, and neither a clear delineation of urban areas.

[15]Ibid.

[16]Sietchiping et al. (2014).

[17]Payments for Environmental Services (PES) "occur when a beneficiary or user or an ecosystem service makes a direct or indirect payment to the provider of that service. The idea is that whoever preserves or maintains an ecosystem service should be paid for doing so" UN-Habitat (2019), p. 43.

[18]KMLPL (2017).

[19]County Government of Kiambu (2018).

3.3 Urban-Rural Linkages from a Food System Perspective in Germany

The discourse on URL in Germany is still relatively young. In 2016, the German Advisory Council on Global Change published its comprehensive survey "Humanity on the move: Unlocking the transformative power of cities",[20] analysing the rapid worldwide urbanization including the impacts on URL from a global environmental, social and political perspective. At political level in Germany, URL played a certain but not prominent role in the preparation process for the UN Conference Habitat III in October 2016 and the following discussion on the implementation of the New Urban Agenda (NUA). Germany belongs to the most densely populated countries in Europe. The influx of population in the city districts proceeds heavily, so that the cities and the metropole regions (e.g. Rhine-Ruhr, Rhine-Main, Berlin-Brandenburg, Munich, Nuremberg) are steadily growing.[21] In 2018, the German government established a Commission for Equivalent Living Conditions in Rural and Urban Areas with the ambition to substantiate the NUA and to improve the URL in Germany.

At the **scientific level**, research on URL and their specific impacts on land use can be traced in several disciplines, such as planning and governance.[22] Research in URL has been promoted by the German Environment Agency (UBA) and the Ministry of the Environment (BMU) since 2015. A particular example is the research project *Rural Urban Nexus: Global sustainable land use and urbanization* (UBA, FKZ 3715 75 1220).[23] The clear focus was on the environmental dimension of URL, especially on resource flows and their management between rural and urban areas, as these two are closely linked through different types of exchange.

Due to the globalization of goods and value chains, changes in economic and consumption patterns, ICT and innovations, the agricultural policies of the industrialized countries (inclusive the common agricultural policy of Europe) changed in the last five decades from domestic traditional to a more and more industrialized production. **The entire food system**, the agricultural production, the processing, distribution, retailing, consumption and management was molded by urbanization and economic development. At the same time, policies for rural and urban development continued to be largely disconnected and rarely reflected from an integrated perspective. Regarding the food system, URL in terms of circular economy have been disconnected, resulting in strong impacts on local to global environment, just to mention impacts as climate gas emissions, soil degradation, water and air pollution or loss of biodiversity.[24] The industrialized food system contributed significantly to

[20]WBGU (2016).

[21]BBSR (2016).

[22]Repp et al. (2012).

[23]Wunder et al. (2019).

[24]UBA (2013).

the increase of the global production volume of agricultural goods and food and to lower product prices for producers and the consumers. Although the industrialized food system externalizes ecological and social costs, the damages caused by the system are not integrated in the product prizes. The damages have to be borne by all tax payers or even by the people of the countries outside Europe.[25]

Another facet of **the broader URL** is the following: Germany is the world's third biggest exporter of agricultural goods (meat, milk, milk products, cereals and processed products) and at the same time also the third biggest importer.[26] Imports of oilseeds and oil seed products and high protein feedstuff, such as soy, are most prominent for the industrialized food production. Especially the intensive animal husbandry is associated with so-called "virtual land imports",[27] also with virtual water imports, imported (virtually) with the goods. This land or water is lost or no longer available for the supply of the population in the producer countries.[28] Imports of bulkfeed (e.g. soy) for animal production effect disruption in global nutrient cycles (e.g. nitrogen, phosphorus) and impact climate additionally by indirect land use change (ILUC)[29] through forest clearance, as well as land grabbing and displacement of indigenous people.[30]

Compared to many countries in the so-called Global South, food availability and access to food in Germany is secure. But, as in many European or other industrialized countries, unhealthy dietary patterns (high consumption of meat-, sugar- and fat) lead increasingly to obesity of the population—54% of German adults are affected by overweight or obesity.[31] The high prevalence of obesity results in various chronical diseases (mainly heart and vascular effects, diabetes) putting increasing pressure on the public household.[32]

New movements for sustainable food systems emerge: For a long time, food policy was shaped only by politics or industry, not by the communities—due to sufficient availability. Nowadays, awareness for sustainable and healthy food is rising; more and more people want to know how their food is produced and where it comes from—indicating a rising preference for regionally and locally produced food. Recently, critical scientists and civil society have been eagerly raising societal

---

[25]Lugschitz et al. (2011) and UBA (2013).

[26]UBA (2018).

[27]The feedstuff for animal husbandry uses agricultural land in other countries (ex. for soy in South America).

[28]UBA (2018).

[29]The concept of indirect land use change assumes that due to the additional demand for raw material from growing sectors (e.g. feedstuff) attention is shifted to other areas on the strength of the assumption that the demand for these products will go down less than the demand for feedstuff. In consequence, the loss of the land used originally to produce the respective commodity is at least compensated by making new cultivation areas elsewhere arable, which leads to land use changes, particular the conversion of forests and grassland to cropland—confer UBA (2013), p. 44.

[30]UBA (2013).

[31]RKI (2017).

[32]Yates et al. (2016).

attention to the ecological, social and health problems of the industrialized food systems and its impacts on the global ecosystem and climate.[33] The Milan Urban Food Policy Pact,[34] signed by more than 180 European cities (April 2019), brought back nutrition on the level of regional politics. In particular, **the interface between cities and surrounding areas in regard to food production and consumption** has been recognized by the civil society in Germany, calling for radical changes to urban food systems. New niche movements, starting mostly from the civil society, supported by science and partly supported by the state,[35] try to develop integrated approaches for a viable relationship between rural and urban areas and to demonstrate linkages to global sustainable land use in the context of urban living. Food Councils, Community Supported Agriculture (CSA), Food Cooperatives, and Food-sharing initiatives are growing more and more all over Germany, as new forms of common organization of food supply in cities and on the countryside.

Food Policy Councils (FPCs) are the most important instruments of regional food policy in the urban-rural space. As food policy in Germany until now is under the responsibility of the federal state and of the European Union, FPCs want to retract the responsibility at local or regional government level. FPCs bring together a cross-disciplinary group of stakeholders from agriculture and food policies, from citizenship and administration to address and improve the urban food system for more local and sustainable food supply. The aim of the FPCs is to develop and implement a masterplan for a **regional food system**. The concept of FPCs is not new; the first council was founded in 1982 in the US. The FPCs' structures, practices and policies are still evolving and may differ from council to council. FPCs enable local, communal, civil and economic actors to exchange, network and pursue common actions.[36] The first Food Councils in Germany have been established in big cities (2016 Cologne, followed by Berlin, Frankfurt, Dresden, Munich). By April 2019, more than 40 councils exist, including also smaller cities and communities and many more are just in process of establishment, aiming at regionalized food systems and resource flows, as well as sustainable and sound diets. Although **regional food strategies** are still young in Germany, their development shows a strong dynamic. A food strategy at federal level does not yet exist, but research on first ideas for elements of a national food strategy are under discussion. Research results show various positive effects of designing and building up regional food policies in the fields of environment protection, health, food sovereignty, participation and governance, education and consumer protection, regional identity and circular economy. The most important precondition for installing regional food strategies is the willingness of all concerned actors to cooperate on all levels of administration.[37]

---

[33] Stierand (2014).

[34] http://www.milanurbanfoodpolicypact.org/text/.

[35] UBA and Ministry of Environment and Nature Protection support several projects researching on niches and community supported agriculture as well as their networks.

[36] INKOTA (2015).

[37] Wunder et al. (2019).

The food system, with focus on regional food sovereignty and ecological products, seems to be an important and very appropriate **entry point to improve URL**, meeting most of the URL-GP. FPCs and the mentioned new movements in Germany can be seen as locally grounded interventions, concerning the URL. They look for an integrated, democratic approach to food systems. Territorial approaches and strengthening of urban–rural linkages is key in developing sustainable food systems. They are environmentally sensitive; land preservation is a key component of urban food security, so they raise awareness for a diligent and responsible use of soil and land. FPCs strengthen URL by local multi-actor platforms for regional nutrition and local food policies. Functional and spatial system-based approaches contribute to circular economy. They are based on balanced partnerships; they provide social protection, as they also demand that regional food systems should be integrated into a national food strategy.

3.4 Urban-Rural Linkages as Research Subject: "Stadt-Land-Plus"

Despite the increasing awareness for food systems, in the mindset of many Germans, cities and countryside are complete opposites. The way of living, working, and consuming is said to differ significantly. However, **cities and countryside are closely interwoven** not only with regard to food as explained above. Regional actors, in particular the municipalities, must collaborate to link the city and surrounding areas. In practice, however, there are often problems between municipalities with mayors not thinking further than they can watch from the tower of their city hall. Only few established or institutionalized processes and structures for the collaboration beyond administrative boarders exist, e.g. in the so-called metropole regions mentioned before. Moreover and more problematic, the knowledge and awareness about the importance and meaning of interdependencies of urban and rural areas is limited. This unawareness is a major stumbling block to a fulfillment of the URL-GP 1-3, locally grounded interventions, integrated governance, functional and spatial system based approaches.

The Germany Federal Ministry of Education and Research (BMBF) has therefore launched the funding initiative "Stadt-Land-Plus" ("Urban-Rural-Plus").[38] As of December 2019, **Stadt-Land-Plus** supports 12 interdisciplinary collaborative projects in three project clusters. A cross-sectional project supports the networking and transfer as well as the processing of cross-project, cross-cutting topics. The Stadt-Land-Plus funding initiative is part of the Zukunftsstadt ("City of the future") flagship initiative within the BMBF framework programme "Research for Sustainable Development - FONA[3]". Stadt-Land-Plus complements the flagship Zukunftsstadt by extending the perspective beyond the borders of the city. Stadt-

[38] Stadt-Land-Plus (2019).

Land-Plus promotes research projects that aim at supporting integrated sustainable development of regions throughout Germany by strengthening urban-rural relations. Taking into account the interests of cities, urban hinterlands and rural areas, a resource-efficient and economically viable land management at regional level is pursued. The BMBF Minister Anja Karliczek commented on the occasion of the inaugural event organized by Stadt-Land-Plus via Twitter: "Good living and working in the whole country - this is a mission of the Federal Government's High-tech Strategy 2025. With the Stadt-Land-Plus funding measure, we are working on the mission by focusing on sustainable land management and equivalent living conditions in the city and country."[39]

The main topics of the "Stadt-Land-Plus" funding measure resonate with the URL-GPs. The aim of a joint sustainable development of urban, peri-urban and rural areas for mutual benefit is linked to both GP 2 "integrated governance" and GP3 "functional and spatial system based approaches". The aim for a development of a sustainable regional circular economy and strengthening regional value creation is connected to GPs 4 "financially inclusive" and 8 "environmentally sensitive"—the latter with consideration of soil as a scare resource and further material stocks and flows as critical indicators of a sustainable regional economy. The goal of improving the common information and knowledge-based decision-making basis of urban and rural actors to pursue innovative solutions for reconciliation of interests echoes with GP 5 "balanced partnership" and also GP 10 "data driven and evidenced based" and again GP 2 "integrated governance".

Within the subject area **regional circular economy and quality of land management,** dependencies and interdependencies between city, urban surroundings and rural area regarding land and material flow management as well as material cycles, are addressed. The aim is to develop options for a resource-efficient, economically viable regional circular economy. The overall purpose is to relieve the burden on land and soils and at the same time achieve a higher regional added value. The nature and quality of land management influence the regional circular economy, e.g. as in the supply and disposal of cities with food, with biogenic raw and secondary raw materials or waste. For this purpose, better knowledge of material flows and their whereabouts are indispensable. Cascade use can support the regional circular economy, increase added value and relieve land use.

Improved reconciliation of interests between city, suburbs and rural area: The growing urban population and urban processes are increasingly effecting the development in regions. Influencing factors on the city, urban surroundings and rural areas include changes in behavior and entitlement to nutrition, multifunctional places to live and work, energy transition, recreation and nature conservation. At the same time, there are opportunities for rural areas, e.g. through the use of

[39]Own translation of "Gut leben und arbeiten im ganzen Land - Das ist eine Mission der neuen Hightech-Strategie 2025 der Bundesregierung. Deshalb fördern wir mit der Maßnahme Stadt-Land-Plus Forschungsprojekte zu nachhaltigem Landmanagement und gleichwertigen Lebensverhältnissen in Stadt und Land."

digitization. The projects are expected to develop medium- to long-term scenarios for sustainable and more effective regional land use structures that consider regionally central trends. It is to be expected that new or intensified conflicting goals as well as an increased appreciation regarding the use of the resource land will be identified.

In addition to the project clusters, the projects work together on **different cross-cutting topics**. These address issues of relevance for several collaborative projects. Given the overarching importance, projects contribute to several of these topics. The findings of the cross-cutting issues are part of the recommendations for action for municipalities and the basis for further scientific discussion.

- Promoting equivalence of living conditions—shaping attractive regions throughout the country
- Pursuing regional sustainability goals
- Forming mechanisms and institutions for urban-regional development processes
- Digitalization—unleashing potentials for urban, rural and connected areas
- Improving internal communication and transfer—achieving lasting results

Each collaborative project in Stadt-Land-Plus must involve urban actors and local and rural stakeholders. In order to be able to adopt a systemic approach, funded projects are required to have a convincing regional relation to urban-rural links. The term "region" refers to spatial units that are particularly closely connected or interdependent with regard to their **functional relationships**. Project regions must include at least one medium or large city (>50,000 inhabitants) involved in the project. A focus of the first call has been on project regions with growing cities or communities, as there is often an increased competition and use interest in the limited resource land.

Moreover, convincing exploitation perspectives were a central concern of this application-oriented funding measure, which is in line with the URL-GP **Framework for Action** pursuing active support for the actual implementation of visions and principles. Solutions and options for action elaborated from the research projects must be exemplary in nature. For the **transferability of the results**, suitable products must be produced within the funded term. The funding period consists of a research and development phase—usually of up to three years—and a period of implementation and stabilization of a maximum of two years. In order to carry out the implementation and perpetuation phase, it is necessary to reach an ambitious milestone at the end of the research and development phase. With the participation of the relevant stakeholders, in particular decision-makers as well as the regional public, concrete steps should be taken towards the scientifically accompanied implementation and perpetuation of project results. It is also encouraged to cooperate with other cities and regions (possibly also in cross-border cooperation) for the transmission and testing of the implementation and dissemination under different framework conditions.

Different actors and scientific disciplines work closely together to achieve this goal throughout Germany. For example, the Bodenseekreis is researching how biological waste from municipal landscape management can be converted into activated carbon in a special process. The aim is to use the product obtained regionally for wastewater treatment in sewage treatment plants. In the Leipzig-Halle metropolitan area, attempts are being made to counteract the disordered growth of residential areas in the suburban areas between the major cities. In a cooperative participation and negotiation process with the surrounding smaller municipalities, an integrated land management and residential area concept is developed. Among other things, this should serve as the basis for recommendations for action in local and regional planning. As in all projects, the results and instruments are also available to other regions throughout Germany.

The synthesis project, coordinated by the German Environment Agency (UBA) in collaboration with Institut Raum & Energie and StadtLand GmbH, promotes interdisciplinary research work and links the joint projects with one another. Another task over the next five years is to supplement research on cross-cutting issues and to feed new findings into the technical dialogue at European level as well. The UBA president, Maria Krautzberger, emphasized in a press release the research focus and the federal political relevance: "One of the tasks of the cross-sectional project is to consider topical issues of relevance to federal politics and to support the research of the associations in this regard. [...] Stadt-Land-Plus is not only intended to add value in the participating regions, but also to improve the framework conditions for regional sustainable spatial development throughout Germany."[40]

4 Commonalities and Specificities of the Cases and Their Specific Link to Soil and Land Management

4.1 Communalities and Specificities of the Cases

The Guiding Principles for strengthening Urban-Rural Linkages have been designed for **universal relevance and contextual adaptation**. The discussion of the cases from Germany, Kenya and Tanzania demonstrates that the ten principles can be linked with each of the cases fruitfully.

The presented cases, despite their heterogeneity, share **obvious communalities**: In none of the examples are the URL fully implemented. This gap manifests with regard to various facets. Neither seems there to be a societal appreciation nor an in-depth understanding of the interdependencies. As a result, it takes no wonder that also institutions and governance approaches—specific regional or functional area

[40]Own translation of: "Das Querschnittsvorhaben hat unter anderem zur Aufgabe, aktuelle bundespolitisch relevante Fragestellungen mitzudenken und die Forschung der Verbünde diesbezüglich zu unterstützen. [...] Stadt-Land-Plus soll nicht nur einen Mehrwert in den beteiligten Regionen leisten, sondern zukünftig die Rahmenbedingungen für eine regionale nachhaltige Raumentwicklung deutschlandweit verbessern."

oriented policies and instruments—are either lacking or priority goals on the research agendas. This must be seen as critical, because governance is important to implement the principles and goals—research priority is a first step in this direction.[41]

Yet, also a **common specific concern is clear: soils and land taken due to urbanization** lead to increased awareness to the URL in all introduced cases. Urbanization demands more land, it takes place on arable land, it demands more resources and endangers natural resources on a global scale. The future of urbanization, but also the health of the population, applies to the relationship between cities and their surrounding rural areas. URL are mainly discussed from the perspective of land use for building and housing as well as for mobility and food production. Sustainability goals and ecological requirements for land use emerging from the Agenda 2030 and the national legislations and respective sustainability standards have not yet been integrated sufficiently in the **actual land use practices** (measured as minimizing land take or increasing ecological farming). This concerns mainly the fields of land use for building and housing, mobility, industrial and food production and consumption. The urban fringe is the area where the URL is most tangible and where conflicts materialize.

We also learn from the cases that participation and willingness for cooperation of all multi-actors and administration is very important in order to realize change. Education and awareness raising is essential and can contribute a lot to reach more sustainable URL.

The presented cases also show **few specificities**. The **status quo of URL consideration is diverging** from nearly not existing in the property market to well established considering the Food Councils in German metropoles—although, as argued before, this is also a movement in the infancy of a broad societal institutional embeddedness.

As an interim result, we can answer to our question "Can the GP at their meta level provide a global framework even though the specificities of the cases are numerous and heterogeneous?", with a clear yes—given the (admittedly restricted but divers) set of cases introduced in this chapter.

4.2 Links of the URL-GP with Soil and Land Management

To answer our second question, "How far are the GP specific to soil and land-use?", we next discuss the relations based on and with reference to the cases and general considerations:

Related to **GP 1 "Locally grounded interventions",** it is obvious that national and subnational commitments to sustainable land and soil management in policy and plans need to incorporate the urban-rural nexus. Recognition of local solutions in

[41] See also Bartke et al. (2018) and Nathanail et al. (2018).

land conservation, improvement of land tenure security and further environmental protection efforts are concrete entry points. Building up regional and community based food system is a commitment to sustainable and fair food production and consumption, which includes sustainable land and soil management in practice. Regional food production and consumption is an important starter and connector for sustainable urban-rural interactions recognizing local land conservation and protection efforts. **GP 2 "Integrated Governance"** is linked to policies, strategies and plans on land and land use that are integrated across spatial scales, public and private sectors and different governance levels (national, regional and local; traditional and public). Regional food strategies integrate multi-sectoral, multilevel and multi stakeholder approaches and in this way, do contribute to integrated governance with potentially positive effects on other policy fields. Also for other areas discussed, only an integrated governance perspective seems promising to answer to specific competing use interests.

GP 3 "Functional and spatial system-based approaches": Policies and plans supporting functional territorial and ecosystem-based inter-linkages between urban and rural land uses are obviously a core of URL as the apparent interdependencies of land use impacts are evident. A prime example are regional food systems, but also other regional products and the logistics needed to exchange the material and product flows must be addressed to support sustainable urban and rural land use in an integrated functional territorial and ecosystem-based approach. Also, **GP 8** emphasizes the aim of an **"Environmentally sensitive"** development. This objective highlights the intrinsic value of soil and environmental health. In URL practice, it requires a priority on protecting and sustaining areas that are significant to biodiversity and ecosystem services—also in land use plans, policies and interventions.

GP 4 refers to a **"Financially inclusive"** development. Sustainable and responsible private and in particular public investments meet the objectives for integrated rural and urban economic, social and environmental development across the rural urban continuum. Public funding of economic development or taxation, e.g. for different land uses, can be a powerful instrument to steer land use.

Sustainable URL seem to be more stable, if following **GP 5** they are based on a **"Balanced partnership"** of stakeholders. Further, land conservation and soil protection efforts call for partnerships, alliances and networks of urban and rural actors and a wide range of stakeholders, including marginalized groups to be involved. In a related way, **GP 6 "Human rights-based"** calls for policies and interventions on soil and land management to respect, promote and fulfil human rights, including access to land and affordable housing. Likewise, the perspective on regional food systems aiming at food sovereignty and stopping hunger or malnutrition are human rights based. Moreover in the societal context, **GP 7 "Do no harm and provide social protection"** is also important in the soil/land context. Policies and initiatives should protect just land tenure systems and a state of soil ecosystem that protects soil functions and the biodiversity. The recognition of cultural diversity and indigenous knowledge in land conservation strengthens tenure security and protection efforts. The protection of land tenure is also important with regard to food sovereignty. **GP**

9 "Participatory engagement" underlines that strategies and interventions in land/soil management should ensure meaningful levels of participation of people, local institutions and communities across the urban-rural continuum in integrated functional area management approaches. Not least for the particular example of establishing regional food systems, this is very important.

GP 10 "Data driven and evidence-based" calls for the use of data systems, like Geographic Information Systems (GIS), in land data collection and land management across national and sub-national levels. Here, the challenges and opportunities of digitalization (e.g. remote sensing) need to be considered. Not least, it expresses that data and evidence (in other words scientifically justified) information should guide URL (and not political short sightedness or populism).

Given the interrelatedness of the URL-GP with land and soil management, it follows that the **Framework of Action** also entails numerous fields of actions for soil/land management. Here, we only give some examples. An achievement of most GP hinges on a sensitive **development of governance, legislation and capacity**. Related to GP2, actions should ensure "whole-of-government" approaches, which could necessitate integrated associations or agencies for land management. In addition, established mechanisms for participatory approaches for reconciling differences to protect the rights of all stakeholders in ongoing land use conflicts are a governance challenge. The Kenyan case study emphasized the need for people driven initiatives in land and soil conservation and management. However, these communities should be empowered through knowledge support. In order to avoid conflicts from early on, **integrated planning across the urban-rural continuum** is a central tool. Instruments that advance cross-sectoral planning at national and subnational levels in a multi-stakeholder setting, addressing the important role of land use planning and management, need to be (further) developed. This includes reviewing, adapting and using locally relevant legal and legislative instruments and methods to develop functional area based development plans, including the integration of land use in the urban-rural continuum. Criteria and approaches for land use planning that enable to monitor and steer integrated management and regulation to address urbanization and rural transformation are needed. The **investment and finance for inclusive urban-rural development** perspective calls for adaptive and innovative models of financing, such as Payment for Environmental Services (PES) to give incentives to promote investment in environmentally motivated soil protection.

Knowledge/data management for dynamic spatial flows of people, products, services and information, development of observatories and GIS-based monitoring systems on land tenure and property transactions, local mapping and GPS data in land and soil management are key areas that require further attention.

Territorial economic development and employment, next to **Infrastructure, technology and communication systems,** are further obvious areas of action that can be linked to soil and land sensitive policy making. Food is mentioned particularly in the UN-HABITAT URL-GP Framework. **Integrated approaches for food security, nutrition, and public health** refers to assessing the quality and performance of territorial food system approaches for linking food security and healthy diets to land. Policies and practices should ensure that food systems provide nutritious and safe foods and address the importance of secure tenure of land or food provision. In this regard, there is an urgent need to preserve regional environments and farmlands to increase access to sustainable and healthy diets.

Notably, **environmental impact and natural resource and land management** are seen as a key field of the URL-Framework, calling for conducting strategic environmental assessments to reduce environmental risks, to addressing land and water tenure and sustainable use of natural resources, to protecting biodiversity and to promoting ecosystem-based production systems and building resilient landscapes. Finance mechanisms and incentives to facilitate fair compensation for ecosystem services between rural and urban areas is one example to protect the world's most fertile and productive lands, which are often found in the path of rapid urbanization. Other areas are the incorporation of integrated landscape management into territorial development plans and the development and implementation of policies and programmes for conservation and sustainable use of natural resources. The critical element also under investigation in the Stadt-Land-Plus projects is the application of systems approaches and circular economy frameworks related to urban-rural flows and streams and to space and land as the basis of production and consumption—and finally, as reuse and remediation of the landscape, including soils.

5 Conclusion, and Recommendations

In this contribution, we addressed land use critically impacting land and soils with a focus on the urban-rural metabolism and relationships, which are claimed to determine the in-/efficiency of land use. We introduced the UN-Habitat "Urban-Rural Linkages: Guiding Principles" as guidance and framework for action to advance integrated territorial development. Different topical cases in the urban-rural continuum in Germany, Kenya and Tanzania were introduced to investigate (1) if the URL-GP at their meta-level do provide a global framework, and (2) in how far the GP are soil and land-use specific.

We answered both questions by concluding that the URL-GP does indeed provide an important contribution to support more sustainable land-use management and policies—as they are relevant in heterogeneous contexts in as much as for land and soil protection.

Overall, the GP do well resonate with the Sustainable Development Goals of the Agenda 2030. This is particularly noteworthy with goals 2 "Zero hunger", 11 "Sustainable cities and communities" and 15 "Life on Land", as the cases clearly demonstrate. For example, the URL-GP are very helpful for building up sustainable regional food systems irrespective of the country. It can help people to improve their living and to protect land and soil and the environment by adequate production and consumption patterns. And it can help to balance the URL.

The ten URL-GP are particularly helpful for building up regional, sustainable food systems. This will work only with a balanced approach, respecting the social and cultural needs and characteristics of all actors concerning the food system, as well as addressing and respecting the local conditions and various interlinkages to all environmental issues (esp. soil, water, biodiversity) of the local and global ecosystem. The URL-GP and the eleven areas of the Framework of Action are all relevant for the case of food systems as the exemplary starting point. As shown by the projects in Stadt-Land-Plus, this also holds true for other areas related to the regional circular economy and creation of value chains in the regional functional areas. Research in this funding programme also—in line with the examples from Tanzania and Kenya—point out the critical role of governance (policy and formal and informal instruments) to steer land use. In particular, the regulation of the property market and the regulation of planning and use restrictions in the urban fringe is a tangible and critical area of in-/efficient land use.

To conclude—the GP-URL, having been elaborated in a broad stakeholder consultation process, are a well-structured support for all local, national and global measures, strategies and programmes. It is important to notice that all dimensions of sustainable development are attentively addressed—including the economic, ecological and social dimensions. As politics is to be implemented locally, it seems to be important to emphasize the social dimension and principles of participation. This needs to be done consciously and responsibly in order to guide a transformation to a socially just, ecological responsible and economical feasible way of living, production and consumption, appreciating the urban and rural areas as different but equally important and valuable inter-connected places.

It was claimed that the URL materialize mostly in the urban fringe. This has both good and bad implications. On the one hand, it is good that at all URL are gaining attention and that a starting point exists from where action can be fathomed. On the other hand, the premise that URL are understood merely as a challenge of the urban fringe and urbanizing areas must be averted. Regions are multi-facet areas with different layers of interdepending relationships, stream of materials, energy, products, services, people, animals, plants, and so forth. The URL-GP enables the appreciation of this broader picture and link it to the SDGs, including sustainable land use.

Annex

Urban-Rural Linkages: Guiding Principles

1. **Locally grounded interventions:** Translate global normative agendas in national and subnational commitments for territorial cohesion and action.
2. **Integrated governance:** Incorporate the urban-rural nexus in multi-sectoral, multi-level and multi-stakeholder approaches to governance integration.
3. **Functional and spatial system-based approaches:** Promote inclusive and systems-based approaches to urban and territorial planning.

4. **Financially Inclusive**: Secure and prioritize sustainable and responsible investments balanced between and conducive to urban-rural linkages formal and informal sectors.
5. **Balanced partnership:** Foster partnerships, alliances and networks that link urban and rural actors and different sectors and are inclusive, participatory and held accountable.
6. **Human rights-based**: Embed human rights-based approaches in all policy instruments and actions across the urban-rural continuum.
7. **Do no harm and provide social protection:** Build urban-rural linkages to recognize cultural differences, overcome conflict, and inequalities in provision of social and health services.
8. **Environmentally sensitive**: Protect, sustain, and expand areas important to biodiversity and ecosystem services in transition to resilient, resource efficient societies.
9. **Participatory engagement:** Create spaces and mechanisms to ensure meaningful participation of people, local institutions and communities across the urban-rural continuum.
10. **Data driven and evidence-based:** Establish or improve knowledge systems for the urban-rural continuum and territorial cohesion.

Framework of Action

A. **Governance, legislation and capacity development**: assess capacity and needs for policy tools; enhance dialogue and cooperation across sectors and planning levels and convene new multi-level, multi-sector and multi-actor governance mechanisms and support inclusion of affected urban and rural populations.
B. **Integrated planning across the urban-rural continuum:** Support localization of national planning; integrate urban-rural linkages and integrated territorial development in National Urban Policies and promote networks and associations of planners in different jurisdictions.
C. **Investment and finance for inclusive urban-rural development:** Address and improve public and private finance and access to credit across the urban-rural continuum, focusing on small and intermediate towns and cities and integrated urban-rural territories. Strengthen financial intermediation services and enable ecosystem service provision in peri-urban and rural areas.
D. **Empower people and communities:** Assess and support inclusive multi-actor participatory processes and partnerships; directly address and compensate for inequities across the urban-rural continuum in order to create balance and inclusion from informal and formal sectors.
E. **Knowledge/data management for dynamic spatial flows of people, products, services and information:** Improve collection and management of spatially and gender disaggregated data and knowledge, including ensuring a balance between inclusion of the formal and informal sectors.

F. **Territorial economic development and employment**: Coordinate urban and rural economic development to enhance synergies; harness the potential of small and medium size towns and apply innovations in participatory credit, finance and enterprise incubation schemes to create new jobs that improve territorial flows of products, services and information.
G. **Coherent approaches to social service provision:** Identify needs and opportunities where stronger urban-rural linkages can impact more spatially and socially equitable service provision; pilot new urban-rural partnerships for health and social services outside cities.
H. **Infrastructure, technology and communication systems:** Perform integrated and inclusive urban-rural infrastructure needs analysis; jointly plan, finance and construct infrastructure for water, sanitation, roads and public transport, electrification and communication, etc.
I. **Integrated approaches for food security, nutrition, and public health:** Mainstream health and well-being by building coherent and linked urban-rural approaches to food, water, energy and health systems with attention to the multiple benefits of the circular economy of the urban-rural nexus, especially when coupled with capacity development and inclusion.
J. **Environmental impact and natural resource and land management:** Conduct strategic environmental assessments to reduce environmental risk from severe flooding, drought, storms, etc.; address land and water tenure and sustainable use of natural resources, protect biodiversity, promote ecosystem-based production systems and build resilient landscapes.
K. **The urban-rural continuum in the face of conflict and disaster:** Assess risks, gaps and vulnerable populations in relation to potential hazards in urban, peri-urban and rural areas; jointly plan for resilience measures and invest across jurisdictions to protect transport, energy, information, health, education, food and water systems, etc.

References

Andreason MH (2013) Population growth and spatial expansion of Dar es Salaam. An analysis of the rate and spatial distribution of recent population growth in Dar es Salaam. RurbanAfrica WP1 Copenhagen: University of Copenhagen

Bartke S, Boekhold S, Brils J, Grimski D, Ferber U, Gorgon J, Guerin V, Makeschin F, Maring L, Nathanail CP, Villeneuve J, Zeyer J, Schröter-Schlaack C (2018) Soil and land management in Europe: lessons learned from INSPIRATION bottom-up strategic research agenda setting. Sci Tot Environ 622–623:1408–1416

BBSR (2016) Metropolregionen-Kooperation und Wettbewerb in Deutschland und Europa. Informationen zur Raumentwicklung – IzR 2016.5

Becker LC (2013) Land sales and the transformation of social relations and landscape in peri-urban Mali. Geoforum 46(May):113–123

County Government of Kiambu (2018) County integrated development plan 2018–2022. County Government of Kiambu, Feb. 2018. https://www.cog.go.ke/downloads/category/106-county-integrated-development-plans-2018-2022?download=308:kiambu-county-integrated-development-plan-2018-2022. Last accessed 17 July 2019

INKOTA (2015) Unser Essen Mitgestalten! Ein Handbuch zum Ernährungsrat. INKOTA-netzwerk e.V. Berlin, Nov. 2015. https://www.inkota.de/uploads/tx_ttproducts/datasheet/handbuch_INKOTA_Ernaehrungsraete_01.pdf. Last accessed 17 July 2019

Kago J, Loose S, Sietchiping R (2019) Implementing the new urban agenda: urban and territorial integration approaches in support of urban food systems. In: Ginzky H et al (eds) International yearbook of soil law and policy 2018. Springer International Publishing, Cham, pp 271–293

Kavilu S (2016) FEATURE-Kenya's rapid urbanisation takes toll on Maasai communal land. Reuters online. https://www.reuters.com/article/kenya-landrights-maasai-idUSL8N1A728G. Last accessed 17 July 2019

Kironde LJM (2000) Understanding land markets in African urban areas:: the case of Dar es Salaam, Tanzania. Habitat Int 24(2):151–165

KMLPL [Republic of Kenya Ministry of Land and Physical Planning] (2017) Sessional Paper No. 1 of 2017 on National Land Use Policy, Oct. 2017. Government Printer, Nairobi

Kombe WJ (2005) Land use dynamics in peri-urban areas and their implications on the urban growth and form: the case of Dar es Salaam, Tanzania. Habitat Int 29(1):113–135

Locke A, Giles H (2016) Urbanisation, land and property rights. ODI report Jan. 2016. Overseas Development Institute, London

Lugschitz B, Bruckner M, Giljum S (2011) Europe's global land demand - a study on the actual land embodied in European imports and exports of agricultural and forestry products. Sustainable Europe Research Institute, Vienna

Macharia JM, Thenya T, Ndiritu GG (2010) Management of highland wetlands in central Kenya: the importance of community education, awareness and eco-tourism in biodiversity conservation. Biodiversity 11(1–2):85–90

Mbiba B, Huchzermeyer M (2002) Contentious development: peri-urban studies in sub-Saharan Africa. Progr Dev Stud 2(2):113–131

Nathanail CP, Boekhold AE, Grimski D, Bartke S (2018) The Europeans' Strategic Research Agenda for Integrated Spatial Planning, Land Use and Soil Management. Final public version of Deliverable D4.3 of the HORIZON 2020 project INSPIRATION. EC Grant agreement no: 642372. UBA, Dessau-Roßlau

Nelson SC (2007) Farming on the fringes: changes in agriculture, land use and livelihoods in peri-urban Dar es Salaam, Tanzania. Geography Honors Projects Paper 10. Macalester College, Minnesota

Repp A, Zscheischler J, Weith T, Strauß C, Gaasch N, Müller K (2012) Urban-Rurale Verflechtungen. Analytische Zugänge und Governance Diskurs. Nachhaltiges Landmanagement Diskussionspapier Nr. 4, Dez. 2012. Müncheberg, ZALF

RKI [Robert Koch Institut] (2017) Journal of Health Monitoring. Fact Sheet 21, Jun 2017. Gesundheitsberichterstattung des Bundes, Robert Koch Institut & Destatis

Sietchiping R, Kago J, Zhang XQ, Augustinus C, Tuts R (2014) Role of urban–rural linkages in promoting sustainable urbanization. Environ Urban Asia 5(2):219–234

Stadt-Land-Plus (2019). https://www.zukunftsstadt-stadtlandplus.de/. Last accessed 17 July 2019

Stierand P (2014) Speiseräume – Die Ernährungswende beginnt in der Stadt. OekomVerlag, München

Tomitho C, Chiombola J (2015) A fact-finding mission report on the dynamics of village lands sale and its implications to rural areas. HakiArdhi, Dar es Salaam

UBA (2013) Sustainable land and biomass resources. Position paper, Federal Environment Agency. Umweltbundesamt, Dessau 2013

UBA (2018) Daten zur Umwelt 2018. Umweltbundesamt, Dessau 2018

UN-Habitat (2019) Urban-Rural Linkages: Guiding Principles - Framework for Action to Advance Integrated Territorial Development. United Nations Human Settlements Programme, Nairobi. https://urbanrurallinkages.files.wordpress.com/2019/04/url-gp.pdf. Last accessed 17 July 2019

WBGU [German Advisory Council on Global Change] (2016) Humanity on the move: unlocking the transformative power of cities. WBGU, Berlin

Wehrmann B (2008) The dynamics of peri-urban land markets in sub-Saharan Africa: adherence to the virtue of common property vs. quest for individual gain. Erdkunde 62(1):75–88

Wunder S et al (2019) Rural Urban Nexus - Globale Landnutzung und Urbanisierung. Integrierte Ansätze für eine nachhaltige Stadt-Land-Entwicklung. UBA Texte 138/2019. Umweltbundesam, Dessau-Rosslau

Yates N, Teuner CM, Hunger M, Holle R, Stark R, Laxy M, Hauner H, Peters A, Wolfenstetter SB (2016) The economic burden of obesity in Germany: results from the population based KORA studies. Obes Facts 9(6):397–409

Aspects of a Legislative and Policy Framework to Manage Soil Carbon Sequestration

Ian Hannam

1 Introduction

Human activities, especially the burning of fossil fuels, land clearing and burning vegetation, have caused a substantial increase in the concentration of carbon dioxide (CO_2) in the atmosphere. This increase in atmospheric CO_2 from about 280 to more than 400 parts per million over the last 250 years is causing measurable global warming.[1] The term "carbon sequestration" is used to describe both natural and deliberate processes by which CO_2 is either removed from the atmosphere or diverted from emission sources and stored in the, terrestrial environment (vegetation, soils, and sediments).[2] Before human-caused CO_2 emissions began, the natural processes that make up the global "carbon cycle" maintained a near balance between the uptake of CO_2 and its release back to the atmosphere. However, existing CO_2

[1] IPCC (2018), p. 6, "This Report responds to the invitation for IPCC '... to provide a Special Report in 2018 on the impacts of global warming of 1.5°C above pre-industrial levels and related global greenhouse gas emission pathways' contained in the Decision [Decision 1/CP.21, paragraph 21] of the 21st Conference of Parties of the United Nations Framework Convention on Climate Change to adopt the Paris Agreement". The Report at A.1 at 6 specifies that "Human activities are estimated to have caused approximately 1.0°C of global warming above pre-industrial levels, with a likely range of 0.8°C to 1.2°C. Global warming is likely to reach 1.5°C between 2030 and 2052 if it continues to increase at the current rate".

[2] United States Geological Survey (2008), p. 2, "Terrestrial sequestration (sometimes termed 'biological sequestration') is typically accomplished through forest and soil conservation practices that enhance the storage of carbon (such as restoring and establishing new forests, wetlands, and grasslands) or reduce CO_2 emissions (such as reducing agricultural tillage and suppressing wildfires)."

I. Hannam (✉)
Australian Centre for Agriculture and Law, School of Law, University of New England, Armidale, NSW, Australia

H. Ginzky et al. (eds.), *International Yearbook of Soil Law and Policy 2019*, International Yearbook of Soil Law and Policy 2019,
https://doi.org/10.1007/978-3-030-52317-6_19

uptake mechanisms, sometimes called CO_2 or carbon "sinks", are insufficient to offset the accelerating pace of emissions related to human activities. It is argued that successful adaptation to climate change impacts will include advances in international and national environmental law particularly in the areas of institutional, technological, education, research, and regulatory practices that concern soil carbon sequestration.[3] Legislative decisions about soil carbon sequestration require consideration of priorities and tradeoffs among multiple resources. For example, converting farmlands to forests or wetlands may increase soil carbon sequestration, enhance wildlife habitat and water quality, and increase flood storage and recreational potential, but the loss of farmlands will decrease crop production.[4] The long-term conversion of grassland and forestland to cropland (and grazing lands) has resulted in historic losses of soil carbon worldwide but there is a major potential for increasing soil carbon through restoration of degraded soils and widespread adoption of soil conservation practices.[5]

1.1 Sustainable Soil Management

Soil is the world's largest terrestrial pool of carbon,[6] and approximately 95% of global food is produced in soil.[7] FAO (2017) states that, "Sustainable Soil Management (SSM) is a valuable tool for climate change adaptation and a pathway for

[3]For an earlier paper on international and national aspects of a framework to manage soil carbon sequestration see Hannam (2004).

[4]IPCC (2018) B.3 at 10, "On land, impacts on biodiversity and ecosystems, including species loss and extinction, are projected to be lower at 1.5°C of global warming compared to 2°C. Limiting global warming to 1.5°C compared to 2°C is projected to lower the impacts on terrestrial, freshwater and coastal ecosystems and to retain more of their services to humans (high confidence)".

[5]FAO Soils Portal http://www.fao.org/soils-portal/soil-management/soil-carbon-sequestration/en/; 'The development of agriculture during the past centuries and particularly in last decades has entailed depletion of substantive soil carbon stocks. Agricultural soils are among the planet's largest reservoirs of carbon and hold potential for expanded carbon sequestration (CS), and thus provide a prospective way of mitigating the increasing atmospheric concentration of CO_2. It is estimated that soils can sequester around 20 Pg C in 25 years, more than 10 % of the anthropogenic emissions'.

[6]FAO (2004), p. 1, "... new strategies and policies within the international framework have been developed for the implementation of agriculture and forestry management practices that enhance carbon sequestration (CS) both in biomass and soils. The importance of these activities is that any action taken to sequester C in biomass and soils will generally increase the organic matter content of soils, which in turn will have a positive impact on environmental, agricultural and biodiversity aspects of ecosystems. The consequences of an increase in soil carbon storage can include increases in soil fertility, land productivity for food production and security, and prevention of land degradation"; http://www.fao.org/3/a-Y5738e.pdf.

[7]FAO (2015a), p. 1, "the most widely recognized function of soil is its support for food production. It is the foundation for agriculture and the medium in which nearly all food-producing plants grow. In fact, it is estimated that 95% of our food is directly or indirectly produced on our soils. Healthy soils supply the essential nutrients, water, oxygen and root support that our food-producing plants

safeguarding key ecosystem services and biodiversity. Due to the incalculable value soils provide to society through ecosystem services, SSM ensures a high return on investment by supporting and increasing these services. Widespread adoption of SSM practices generates multiple socioeconomic benefits, especially for smallholder farmers and large scale agricultural producers worldwide whose livelihoods directly depend on their soil resources. However, evidence recently provided in the Status of the World's Soil Resources (SWSR) report and other studies shows that about 33% of global soils are moderately or highly degraded,[8] i.e. due to unsustainable management practices. On a global scale an annual loss of 75 billion tons of soil from arable land is estimated to cost about USD400 billion each year in lost agricultural production. This loss also significantly reduces the soil's ability to store and cycle carbon, nutrients, and water. Annual cereal production losses due to erosion have been estimated at 7.6 million tones".[9]

1.2 *Instruments*

The primary international legal instrument concerning atmospheric conditions is the United Nations Framework Convention on Climate Change (UNFCCC). Its main aim is "... to achieve the stabilisation of greenhouse gases in the atmosphere at a level that would prevent dangerous anthropogenic interference with the climate system. Such a level should be achieved within a time-frame sufficient to allow ecosystems to adapt naturally to climate change, to ensure that food production is not threatened and to enable economic development to proceed in a sustainable manner."[10] The objective and role of the UNFCCC is strengthened and broadened by the Kyoto Protocol,[11] the Marrakech Accords,[12] and more recently, the 2015 Paris Agreement.[13] The Protocol sets targets for allowable greenhouse gas emissions

need to grow and flourish. Soils also serve as a buffer to protect delicate plant roots from drastic fluctuations in temperature"; http://www.fao.org/3/a-i4405e.pdf.

[8]FAO (2015b) Status of the World's Soil Resources; FAO (2017), p. 1.

[9]FAO (2017), p. 1; FAO (2017) further reports that "Growing concerns about the state of global soils resulted, amongst others, in the establishment of the Global Soil Partnership, the proclamation of the International Year of Soils (2015) by the UN General Assembly and the adoption of the revised World Soil Charter by the FAO Conference. In a broader context, the 2030 Agenda for Sustainable Development adopted a number of related targets in 2015, i.e. those aimed at restoring degraded soil, striving to achieve a land degradation-neutral world and implementing resilient agricultural practices that progressively improve soil quality and minimize soil contamination".

[10]UNEP (1995b) Article 2.

[11]Kyoto (1997) The Kyoto Protocol entered into force generally on 16 February 2005.

[12]UNFCCC (2002a).

[13]Paris Agreement, available at: http://unfccc.int/paris_agreement/items/9485.php; adopted by consensus on 12 December 2015; Article 2(a), 'Holding the increase in the global average temperature to well below 2°C above pre-industrial levels and pursuing efforts to limit the

from developed countries, and recognises the role of forests and land use practices in controlling greenhouse emissions. The Marrakech Accords describe the rules for implementing the Kyoto Protocol as well as important criteria to implement the UNFCCC. The Paris Agreement enhances the implementation of the Convention, including its objective, and aims to strengthen the global response to the threat of climate change, in the context of sustainable development and efforts to eradicate poverty (Article 2).

While these four instruments are the principal legislative instruments for the management of greenhouse issues, they, in conjunction with other international and regional multilateral treaties and strategies for land use and ecosystem management, form a comprehensive framework of international law to deal with soil carbon sequestration. Other instruments in the 'synergy' include the Rio Declaration[14] and Agenda 21,[15] various documents of the World Commission on Sustainable Development 2002,[16] the Marrakech Ministerial Declaration 2002,[17] the Convention on Biological Diversity (CBD),[18] United Nations Convention to Combat Desertification (UNCCD),[19] and the United Nations Sustainable Development Goals (SDGs).[20]

Relevant regional instruments include the Revised African Convention for the Conservation of Nature and Natural Resources,[21] the 1985 ASEAN Agreement on the Conservation of Nature and Natural Resources (not in force), and the Protocol for the Implementation of the 1991 Alpine Convention in the Field of Soil Protection.[22] At the national level, there has been a significant amount of legislative activity in recent years to improve the management of soil as a direct measure against the effects of climate change and land degradation.[23] The approaches adopted and the reasons for environmental law development vary between countries and regions.

---

temperature increase to 1.5°C above pre-industrial levels, recognizing that this would significantly reduce the risks and impacts of climate change'.

[14]United Nations (1992a).

[15]United Nations (1992b).

[16]WSSD (2002a, b, c).

[17]UNFCCC (2002b).

[18]UNEP (1995a).

[19]United Nations (1995).

[20]Transforming our World: the 2030 Agenda for Sustainable Development, UN A/RES/70/1, https://sustainabledevelopment.un.org/post2015/transformingourworld; for a discussion on how soil scientists can help to reach the recently adopted UN Sustainable Development Goals (SDGs) in the most effective manner see—Keesstra et al. (2016).

[21]Adopted in March 2017, https://au.int/en/treaties/african-convention-conservation-nature-and-natural-resourcesrevised-version.

[22]The Protocol for the Implementation of the Alpine Convention of 1991 in the Field of Soil Protection 1991 Official Journal of the European Union, L 337/29, available at: http://www.alpconv.org/en/convention/protocols/Documents/SoilProtocolEN.pdf.

[23]Hannam and Boer (2002) and Boer and Hannam (2019).

2 Legal and Institutional Frameworks

Understanding the basic physical and human related land use issues associated with the process of soil carbon sequestration provides a good guide to the most appropriate legislative framework and the legislative system needed to manage the soil environment. These issues involve how to effectively adapt to climate change impacts, the problems associated with agricultural land use and food security, maintaining and restoring biological and ecosystem diversity, and the control and prevention of land degradation.[24] A comprehensive understanding of these issues points towards the actual legislative and institutional elements that are essential within a national legal and institutional system to effectively manage soil carbon. It also identifies the specific legal and institutional elements necessary within a national legislative system to deal with the land management issues associated with the process of soil carbon sequestration.[25] A legal framework provides law and policy-makers with a practical method and guideline for identifying, developing, or strengthening a legal and institutional system that is concerned with aspects of the environment.[26] An important characteristic of a legal framework that would provide the legal means to manage soil carbon processes is the need for cooperation and coordination between the international and national levels so that the legal issues concerned can be effectively addressed.

3 Soil Issues Associated with Soil Carbon and Climate Change Impacts

3.1 *Physical Processes and Consequences, Changes in Land Use*

The conversion of terrestrial ecosystems for agriculture leads to emissions of carbon through decomposition of terrestrial biomass and from mineralisation or solubilisation of soil carbon. Prominent activities include deforestation, biomass burning, ploughing, and use of organic manure, applying nitrogenous fertilisers, rice cultivation, livestock production, and draining wetlands. These activities lead to land degradation that, in turn, exacerbates the emission of radiatively active gases from terrestrial ecosystems to the atmosphere.[27] The development of agriculture during the past centuries and particularly in last decades has entailed depletion of substantive soil carbon stocks. Agricultural soils are among the planet's largest

[24]Boer and Hannam (2019) and Hurni and Meyer (2002).

[25]Hannam (2004); Hannam and Boer (2004) section IV at 37.

[26]Ibid section 111 at 29.

[27]Lal (2010), p. 7.

reservoirs of carbon and hold potential for expanded carbon sequestration, and thus provide a prospective way of mitigating the increasing atmospheric concentration of CO_2. It is estimated that soils can sequester around 20 Pg C in 25 years, more than 10% of the anthropogenic emissions.[28] At the same time, this process provides other important benefits for soil, crop and environmental quality, prevention of soil erosion and desertification and for the enhancement of biodiversity. Land degradation reduces crop yields and the carbon content of agro-ecosystems.

Effective legal and institutional systems can restore degraded land and lead to the restoration of the carbon sinks.[29] A clear understanding of the principal processes and consequences of land degradation provides direction as to how environmental legislation can best be framed to support a strategy for successful management and control of soil carbon sequestration. For example, soil erosion involves preferential removal of soil clay and soil organic carbon. Displaced organic matter is redistributed over the landscape, becoming more prone to mineralisation. Although some of the carbon displaced is accumulated in depressions and aquatic ecosystems and may become sequestered, soil erosion may exacerbate carbon emissions due to other processes, including the breakdown of soil aggregates. Many changes in land use and land use practices that lead to the emission of greenhouse gases are also those which can benefit other aspects of environmental management, including biodiversity conservation. Of particular importance for the law is the recognition that carbon dioxide is a dangerous greenhouse gas and that the same carbon atom if located in the soil becomes organic matter which is the key to soil fertility, soil stability and general ecosystem health.[30] The mitigation of land degradation by the inclusion of procedures for changing land use must be a significant feature of any national legal and institutional strategy to reduce greenhouse gasses.[31] It is therefore important to identify what important synergies can be found in the area of soil carbon sequestration between the three UN conventions: UNFCC, UNCCD and UNCBD.[32]

[28]FAO Soil Portal http://www.fao.org/soils-portal/soil-management/soil-carbon-sequestration/en/.

[29]Severinsen (2014), p. 6631; WSSD (2002a, b, c).

[30]Lal (2010), p. 14.

[31]Severinsen (2014), p. 6632.

[32]FAO Soils Portal; Lehner and Rosenberg (2017), p. 10845 "Agriculture is both a source and a sink for greenhouse gases. Decision makers can take full advantage of agriculture's potential to slow climate change only by acknowledging the sector's dual role in decarbonizing the economy, and seeking both to minimize agricultural greenhouse gas emissions and to maximize carbon storage".

4 International Treaties and Strategies

A group of closely related international and regional environmental treaties and strategies with a role in mitigating the effects of climate change contain detailed rules for the protection of the terrestrial environment and they have an important role in the management of soil carbon.[33]

4.1 *Framework Convention on Climate Change*

The UNFCCC sets the overall framework for intergovernmental efforts to address climate change. The UNFCCC emphasises that human activities have been substantially increasing the atmospheric concentrations of greenhouse gases, and recognises the role and importance of terrestrial ecosystems as a sink and reservoir of its potential greenhouse gases.[34] The ultimate objective of the UNFCCC (Article 2) is 'to achieve stabilisation of atmospheric concentrations of greenhouse gases at levels that would prevent dangerous anthropogenic (human-induced) interference with the climate system ...' Its key Principles (Article 3) encourage Parties to protect the climate system for the benefit of present and future generations, take precautionary measures to minimise the causes of climate change and mitigate its effects, and promote sustainable development.

The UNFCCC identifies the principal sources of greenhouse gases, including changes in land cover and land use. The main agricultural activities that contribute to emissions of greenhouse gases are deforestation, biomass burning, and cultivation, using organic manure, applying nitrogenous fertilisers and the keeping of livestock.[35] It is argued that, under Article 4, nations have the mandate to develop and improve their legal and institutional systems and responsibilities to protect the climate system for the benefit of present and future generations, to take precautionary measures to prevent or minimise the causes of climate change, and to mitigate its adverse effects.

[33]Hannam (2004); IUCN (2015), p. 2, the Covenant responds "to the call for an integrated international legal framework to provide a consolidated ecological and ethical foundation for present and future international and national policies and laws on environment and development".

[34]German Federal Government (1998), p. 14.

[35]Boer and Hannam (2019), p. 447; 'The UNFCCC promotes sustainable management, and the conservation and enhancement of sinks and reservoirs of greenhouse gases not controlled by the Montreal Protocol, including biomass, forests, and oceans as well as other terrestrial, coastal, and marine ecosystems. However, while its focus is on the reduction in emissions in all sectors, it cannot be considered as a sufficiently specific legal vehicle to address land degradation'.

4.1.1 General Commitments

Developing countries, for their part, are more vulnerable to the adverse effects of climate change and their technological, economic and institutional capacity to respond is generally lower than those in industrialised countries.[36] Parties to the Convention are subject to a set of general commitments which place a fundamental obligation on both industrialised and developing countries to respond to climate change. These include preparation and regular update of national climate mitigation and adaptation programs, measures to address sources of greenhouse gas emissions and to protect carbon sinks and reservoirs, including the forests and other natural systems that remove carbon from the atmosphere. They must also take climate change considerations into account in their other relevant social, economic and environmental policies. To create or improve the capability to meet the obligations of both industrialised countries and developing countries includes an obligation to make better use of the international and the national legislative and institutional frameworks as an essential procedure to reduce emissions and control soil carbon sequestration. It has been recognised that the legislative capacity of individual nations to implement international instruments for land management can vary enormously, particularly in the developing nations where the general capacity of national legislation to manage land and water problems can be inadequate.[37]

4.1.2 Kyoto Protocol

The Kyoto Protocol[38] under the UNFCCC, adopted in 1997, contains a responsibility to promote sustainable forms of agriculture in the light of climate change considerations (Article 2(a) (iii)). It recognizes the need to expand and preserve soil carbon sinks and improve agricultural practices in countries where a significant proportion of the emissions are related to the clearing of vegetation for agriculture (Articles 3 and 5). The Marrakech Accords provided rules for the implementation of the Protocol.[39]

The Protocol supplements and strengthens the UNFCCC, and it shares the same objectives and principles. Nations could implement the general procedures of the Protocol in conjunction with the rules of the Marrakech Accords to frame or reform

---

[36]Climate Change Secretariat (2002), p. 8; Marrakech Accords, Decisions I, A and B, and VII—'Matters relating to Least Developed Countries'.

[37]Hannam (2002), p. 2.

[38]Kyoto Protocol to the Framework Convention on Climate Change, Bonn (1998) 37 ILM 22; The Kyoto Protocol was adopted in Kyoto, Japan on 11 December 1997 and entered into force on 16 February 2005. There are currently 192 parties (Canada withdrew from the protocol, effective December 2012) [41] to the Protocol.

[39]Hannam (2004); Marrakech Accords at: http://unfccc.int/land_use_and_climate_change/lulucf/items/3063.php.

national laws to more effectively manage soil carbon processes.[40] The additional commitments it makes to the UNFCCC are particularly significant to the development of specialised legislative and policy responses to manage the impacts of climate change and greenhouse gasses.[41] To this extent, Article 2 could be adopted as a legislative guideline. This Article promotes sustainable development and calls for each Party to implement the policies and measures to protect and enhance sinks and reservoirs of greenhouses gases, taking into account their commitments under various international environmental agreements, sustainable forest management practices, afforestation and reforestation (see Article 2(i)-(iv)). There is also the responsibility to promote sustainable forms of agriculture in light of climate change characteristics.

It is further argued that the main elements of the Kyoto Protocol and its rulebook set out in the Marrakech Accords provide a valuable tool to aid in the development of national legal and institutional frameworks to manage the processes associated with soil carbon sequestration for both developing and developed countries.

4.1.3 Paris Agreement

The 2015 Paris Agreement under the UNFCCC deals with mitigation of greenhouse gas emissions, adaptation, and finance.[42] Its Preamble recognizes "the importance of the conservation and enhancement, as appropriate, of sinks and reservoirs of the greenhouse gases referred to in the Convention". Under the Paris Agreement, each country determines plans and regularly reports on the contribution it should make to mitigate global warming. There is no mechanism to force a country to set a specific target by a particular date, but each target should go beyond previously set targets, an approach that is now recognized as the principle of progression,[43] as expressed in UNFCCC Article 3: "the efforts of all Parties will represent a progression over time". The aim of the Paris Agreement is to enhance the implementation of the UNFCCC through: "(a) Holding the increase in the global average temperature to well below 2°C above pre-industrial levels and to pursue efforts to limit the temperature increase to 1.5°C above pre-industrial levels, recognizing that this would significantly reduce

[40]Climate Change Secretariat (2002), p. 8, "The landmark Marrakesh Accords adopted at COP 7 (Marrakesh, October/November 2001) were especially important in elaborating the Convention's rulebook on issues of particular concern to developing countries ...[t]he Accords also set out a detailed rulebook for the implementation of the Kyoto Protocol".

[41]For example, Verschuuren (2017), p. 1 argues for a new approach to the use of legal instruments associated to the common agricultural policy should be used in the EU.

[42]Paris Agreement, available at: http://unfccc.int/paris_agreement/items/9485.php; adopted by consensus on 12 December 2015.

[43]Voigt and Ferreira (2016), p. 285; see also Principle 13, IUCN World Declaration on the Environmental Rule of Law, IUCN World Commission on Environmental Law, available at: https://www.iucn.org/sites/dev/files/content/documents/world_declaration_on_the_environmental_rule_of_law_final_2017-3-17.pdf.

the risks and impacts of climate change and (b) Increasing the ability to adapt to the adverse impacts of climate change and foster climate resilience and low greenhouse gas emissions development, in a manner that does not threaten food production" (Article 2).These actions will likely include a wide range of national sustainable land management actions to control or prevent land degradation, including limitations on forest clearing, and improving cover of agricultural land by employing conservation farming techniques and preventing overgrazing of pastures.[44]

In general, many of the Articles in the Paris Agreement provide a good framework for the development of specific elements of national laws to manage the processes associated with soil carbon sequestration in particular.[45]

4.2 Convention on Biological Diversity (CBD)

The objective of the CBD is relevant to soil carbon sequestration as it provides for the conservation of biological diversity and encourages sustainable use of its components. It takes into account various rights over those resources (Article 1). Fundamental to the CBD is the concern that biological diversity is being significantly reduced by human activities which includes the destruction of sinks through the different processes of land degradation. The avoidance of land degradation is implicit in the definition of 'sustainable use' under the CBD, defining the term as 'the use of components of biological diversity in a way and at a rate that does not lead to the long-term decline of biological diversity, thereby maintaining its potential to meet the needs and aspirations of present and future generations' (Article 2).

The CBD and UNCCD Secretariats are actively developing a strategy for land degradation in the context of biodiversity conservation and as part of the effort to 'monitor and report on progress towards target 15.3 of the Sustainable Development Goals, the strategic objectives of the Rio Conventions and other relevant targets and commitments'.[46] The purpose of the Framework is 'to provide consistent definitions and the best available methodologies as well as global/regional data options for three

[44]Neely et al. (2009) see generally 5–11; The Katowice Work Program adopted at COP 24 sets out how countries will provide information about their national climate action plans, including the reduction of greenhouse gas emissions as well as mitigation and adaptation measures.

[45]The comment against each Article is an extract from the respective Article as it appears in the Paris Agreement.

[46]As reported in Boer and Hannam (2019), p. 446; UNCCD, CBD, FAO, Global Environment Facility and UNEP (2016) Draft for Consultation, Framework and Guiding Principles for a Land Degradation Indicator to Monitor and report on progress towards Target 15.3 of the Sustainable Development Goals, the strategic objectives of the Rio Conventions and other relevant targets and commitments; https://www.cbd.int/doc/meetings/sbstta/sbstta-20/information/; under 3.3 "Carbon stocks reflect the integration of multiple processes affecting plant growth and the losses/gains from terrestrial organic matter pools. These changes also reflect trends in soil health and ecosystem functioning as well as land use and management practices. Current carbon stocks are much larger in soils than in vegetation, particularly in non-forested ecosystems in the middle and high latitudes".

sub-indicators'.[47] These include land cover and land cover change, land productivity, and carbon stocks above and below ground 'that could be used to derive the indicator for monitoring and reporting progress towards SDG Target 15.3 as well as other relevant targets and commitments.'[48] It is contended here that for the CBD to take on an expanded, more precise role in addressing land degradation and encouraging the sustainable use of land, it would be desirable for technical guidelines to be drafted on sustainable land management, reinforced by an extra protocol to the Convention.[49] Such provisions could be drawn upon in framing legal elements to protect carbon sinks in national legislation.

Many Articles of the CBD are relevant to the object of managing soil carbon particularly the concern that biological diversity is being significantly reduced by human activities, mostly through habitat loss and soil degradation activities. In particular, the environmental, economic and social benefits of conserving biodiversity outlined under Articles 5–10 of the CBD could assist when framing domestic legislation to manage soil carbon sequestration.

4.2.1 Other Strategies

4.2.1.1 Addis Ababa Principles and Guidelines for the Sustainable Use of Biodiversity

These Principles and Guidelines are based on the assumption that it is possible to use biodiversity in a manner in which ecological processes, species and genetic variability remain above the thresholds needed for long-term viability, and that therefore all resource managers and users have the responsibility to ensure that that use does not exceed these capacities.[50] It is crucial that the biodiversity in ecosystems is maintained, or in some cases recovered, to ensure that those ecosystems are capable to sustain the ecological services on which both biodiversity and people depend and in this way they would be effective as carbon sinks. In this context, the Guidelines are yet another tangible tool in keeping with the commitment of Parties to the Convention to focus on its practical implementation and the target to achieve a significant reduction in the current rate of biodiversity loss, which is at the heart of the Convention's strategic plan.

Because of their interdependence, the fourteen Principles should be considered together and the sustainability of use of biodiversity components will be enhanced,

---

[47]Ibid at 3.

[48]Ibid.

[49]As reported in Boer and Hannam (2019), p. 446; Article 28, Adoption of Protocols CBD.

[50]Secretariat of the Convention for Biological Diversity (2004), p. 3, https://www.cbd.int/doc/publications/addis-gdl-en.pdf; "the practical principles in most instances apply to both consumptive and non-consumptive uses of biodiversity components. They take into account requirements related to: (i) policies, laws, and regulations; (ii) management of biological diversity; (iii) socio-economic conditions; and (iv) information, research and education".

including sink enhancement, if the practical principles and related operational guidelines are applied.[51]

4.2.1.2 Strategic Plan for Biodiversity 2011–2020, Including the Aichi Biodiversity Targets

The tenth meeting of the Conference of the Parties called upon countries to implement the 2011–2020 Strategic Plan for Biodiversity, including the Aichi Biodiversity Targets,[52] with attention given to a number of activities many of which have a direct relationship to the improvement of carbon levels in soils. Some of the more relevant activities include the development of national and regional targets, using the Strategic Plan and its Aichi targets as a flexible framework, review, revision and update of national biodiversity strategies and action plans (NBSAPs) in line with the Strategic Plan, integration of national targets into revised and updated NBSAPs, adopted as a policy instrument, monitoring and review of NBSAP implementation in accordance with the Strategic Plan and national targets, making use of the set of indicators developed for the Strategic Plan as a flexible framework, reporting on progress achieved towards implementation of the Strategic Plan and Aichi Biodiversity Targets. The most relevant Aichi Targets to increasing soil carbon sequestration include,[53] Target 7, Target 11 and Target 15 (by 2020, ecosystem resilience and the contribution of biodiversity to carbon stocks has been enhanced, through conservation and restoration, including restoration of at least 15 per cent of degraded ecosystems, thereby contributing to climate change mitigation and adaptation and to combating desertification).

4.3 Convention to Combat Desertification

It has been effectively pointed out that desertification and drought (as defined under Article 1, UNCCD) is a problem of global dimension, affecting all regions of the world.[54] The UNCCD acknowledges that arid, semi-arid, and dry sub-humid areas together account for a significant proportion of the earth's land area and form the

---

[51]Ibid at 11.

[52]COP 10 Decision X/2 Strategic Plan for Biodiversity 2011–2020 https://www.cbd.int/sp/implementation/default.shtml; 'The purpose of the Strategic Plan for Biodiversity 2011-2020 is to promote effective implementation of the Convention through a strategic approach, comprising a shared vision, a mission, and strategic goals and targets ("the Aichi Biodiversity Targets"), that will inspire broad-based action by all Parties and stakeholders'.

[53]Secretariat for the Convention on Biological Diversity https://www.cbd.int/sp/targets/default.shtml.

[54]UNCCD (2017) Global Land Outlook, at 44–50; https://www.unccd.int/actions/global-land-outlook-glo.

habitat and source of livelihood for a large proportion of its population. The UNCCD's geographic focus of 'desertification' to arid, semi-arid, and dry sub-humid areas excludes other climatic regions of the world that experience severe land degradation processes. While the addition of annexes enables the UNCCD to be applied more specifically at the regional level, taking into account different climatic and geographical conditions,[55] the annexes do not operate to extend the reach of the convention beyond arid, semi-arid, and dry sub-humid areas in the regions.

Despite its limitations, the objective of the UNCCD is fundamental to soil carbon management in that it promotes reduction of land degradation, rehabilitation of degraded land and reclamation of desertified land (Article 2).[56] Joint action is needed in the international community to combat desertification and mitigate the effects of drought, and this creates an important synergistic relationship between the obligations of the UNCCD and obligations of the UNFCCC that can directly benefit and improve soil carbon sequestration management. Techniques used to combat desertification are conducive to the improvement of soil carbon levels. The planning and decision-making mechanisms of the UNCCD (e.g., formation of National Action Programs and Subregional Action Programs, Part III), and its requirements for information collection (Article 16) research and development (Article 17), capacity building and community participation (Article 19), are ideal vehicles for the implementation of soil carbon sequestration management procedures and functions.

4.4 The UN Sustainable Development Goals 2015 and Land Degradation Neutrality

In 2015, the United Nations completed negotiations for the new UN Sustainable Development Agenda.[57] The seventeen Sustainable Development Goals (SDGs), which form part of the 2030 Agenda, came into effect in January 2016. Over the next fifteen years, the SDGs are intended to mobilize efforts to end poverty, fight inequality, and address a wide range of environmental issues, including climate change, conservation of the oceans, desertification, land degradation, and the loss of biodiversity.[58]

---

[55]Annex I: Africa; Annex II: Asia; Annex III: Latin America and the Caribbean (LAC); Annex IV: Northern Mediterranean; Annex V: Central and Eastern Europe (CEE), at http://www2.unccd.int/convention/regions.

[56]FAO (2004), p. xi, 'The case studies presented here assess the effect of different management practices on soil carbon stocks in various dryland ecosystems. The effect of climate and/or land-use change can be predicted only through the use of accurate dynamic models'.

[57]See Boer and Hannam (2019), p. 451; Transforming our World: the 2030 Agenda for Sustainable Development, UN A/RES/70/1, https://sustainabledevelopment.un.org/post2015/transformingourworld.

[58]United Nations, The Sustainable Development Agenda, available at: http://www.un.org/sustainabledevelopment/development-agenda/.

With regard to land degradation control, and soil carbon management specifically, the general aim of SDG 15 is to "conserve and restore the use of terrestrial ecosystems such as forests, wetlands, drylands and mountains by 2020". The UNCCD Global Land Outlook observes that 'SDG 15 ... puts a strong emphasis on the need to scale up transformative management practices with the goal to "Protect, restore and promote sustainable use of terrestrial ecosystems, sustainably manage forests, combat desertification, halt and reverse land degradation, and halt biodiversity loss"'.[59] Land Degradation Neutrality (LDN) is seen as 'a new paradigm in environmental politics for Sustainable Development Goal (SDG) 15'.[60] SDG Target 15.3 ambitiously aims, by 2030, 'to combat desertification, restore degraded land and soil, including land affected by desertification, drought and floods, and strive to achieve a land-degradation neutral world'.[61] In this regard achieving the objective of SDG 15.3 is one of the most important strategies for increasing soil carbon sequestration processes.

A 2016 Science-Policy Brief defines LDN as "a state whereby the amount and quality of land resources necessary to support ecosystem functions and services and enhance food security remain stable or increase within specified temporal and spatial scales and ecosystems".[62] LDN is a new initiative intended to halt the ongoing loss of healthy land through land degradation. Unlike past approaches, LDN creates a target for land degradation management, promoting a two-pronged approach: to avoid or reduce degradation of land combined with measures to reverse past degradation. The objective is that losses are balanced by gains, to achieve a position of no net loss of healthy and productive land.[63]

The UNCCD 2018-30 Strategic Framework[64] that emerged at COP 13 was described as 'a new global roadmap to address land degradation' and 'the most comprehensive global commitment to achieve Land Degradation Neutrality (LDN) in order to restore the productivity of vast swathes of degraded land, improve the livelihoods of more than 1.3 billion people, and to reduce the impacts of drought on

[59]UNCCD (2017) Global Land Outlook, at 36.

[60]See Boer and Hannam (2019), p. 452; Metternich and Cowie (2017) http://sdg.iisd.org/commentary/guest-articles/kill-not-the-goose-that-lays-the-goldenegg-striving-for-land-degradation-neutrality/, fn 77 at 452 in Boer and Hannam (2019).

[61]The UNCCD 2018–30 Strategic Framework adopted by the 2017 UNCCD Conference of the Parties links the UNCCD's aims with SDG 15.3. It includes a 'Vision' which emphasizes the achievement of 'a land degradation-neutral world consistent with the 2030 Agenda for Sustainable Development, within the scope of the Convention'.

[62]Boer and Hannam (2019), p. 452; UNCCD Science-Policy Brief 02—September 2016; https://www.unccd.int/sites/default/files/documents/18102016_Spi_pb_multipage_ENG_1.pdf.

[63]See UNCCD Policy Brief, Zero Net Land Degradation: A Sustainable Development Goal for Rio +20 (2012); http://www.unccd.int/Lists/SiteDocumentLibrary/Rio+w.unccd.int/Lists/SiteDocumentLibrary/Rion.pdfR.

[64]Ibid, UNCCD 2018–30 Strategic Framework.

vulnerable populations'.[65] In this context it is a vitally important framework for global soil carbon sink protection and restoration. The Ordos Declaration that was also issued at COP 13 committed to facilitate, 'for Parties that wish to do so, the voluntary land degradation neutrality target-setting process, and to provide sufficient support to national efforts to turn defined land degradation neutrality targets into effective projects and equitable action'.[66] The challenge to translate Goal 15 and Target 3, together with the aims of the Strategic Framework into implementable legislation and policies that will deliver LDN at a national level in all countries suffering from land degradation, is potentially one of the most important global strategies for soil carbon sink protection and restoration.

The Secretariat and the Global Mechanism of the UNCCD are supporting interested countries in the national LDN target setting process, including the definition of national baselines, targets and associated measures to achieve LDN by 2030 through the LDN Target Setting Program (TSP).[67] Target 15.3 has therefore become an important vehicle for driving UNCCD implementation, while at the same time contributing to other SDGs, including those relating to climate change mitigation and adaptation, biodiversity conservation, ecosystem restoration, food and water security, disaster risk reduction, and poverty.[68] To address the multiple Sustainable Development Goals in a synergistic and cost effective manner, countries can now formulate voluntary targets to achieve LDN, according to their specific national context and development priorities. The targets will also further strengthen the implementation of the countries' UNCCD National Action Program. Under the LDN Target Setting Program, participating country Parties receive a number of benefits that will assist a carbon sequestration objective.[69]

---

[65]Boer and Hannam (2019), p. 453; 'Countries agree on a landmark 2030 strategy to save fertile lands', see http://www2.unccd.int/news-events/countries-agree-landmark-2030-strategy-save-fertile-lands; under Clause 5 of the Ordos Declaration—"Pledge to further integrate sustainable land management objectives and practices into relevant policies, plans and investments in order to contribute to achieving land degradation neutrality".

[66]Boer and Hannam (2019), p. 453; The Ordos Declaration, para. 1.

[67]At the 12th session of the Conference of Parties of the UNCCD held in October 2015 in Ankara, Turkey, country Parties reached a breakthrough agreement to endorse this vision and link the implementation of the Convention to the SDGs in general, and target 15.3 in particular—https://www.unccd.int/actions/ldn-target-setting-programme.

[68]To date, 120 countries have participated in the LDN TSP.

[69]The countries with voluntary LDN Targets are listed at https://knowledge.unccd.int/home/country-information/countries-with-voluntary-ldn-targets.

4.5 United Nations Conference on Environment and Development 1992

The 1992 United Nations Conference on Environment and Development (UNCED) produced two strategies for ecologically sustainable development that have significance for goal of soil carbon management. The Rio Declaration on Environment and Development[70] established the goal of a new and equitable global partnership, through the creation of new levels of cooperation among States, key sectors of society, and individuals. It calls for the international community to work towards and develop international instruments and strategies to protect the integrity of the global environmental system. Specifically, Principle 11 calls for States to enact effective environmental legislation. Various chapters of Agenda 21 provide impetus for the preparation of environmental law and policy to manage land degradation and therefore have a direct connection to soil carbon management objectives. A number of its chapters are important to the management and control of soil carbon sequestration and outline an international environmental law process that can help achieve this goal. Chapter 8 outlines the essential environmental management requirements which can be integrated within policies and practices to develop and implement enforceable and effective sustainable land use laws and regulations. Chapter 38 emphasises the development of international law on the basis of 'sustainability'.[71]

4.6 The World Summit on Sustainable Development 2002

The World Summit on Sustainable Development (WSSD) reaffirmed sustainable development as a central element of the international agenda and stressed the need for global action to protect the environment.[72] The WSSD reaffirmed a wide range of concrete commitments and targets for action to achieve more effective implementation of sustainable development objectives. Of particular importance to an objective of soil carbon management is the recognition of the need to increase protection of the land as a major strategy and reduce the loss of fertile soil. This objective is reinforced in the Johannesburg Declaration on Sustainable Development (see paras 16–30) and WSSD Plan of Implementation (see paras 24–46; 81–136; 137–170), both of which call for the development of integrated land management plans, laws and policies to improve land productivity.

---

[70]United Nations (1992a).

[71]United Nations (1992b); see Foreword, Article 17 on Climate Change, and Article 21 on "Soil" of IUCN (2010).

[72]WSSD (2002a, b, c).

4.7 United Nations Environment Program for Development and Periodic Review of Environmental Law

From 1982 to date, UN Environment's environmental law activities have been organized and coordinated through a series of 10-year programmes for the development and periodic review of environmental law (known as the Montevideo Programme), adopted by UN Environment's Governing Council. This Programme has consistently set the strategy for UN Environment's engagement in the field of environmental law, and the latest version, Montevideo IV, was also intended to provide a broad strategy for the international community at large. Montevideo IV was adopted in 2009 by Governing Council decision 25/11, as a broad strategy for the international law community and UN Environment in the formulation of activities in the field of environmental law for the decade beginning in 2010.[73] As such it does provide a very comprehensive framework of activities that address the whole spectrum of environmental law challenges and the specific functional challenges that countries face in the actual development, implementation and enforcement of environmental law.

Many of the program areas are directly relevant to a goal of improved management of carbon sequestration processes.[74] The objective of the Soil Program (in Pillar 2) is "to improve national and international principles and standards and to support efforts under the United Nations Convention to Combat Desertification for the further development of legal approaches for the conservation, restoration and sustainable use of soils".[75] The strategy is to promote the development, dissemination and implementation of laws and policies that aim to enhance the conservation, sustainable use, control and reduction of soil degradation and, where appropriate, restoration of soils, including in support of work conducted by relevant bodies such as the Conference of Parties, its subsidiary bodies and the secretariat of the UNCCD.[76]

4.8 IUCN Covenant on Environment and Development

With respect to global initiatives, the provisions of the IUCN Covenant on Environment and Development should be noted.[77] The Covenant takes a holistic approach to

---

[73] UNEP (2018) Final Assessment of the Fourth Program for the Development and Periodic Review of Environmental Law (Montevideo IV), Evaluation Office of UN Environment.

[74] The Program consists of Four Pillar Areas (Effectiveness of Environmental Law, Conservation, management and sustainable use of natural resources, Challenges for Environmental Law, Relationships with other Fields) and there are numerous programs within each of these Pillars.

[75] Ibid at 76.

[76] Ibid.

[77] IUCN (2015).

the protection of the environment and conservation of natural resources. Its commentary recognizes that protection and restoration of soils are essential to many natural systems and resources, as well as to biological diversity'.[78] Article 23 states: "Parties shall take all appropriate measures to ensure the conservation and where necessary the regeneration of soils for living systems by taking effective measures to prevent large-scale conversion and soil degradation and loss, to combat desertification, to safeguard the processes of organic decomposition and to promote the continuing fertility of soils". Many other fundamental principles of the Covenant can be used in the development of legal elements when framing national law or strategies to manage soil carbon sequestration.[79]

4.9 The Revised World Soil Charter 2014

The non-legally binding Revised World Soil Charter of 2014 specifies that '[s]oils are a key enabling resource, central to the creation of a host of goods and services integral to ecosystems and human well-being. The maintenance or enhancement of global soil resources is essential if humanity's overarching need for food, water, and energy security is to be met.'[80] The Charter includes guidelines for actions by government, including to: 'incorporate the principles and practices of sustainable soil management into policy guidance and legislation at all levels of government, ideally leading to the development of a national soil policy'.[81]

4.10 FAO Voluntary Guidelines for Sustainable Soil Management

The Voluntary Guidelines for Sustainable Soil Management (VGSSM),[82] endorsed by the FAO Council in December 2016 at its 155th Session complement the Revised World Soil Charter by further elaborating principles and practices for incorporation

[78]Ibid at 83.

[79]Ibid, see Fundamental Principles 1–79.

[80]Principle 3 at 4; Global Soil Partnership and FAO, Revised World Soil Charter (2014), available at: http://www.fao.org/fileadmin/user_upload/GSP/docs/ITPS_Pillars/annexVII_WSC.pdf; see Action I at 6, "All individuals using or managing soil must act as stewards of the soil to ensure that this essential natural resource is managed sustainably to safeguard it for future generations".

[81]Ibid. at 6, 'Actions by Governments'; in particular Action VI specifies "Explicitly consider the role of soil management practices in planning for adaptation to and mitigation of climate change and maintaining biodiversity".

[82]FAO (2017) Voluntary Guidelines for Sustainable Soil Management, Rome; http://www.fao.org/3/a-bl813e.pdf.

into policies and decision-making.[83] The VGSSM are of voluntary nature and are not legally binding. They elaborate the principles outlined in the revised World Soil Charter, taking into account the evidence provided in the Status of the World Soil Resources (SWSR) report of 2015.[84] The guidelines address technical aspects of SSM including core characteristics of sustainably managed soils, key challenges and potential solutions to address them and for these reasons they are applicable to meet the goal of soil carbon management. The VGSSM focus mostly on agriculture which is broadly defined as the production of food, fibre, feed, timber and fuel, although many of the principles described have a significant influence on ecosystem services provided by managed and unmanaged soil systems. The basis of the VSGSSM is SSM is defined under Principle 3 in the Revised World Soil Charter:[85]

> Soil management is sustainable if the supporting, provisioning, regulating, and cultural services provided by soil are maintained or enhanced without significantly impairing either the soil functions that enable those services or biodiversity. The balance between the supporting and provisioning services for plant production and the regulating services the soil provides for water quality and availability and for atmospheric greenhouse gas composition is a particular concern.

In particular, Guideline 3.2 of the VGSSM states that "Soil organic matter (SOM) plays a central role in maintaining soil functions and preventing soil degradation". It further states that "Soils constitute the largest organic carbon pool on the Earth and play a critical role in regulating climate and mitigating climate change through trade-offs between greenhouse gas emission and carbon sequestration. For this reason, SOM is strategic for climate change adaptation and mitigation, and global stores of SOM should be stabilized or increased. A loss of soil organic carbon (SOC) due to inappropriate land use or the use of poor soil management or cropping practices can cause a decline in soil quality and soil structure, and increase soil erosion, potentially leading to emissions of carbon into the atmosphere. On the other hand, appropriate land use and soil management can lead to increased SOC and improved soil quality that can partially mitigate the rise of atmospheric CO_2".[86] In this context, the elements of Guideline 3.2 can be used as a basis for framing specific legal elements in the development of a soil carbon management law, or alternatively, the elements based on these factors can be inserted within an existing appropriate land management law.[87]

---

[83]Ibid at 1, "the VGSSM were developed through an inclusive process within the framework of the Global Soil Partnership (GSP). They aim to be a reference providing general technical and policy recommendations on sustainable soil management (SSM) for a wide range of committed stakeholders. The guidelines were adopted by the 4th GSP Plenary Assembly (Rome, 25 May 2016), approved by the 25th session of the FAO Committee on Agriculture (Rome, 28 September 2016) and finally endorsed by the 155th session of the FAO Council (Rome, 5 December 2016)".

[84]See 2.1.4 at 56 "Factors influencing soil C storage; https://www.researchgate.net/publication/289249254_Status_of_the_World's_Soil_Resources_SWSR_-_Main_Report.

[85]FAO (2017), p. 3.

[86]Ibid at 8.

[87]Ibid.

5 Regional Treaties

Many regional treaties contribute to the management of soil carbon.[88] The following is a selection for discussion purposes. Attempts to address land degradation prior to the 1992 UN Conference on Environment and Development were predominantly regional in nature, and did not establish specific rules for sustainable land use, or particularly, for soil carbon management.[89]

5.1 Protocol for the Implementation of the Alpine Convention of 1991 in the Field of Soil Protection 1991

Only one regional instrument is a specific soil instrument, made under the 1991 Alpine Convention,[90] being the Protocol for the Implementation of the Alpine Convention in the Field of Soil Protection.[91] This is the only legally binding regional instrument in the world specifically directed to the protection of soil. It aims 'to reduce quantitative and qualitative soil impairments, in particular by applying agricultural and silvicultural production processes which have a minimal detrimental impact on the soil, by using land economically, controlling erosion and restricting soil sealing' (preamble). Among its objectives it provides,[92] "In particular, the ecological functions of soil, which are essential elements of the ecological balance, shall be safeguarded and preserved both qualitatively and quantitatively on a long-term basis. The restoration of impaired soils shall be promoted".

---

[88]Hannam (2004); IUCN (2015) see discussion under Article 21 at 77–78.

[89]E.g. the 1968 African Convention for the Conservation of Nature and Natural Resources 1001 UNTS 4 (note that this instrument was revised in 2017 to include land degradation); the 1974 Convention Establishing a Permanent Inter-State Drought Control Committee for the Sahel; also Art. 7 1985 ASEAN Agreement on the Conservation of Nature and Natural Resources (not in force), available at: http://sunsite.nus.edu.sg/apcel/kltreaty.html.

[90]The 1991 Alpine Convention is focused on pursuing 'a comprehensive policy for the preservation and protection of the Alps by applying the principles of prevention, payment by the polluter (the 'polluter pays' principle) and cooperation, after careful consideration of the interests of all the Alpine States, their Alpine regions and the European Economic Community, and through the prudent and sustained use of resources . . .' (Art. 2(1)). The Convention includes soil conservation under its 'General Obligations' in Art. 2(2) (d).

[91]The Protocol for the Implementation of the Alpine Convention of 1991 in the Field of Soil Protection 1991 Official Journal of the European Union, L 337/29, available at: http://www.alpconv.org/en/convention/protocols/Documents/SoilProtocolEN.pdf.

[92]Final paragraph to Article 1(2) Protocol to Alpine Convention.

5.2 *African Convention for the Conservation of Nature and Natural Resources*

In the African region, the Revised African Convention on the Conservation of Nature and Natural Resources,[93] adopted in March 2017, specifically mentions soil under its definition of 'natural resources' (Article V 1). It is one of few regional treaties that play a significant role in addressing land degradation and therefore provide benefits for soil carbon sequestration. Its objectives are to enhance environmental protection, foster the conservation and sustainable use of natural resources, and to harmonize and coordinate policies in these fields with a view to achieving ecologically and socially acceptable development policies and programs (Article II). These provisions are critical for the prevention and control of land degradation. The parties are guided by three important principles that are relevant for soil carbon management, including: 'the right of all peoples to a satisfactory environment favourable to their development, the duty of States, individually and collectively to ensure the enjoyment of the right to development, and the duty of States to ensure that developmental and environmental needs are met in a sustainable, fair and equitable manner' (Article III). Article VI includes specific measures on land and soil.

5.3 *The ASEAN Agreement on the Conservation of Nature and Natural Resources*

The ASEAN Agreement was adopted in 1985, but is not yet in force.[94] However it includes provisions that can be adopted in the development of legal instruments for soil carbon sequestration management. Its fundamental principle (Article 1 (1)) states that 'Parties are to adopt measures necessary within their legislative systems to maintain essential ecological processes and life-support systems, and preserve genetic diversity. Parties could develop and co-ordinate a national soil carbon sequestration management strategy under the framework of a regional conservation strategy'. Further, Article 7 for 'Soil' requires 'Parties to observe the role of soil in the functioning of natural ecosystems, undertake soil conservation measures to rehabilitate eroded and degraded soils, establish soil policies, control soil erosion and improve soil fertility'. Article 10 is also relevant to the management of soil carbon as it provides for the Parties to 'maintain the proper functioning of ecosystems and to prevent the degradation of the environment'.

---

[93]https://au.int/en/treaties/african-convention-conservation-nature-and-natural-resourcesrevised-version.

[94]http://sedac.ciesin.org/entri/texts/asean.natural.resources.1985.html.

6 National Environmental Law

Legislation has been used for many years in a piece-meal fashion to manage soil problems (e.g., soil erosion), control land use activities which directly cause land degradation (e.g., over-grazing of cattle and sheep), and to indirectly control soil management problems (e.g., through environmental planning and land use allocation). An examination of national legal and institutional frameworks from over 100 countries indicated that soil management is approached in a fragmented way. Few countries have developed a law specifically to manage soil carbon processes. Some jurisdictions have a group of legislation that cover a broad range of land management functions including, land planning, farming practices, management of sensitive areas, and soil restoration.[95] The types of legal mechanisms that have been used over many decades to generally protect and manage soil, including acts, decrees, resolutions, ordinances, codes, regulations, circulars, decisions, orders and by-laws, could be more effectively applied to the management of the processes that lead to the sequestration of soil carbon.[96]

6.1 Role of National Environmental Law in the Management of Soil Carbon Sequestration

National environmental law has a central role in the maintenance and return to the soil of its organic carbon. Individual national laws and legal instruments provide mechanisms to directly or indirectly manage processes related to soil carbon sequestration and environmental law relating to the use and conservation of agricultural land has a special role in this regard.[97] The following discussion of different areas of national law overviews the role of these laws to manage soil carbon sequestration. References are made to examples of specific laws where it is regarded they contain legal mechanisms suitable to recognize a role in soil carbon management and to implement procedures for soil carbon sequestration management.

---

[95] See Hannam (2012); The principal aim of this analysis is to provide information on the capacity of three selected PRC environmental laws relating to the management of carbon associated with agricultural land use activity in the grassland areas.

[96] Hannam and Boer (2002) Section 3, 'National Soil Regimes'; Boer and Hannam (2019).

[97] Hannam (2004); Johnson (2018) Chapter 4 'Soils'.

6.1.1 Carbon Laws[98]

The enactment of carbon rights legislation to recognize rights associated with carbon sequestration by trees and forests has been recognized for some time now.[99] Carbon rights law enables acquisition and trading in such rights through a covenant that gives access to or the maintenance of trees or forest of any sequestered carbon. The main purpose of the legislation is to encourage investment in carbon sinks, a legal concept that could be applied readily to sequester soil carbon. A carbon sequestration right in relation to land may mean a right that is conferred on a person by a legal agreement, to the legal, commercial or other benefit of carbon sequestration by any existing or future use of the land. This area of law opens the way for a market in stored carbon and ultimately the future creation of carbon credit schemes for soil carbon.[100]

The following discussion of three laws from the Australian jurisdiction, one at the national level and two from State level, provides examples of different ways of applying the law to manage soil carbon.

6.1.1.1 Commonwealth of Australia Carbon Credits (Carbon Farming Initiative) Act 2011

The Carbon Farming Initiative is a carbon offsets scheme that provides new economic opportunities for farmers, forest growers and landholders to help the Australian environment by reducing carbon pollution.[101] Farmers and land managers are able to generate credits that could then be sold to other businesses wanting to offset their own carbon pollution. In particular, the Carbon Farming Initiative enables land managers to earn credits for various land management actions including: reforestation and revegetation; reduced methane emissions from livestock digestion; reduced fertilizer pollution; reduced pollution or increased carbon storage in agricultural soils (soil carbon); savannah fire management; native forest protection; forest management; reduced pollution from rice cultivation; reduced pollution from legacy landfill waste.

Under Section 3 of the Act the first object is "to remove greenhouse gases from the atmosphere, and avoid emissions of greenhouse gases, in order to meet Australia's obligations under the Climate Change Convention, the Kyoto Protocol and an international agreement (if any) that is the successor (whether immediate or otherwise) to the Kyoto Protocol". The second object of the Act is "to create incentives for people to carry on certain offsets projects". The third object of the

[98]For a discussion on the history of legislative aspects of carbon in Australia see Guglyuvatyy and Stoianoff (2016) http://www5.austlii.edu.au/au/journals/UTSLRS/2016/25.html.

[99]Bonnie et al. (2002).

[100]Jeffery and Baird (2001), p. 10.

[101]Power (2011), p. 59; Macintosh (2012), p. 28.

Act is "to increase carbon abatement[102] in a manner that is consistent with the protection of Australia's natural environment, and improves resilience to the effects of climate change". The fourth object of the Act is "to authorize the purchase by the Commonwealth of units that represent carbon abatement".

Under Section 5 of the Act "eligible carbon abatement" from an offsets project means "carbon abatement that: (a) results from the carrying out of the project; and (b) is able to be used to meet Australia's climate change targets under: (i) the Kyoto Protocol; or (ii) an international agreement (if any) that is the successor (whether immediate or otherwise) to the Kyoto Protocol". An "Offsets project" means: "(a) a sequestration offsets project; (b) an emissions avoidance offsets project". Under Section 54 "a project is a sequestration offsets project if it is a project: (a) to remove carbon dioxide from the atmosphere by sequestering carbon in one or more of the following - (i) living biomass; (ii) dead organic matter; (iii) soil; or (b) to remove carbon dioxide from the atmosphere by sequestering carbon in, and to avoid emissions of greenhouses gases from, one or more of the following: (i) living biomass; (ii) dead organic matter; (iii) soil".

6.1.1.2 Victoria, Australia Climate Change Act 2017[103]

The Purpose of this Act under Section 1 is "to set a long-term greenhouse gas emissions reduction target; and to provide for the setting of 5-yearly interim greenhouse gas emissions reduction targets in order to reach the long-term greenhouse gas emissions reduction target".[104] It also facilitates the consideration of climate change issues in specified areas of decision making of the Government of Victoria and sets policy objectives and guiding principles to inform decision-making under this Act and the development of government policy in the State. Importantly, the Act provides for a strategic response to climate change through a climate change strategy,[105] adaptation action plans[106] and emissions reduction pledges and provides for greater clarity and accountability through information collection and reporting. It also provides for the State's contribution to national and international carbon sequestration efforts, for the creation of forestry rights, carbon sequestration rights and soil carbon

[102]Under Section 5, "*carbon abatement*" means: (a) the removal of one or more greenhouse gases from the atmosphere; or (b) the avoidance of emissions of one or more greenhouse gases.

[103]Under Section 3, "climate change" means 'a change of climate which is attributed directly or indirectly to human activity that alters the composition of the global atmosphere and which is in addition to natural climate variability observed over comparable time periods'.

[104]Under Section 3, "carbon sequestration right" means 'an exclusive right to the economic benefits associated with carbon sequestered by vegetation other than vegetation that has been harvested, lopped or felled'.

[105]Section 30 (1)-(6) Content of a Climate Change Strategy.

[106]Section 35 (1)-(3) Content Adaption Action Plan.

rights, and provides for Forestry and Carbon Management Agreements in relation to private land and Carbon Sequestration Agreements in relation to Crown land.[107]

6.1.1.3 Western Australia Carbon Rights Act 2003

This Act provides for the creation and effect of certain interests in land in relation to the effects of carbon sequestration from, and carbon release to, the atmosphere, and for related matters. Under Section 3 carbon sequestration means "(a) the absorption from the atmosphere of carbon dioxide by land or anything on land; and (b) the storage of carbon in land or in anything on land". Under Section 5 (1) "A person may lodge with the Registrar, for registration, a form for the creation of a carbon right in respect of freehold land or Crown land". Under Section 8 (1) "A proprietor of a carbon right has the legal and commercial benefits and risks arising from changes to the atmosphere that are caused by carbon sequestration and carbon release occurring in or on land in respect of which the carbon right is registered".

In the following sections (6.1.2–6.1.8), examples of specific laws are provided within particular areas of law, where various procedures within the laws that can provide a soil carbon sequestration benefit are outlined.

6.1.2 Soil

Soil legislation has been the principal area of law concerned with land degradation control and soil conservation.[108] In general, soil laws provide for many activities that can either lead to the conservation of existing levels of soil carbon, or to its return to the soil (e.g., Nepal, Soil and Water Conservation Act 2039 (1982)[109]; 1991 Soil and Water Conservation Law of the Peoples Republic of China; the Kyrgyzstan Law on Soil Fertility Protection of Agricultural Lands 2012,[110] and the Bulgarian Law for the Preservation of Agricultural Lands 1996[111]). This type of law generally has jurisdiction over large areas of agricultural land but the evidence indicates[112] that it

[107]Under Section 4(1) "forest carbon right" means (a) a carbon sequestration right; or (b) a forestry right; or (c) a soil carbon right.

[108]Hannam and Boer (2002), p. 35; Hannam and Boer (2004) 1.2 at 5, 'What is wrong with the national legislation in many jurisdictions?'

[109]Section 10 prescribes 'Prohibited acts in land vulnerable or likely to be vulnerable to natural calamity', which includes (c) 'To cut or otherwise destroy any existing forests, trees, plants, weeds, grasses or other natural vegetations'.

[110]Anarbaev (2017), p. 70, "This law was introduced for the purpose of protecting the soil and its fertility, the preservation of soil quality and protection of soil from degradation, in relation to the use of agricultural land. In particular, it aims at regulating activities and competence in the area of soil protection and ensuring the fertility of agricultural land is maintained or improved."

[111]See Article 2 (1).

[112]FAO (2015b) Status of the World's Soil Resources at 5, "Soils are Endangered".

has not been used effectively to control the land use activities that contribute directly to the loss of soil carbon. Such activities have included, for example, vegetation depletion from overgrazing, cultivation of land not physically suitable for cultivation, and failure to rehabilitate degraded land. Some countries have individual statutes, decrees, regulations, ordinances, and codes specifically to protect and manage agricultural soil.[113] Soil legislation is in urgent need of reform worldwide, where the emphasis should be on managing the soil from an ecosystem perspective.[114]

6.1.3 Vegetation

There are at least three categories of vegetation law that offer a significant role in the management of soil carbon. The first is 'forest' law, which is used to manage State-owned forests (e.g., United States National Forest Management Act 1976), forest on private land (e.g., Denmark Forestry Act 1989), and multiple functions of forests. These laws generally enable the preparation of land management guidelines and plans that include soil management and protection measures, and therefore have a significant role in soil carbon sequestration (e.g., People's Republic of China, Forestry Law 1985[115]). Expanding the area of forest is one of the most effective ways recognised under the UNFCCC regime to return and maintain carbon in the soil and reduce greenhouse gases. The second category involves national laws for the conservation and retention of native vegetation and control of land clearing. Some laws in this category promote management of indigenous vegetation in consideration of social, economic and environmental parameters (e.g., Albania Pastures and Meadows Law 1995[116]) and others protect vegetation of a high conservation value (e.g., Australia, New South Wales Biodiversity Conservation Act 2016[117]). Generally, these laws offer a significant benefit for soil carbon sequestration by setting rules and limits for clearing vegetation and ecological assessment of land. There are also laws to protect vegetation as an important landscape element, and the laws that manage large areas of grassland, (e.g., Turkey Rangeland Law 1998; Australia, South Australian Pastoral Land Management and Conservation Act 1989[118]). A third category includes legislation to establish plantations and enable reafforestation, which is crucial to long term improvement and management of soil carbon, especially carbon sinks (e.g., Norway Forestry Act

---

[113]Hannam (2020) indicates that there are around 190 individual laws in Australia that contribute in some way to the activity of soil conservation.

[114]Hannam and Boer (2002), p. 17; Boer and Hannam (2019), p. 453.

[115]See Article 1.

[116]See Article 1.

[117]See Section 1.3.

[118]See Section 4.

1965,[119] Australia, New South Wales Plantations and Reafforestation Act 1999[120]). Plantations protect environmental values critical to improving soil carbon stocks, including, reducing soil salinity, controlling soil erosion, and managing biodiversity conservation. Codes of practice are used to conserve and manage native vegetation, prevent soil erosion, conserve habitat, and manage fire, being activities that conserve soil carbon.

6.1.4 Biodiversity

At the national level biodiversity law provides the legal mechanisms to conserve ecosystems, protect landscapes and protect habitat, which, in turn, aids the management of soil carbon processes (e.g., United States Endangered Species Act 1973; Danish Nature Protection Act 1992[121]; Commonwealth of Australia Environment Protection and Biodiversity Act 1999[122]). Biodiversity laws generally do not apply directly to agricultural activities, but rather to conserve and manage protected and reserved areas within agricultural landscapes. In this regard they play a very important role in soil carbon sequestration management in agricultural environments. The legal mechanisms used to identify biodiversity areas, make bioregional plans, protect ecological communities, and conduct environmental audits, greatly assist the management of soil carbon sequestration by retaining vegetative cover and therefore preventing soil degradation.[123]

6.1.5 Protected Areas

The IUCN advocates that protected areas must be planned and managed using an ecosystem approach.[124] Protected areas are essential for biodiversity conservation, often providing habitat and protection from hunting for threatened and endangered species. In this context they are important as a carbon storage sink. Protection helps maintain ecological processes that cannot survive in most intensely managed landscapes. The range of natural resources that any one protected area may guard is vast. Many will be allocated primarily for species conservation whether it is flora or fauna or the relationship between them, but protected areas are similarly important for

[119]Chapter 1, Section 1 'The purpose of this Act is to promote sustainable management of forest resources in Norway with a view to promotion of local and national economic development, and to secure biological diversity, consideration for the landscape, outdoor recreation and the cultural values associated with the forest'.

[120]Section 3.

[121]The Nature Protection Act includes a general protection of different nature types (e.g. heath, meadows, lakes) and rules for public access.

[122]See Section 3 (1).

[123]Ibid.

[124]Lausche (2011), p. 22.

conserving sites of cultural importance and considerable reserves of natural resources.[125] Today, in addition to the state-owned or state controlled protected area, new governance types offer important management options.[126] This is especially relevant in countries where possibilities exist to recognize voluntarily conserved areas as part of the formal protected areas system. The IUCN guidelines identify new or strengthened legal elements that countries should take into account in their protected area legal frameworks. These elements have broad applicability and general value because in many aspects of protected areas law the legal approach is similar, if not identical, across countries and jurisdictions.[127]

6.1.6 Environmental Planning and Assessment

Environmental planning and assessment legislation is one of the more important areas of law to manage soil carbon sequestration processes as it provides for the formulation of land use policy, development of national and local land use plans, and the control of land development (e.g., New Zealand Resource Management Act 1991[128]). The national plans can be used to set rules, criteria and policies critical to soil carbon management by zoning land for specific uses, prescribing conservation and environmental protection measures, and outlining the land use decision-making procedures.[129]

6.1.7 Protection of the Environment

Environment protection legislation aims to protect and enhance environmental quality and in this regard it provides substantial benefit to the management of soil carbon sequestration. The legislation generally focuses on preventing environmental degradation from 'pollution' activities and the formation of strategies that include environment protection standards, environment protection guidelines, and environment protection protocols (e.g., Australia, New South Wales Environmental Planning and Assessment Act 1979[130]). These laws are generally managed by a regulatory authority with powers to conduct environmental audit, remedy and

[125]Of all global terrestrial carbon stock, 15.2% is contained within protected areas. Protected areas in South America hold 27% of the world's carbon stock, which is the highest percentage of any country in both absolute terms and as a proportion of the total stock, https://en.wikipedia.org/wiki/Protected_area.

[126]For a comprehensive coverage of generic elements of protected area legislation see Lausche (2011), pp. 107–207.

[127]Ibid at 4.

[128]See Part 1 Section 2.

[129]E.g., Severinsen (2010), p. 334, IV.

[130]Section 1.3.

restrain harm to the environment and undertake criminal prosecution (India Environment Protection Act 1986).

6.1.8 Land Administration

This area of law is important for soil carbon management as the legal manner in which land is held and the security of access to land will often determine the capacity or willingness of a land user to adopt good soil and vegetation management practices.[131] It includes a wide variety of land tenancy and leasehold laws where the general principles cover the assessment of State land, regulation of conditions of occupancy, use, sale, lease and reservation (e.g., Norway Land Act 1955[132]; Australia, Queensland Land Act 1994[133]).

7 Framing Legislation to Manage Soil Carbon Sequestration

There are many possible ways for framing legislation to control the impacts of land use on soil carbon management.[134] However, the ability of legislation to achieve effective soil carbon management will depend on the legal and institutional elements that protect physical processes associated with soil carbon sequestration and the establishment, maintenance and protection of carbon sinks and reservoirs. Such elements should be incorporated within procedures that regulate and manage the land use activities that cause the loss of soil carbon, lead to land degradation and contribute to the atmospheric CO_2 and global warming. Legislation and policy reform should be approached from a sound conceptual basis, preferably with an overall societal goal of sink management and enhancement. A change in attitude to create legislation for soil carbon sequestration management will depend on the willingness of society to accept new values in a legal system for land management.[135] Some of the important conceptual and ethical elements for land

[131]Claassen and Morehart (2009), p. 1.

[132]See Chapter 1, Section 1.

[133]Section 4 'In the administration of this Act, land to which this Act applies must be managed for the benefit of the people of Queensland by having regard to various principles, including - Sustainability'.

[134]Hannam (2004); Hannam and Boer (2004) Section IV at 37.

[135]See specifically IUCN (2015) for examples of fundamental principles, to achieve the Objectives of the Covenant (Article 1) which "provides a comprehensive legal framework with the aim of achieving environmental conservation as an indispensible foundation for sustainable development".

management include the concept of natural rights for soil, to provide for the public interest, to apply the precautionary principle, and to conserve biological diversity.[136]

7.1 *Legal and Institutional Elements to Manage Soil Carbon*

Important physical issues which need to be enshrined within the legal and institutional frameworks to successfully deal with soil carbon sequestration include soil characteristics, soil degradation processes, extent and impact of soil degradation, soil degradation causative agents, and methods to control the rate of land degradation and achieve sustainable land management including ecosystem protection. The legislation should include goals and objectives that establish the mandate to control and manage soil carbon sequestration and this should preferably be embodied within an ecologically sustainable approach to soil management. Principles and elements can be adapted from the procedural material of the UNFCCC, Kyoto Protocol, Marrakech Accords, Paris Agreement, and other relevant international treaties, strategies and policies concerning ecology, the conservation of nature, biodiversity, and sustainable land management.[137] Specific legal elements should be formulated to enable governments, landowners and land managers, and the community to share responsibility in soil carbon management and to enable development of policy, guidelines and ecological standards for soil carbon management, including procedures for policy implementation, development of special codes of practice, land management indicators and the physical and ecological limits of land use.[138]

7.2 *National Strategy for 'Soil Carbon Sinks'*

Nations should be innovative in their approach to manage the soil carbon sequestration aspects of climate change and this opportunity presents with the implementation of the 2015 Paris Agreement. In this regard, the development of a national 'soil carbon management' strategy would provide an overall policy position and guidelines for the management of soil carbon at the national level. Such a strategy, based on the concept of sustainable soil management', for example, could draw on the objectives, principle elements and procedural materials of relevant international

[136]Hannam and Boer (2002) Section 4(4) at 41.

[137]Ibid Hannam and Boer (2004); Hannam (2004), FAO (2017) Section 3 at 7–13, 'Guidelines for Sustainable Soil Management, constitute[s] technical guidelines to address soil threats that hamper SSM. They should not be viewed as a full list of good practices, but rather a technical reference to be applied on a context specific basis. Specific technical manuals may be developed later to provide complementary tools'.

[138]Ibid Hannam and Boer (2004) numerous legal elements are set out in Section IV at 37–80; see also Durrant (2008), p. 906.

treaties and strategies (UNFCCC, Kyoto Protocol, Paris Agreement, CBD and the UNCCD) and other international guidelines,[139] and integrating these with national protocols, land use standards and national environmental policy. It could be a 'stand alone' instrument, or the strategy could be embodied within an existing land management law or enshrined within a new legal and institutional framework. The challenge will be to balance the required environmental outcomes of the strategy with national economic objectives and marketing strategies.[140] In particular, a national soil carbon management strategy should aim to set out the relationship between the strategy and the UNFCCC, Kyoto Protocol, Marrakech Accords and the Paris Agreement.

7.3 *Policies to Encourage Development and Protection of Carbon Sinks*

It is essential that policy elements be aimed directly at the specific activities and processes which cause loss of soil carbon, so as to identify the principal ways and means of enhancing existing sinks and increasing the total area of sinks.[141] Specific policy elements should promote methods to identify existing terrestrial sinks and new sinks and outline procedures to identify soil and land degradation 'hot spots', as they relate to identification of sinks. They could also outline procedures to implement sustainable soil and vegetation management (e.g., land clearing controls). These could focus on national policies for conservation farming, organic farming systems, agroforestry, plantation establishment, and regional land use agreements.

[139]IPCC (2018), e.g., Section D at 20–26, "Strengthening the Global Response in the Context of Sustainable Development and Efforts to Eradicate Poverty"; D.2.1 states 'Climate change impacts and responses are closely linked to sustainable development which balances social well-being, economic prosperity and environmental protection. The United Nations Sustainable Development Goals (SDGs), adopted in 2015, provide an established framework for assessing the links between global warming of 1.5°C or 2°C and development goals that include poverty eradication, reducing inequalities, and climate action'.

[140]Including the right to quality environment, see Glazebrook (2009).

[141]See Lal (2010), p. 25.

8 National Legal and Institutional Frameworks for Managing Soil Carbon Sequestration

There is a variety of ways to design an appropriate legal and institutional system for management of soil carbon processes, but two principal approaches can be considered,[142] depending on respective national physical, sociological and economic conditions.[143]

Non-Regulatory Strategy[144]

This strategy is characterised by elements that focus on:

- Extension, education, and awareness programs for soil carbon management.
- Ecosystem research, assessment, and monitoring of soil carbon stocks.
- Financial support to research soil carbon processes and sink protection.
- Extensive use of community participatory facilities in sink education.
- Development of land use practices that minimise the loss of soil carbon.
- Development of soil carbon management, protection, and incentive-based programs.

Regulatory Strategy[145]

This strategy is characterised by elements that focus on:

- Development of statutory land use plans that prescribe legal limits and targets of soil and land use to reduce the loss of soil carbon (e.g., cultivation practices, vegetation retention levels and water quality levels).
- Issue of licenses or permits to control soil use. These would prescribe use entitlements relating to soil restoration, management of sinks, habitat protection, organic matter levels etc.
- Land use agreements between the State and individuals that set binding land use standards.
- The use of restraining notices where sustainable use limits are exceeded.
- Prosecution for failure to follow prescribed standards for managing soil carbon sinks and reservoirs.

---

[142] Hannam (2004); Hannam and Boer (2004) Section III, 2.1 at 30 'Regulatory and Non-Regulatory Strategies'.

[143] Ibid, 'Some States may prefer to develop sustainable soil strategies with a minimum of legal regulation, whereas others may prefer a stronger regulatory-based law. In the short term a comprehensive soil policy may be regarded as being more beneficial or politically convenient in helping to achieve a sustainable use of soil objective than a strictly legal approach'.

[144] Power (2011), p. 59.

[145] See Hannam (2012), p. 62.

9 Conclusions

There is currently a broad range of international and national environmental law instruments that constitute a comprehensive framework to manage soil carbon processes. The climate change process has evolved rapidly since the UNFCCC was adopted over two decades ago and the most high profile development has undoubtedly been the adoption of the Paris Agreement. These instruments provide useful guideline for nations to review and reform their legislative system for soil carbon management. The Paris Agreement process provides the world with an opportunity to approach the effects of climate change with a new purpose. A major policy decision of governments should be to reduce the build up of carbon dioxide in the atmosphere by gaseous emissions and at the same time increase the build up of carbon in the terrestrial environment through sink development and enhancement. This will have the dual benefit of minimising the build up of carbon in the atmosphere and the land rehabilitation measures will reduce the emissions and help achieve Target 15.3 of the SDGs in relation to land degradation neutrality. There must be greater effort on the part of the community to improve the international and national environmental law to manage soil carbon. The UNFCCC, Kyoto Protocol and Paris Agreement pay special attention to the concerns of developing countries.[146] This chapter has outlined ways that individual nations may approach the task of developing a legal and policy framework to more effectively manage these processes.

References

Anarbaev M (2017) Soil legislation and policy in the Kyrgyz Republic on the development of the law "on soil fertility protection of agricultural lands". In: Ginzky H et al (eds) International yearbook of soil law and policy 2017. Springer International Publishing

Boer B, Hannam I (2019) Land degradation. In: Lees E, Vinuales J (eds) Oxford handbook of comparative environmental law. Oxford University Press, UK, pp 438–459

Bonnie R et al (2002) Protecting terrestrial ecosystems and the climate through a global carbon market. Philos Trans R Soc Lond 360:1853–1873

Claassen R, Morehart M (2009) Agriculture land tenure and carbon offsets. United States Department of Agriculture Economic Research Service, Economic Brief 14

Climate Change Secretariat (2002) A guide to the climate change convention and its Kyoto Protocol, Preliminary Version, Bonn

Durrant N (2008) Legal issues in biosequestration: carbon sinks, carbon rights and carbon trading. Univ N S W Law J 49; 31(3)

FAO (Food and Agriculture Organization of the United Nations) (2004) Carbon sequestration in dryland soils, World Soil Resources Reports 102, Rome

FAO (Food and Agriculture Organization of the United Nations) (2015a) Healthy soils are the basis for food production, Rome

[146] IPCC (2018), p. 20.

FAO (Food and Agriculture Organization of the United Nations) (2015b) Status of the worlds soil resources, Rome
FAO (Food and Agriculture Organization of the United Nations) (2017) Voluntary guidelines for sustainable soil management, Rome
German Federal Government (1998) The accounting of biological sinks and sources under the Kyoto Protocol - a step forward or backwards for global environmental protection? Special Report, German Advisory Council on Global Change
Glazebrook S (2009) Human rights and the environment. Victoria Univ Wellington Law Rev 40 (1):293
Global Soil Partnership and FAO (2014) Revised World Soil Charter, Rome
Guglyuvatyy E, Stoianoff N (2016) Carbon policy in Australia- a political history. UTSLRS 25
Hannam I (2002) Legal and institutional frameworks for water and land management with particular reference to marginal areas in selected countries in South and South East Asia and the People's Republic of China. International Water Management Institute, Colombo, Sri Lanka
Hannam I (2004) International and national aspects of a legislative framework to manage soil carbon sequestration. Clim Change 65(3):365–387
Hannam I (2012) Legal framework for improving carbon management in grasslands in the People's Republic of China, a preliminary analysis of three environmental laws in relation to carbon management, ADB R-CDTA 7534: Strengthening Carbon Financing for Regional Grassland Management in Northeast Asia, Manila
Hannam I (2020) Soil legislation in Australia. In: Yahyah H et al (eds) Legal instruments for sustainable soil management in Africa. International yearbook of soil law and policy. Springer, Cham
Hannam I, Boer B (2002) Legal and institutional frameworks for sustainable soils. IUCN, Gland, Switzerland and Cambridge, UK
Hannam I, Boer B (2004) A guide to drafting soil legislation. IUCN, Gland, Switzerland and Cambridge, UK
Hurni H, Meyer K (eds) (2002) A world soils agenda, discussing international actions for the sustainable use of soils, prepared with the support of an international group of specialists of the IASUS Working Group of the International Union of Soil Sciences (IUSS). Centre for Development and Environment, Berne
IPCC (Intergovernmental Panel on Climate Change) (2018) Global warming of 1.5C, An IPCC Special Report on the impacts of global warming of 1.5°C above pre-industrial levels and related global greenhouse gas emission pathways, in the context of strengthening the global response to the threat of climate change, sustainable development, and efforts to eradicate poverty, Summary for Policymakers, Switzerland
IUCN Commission on Environmental Law (2015) Draft International Covenant on Environment and Development: implementing sustainability. Fifth edition, updated text, prepared in cooperation with the International Council of Environmental Law, IUCN, Gland, Switzerland xxxii + 201
Jeffery M, Baird W (2001) Using market-based incentives to curtail greenhouse gas emissions: factors to consider in the design of the clean development mechanism, joint implementation and emissions trading. Asian Pac J Environ Law: 10–19
Johnson H (2018) International agricultural law and policy: a rights-based approach to food security. Edward Elgar, 357 p
Keesstra et al (2016) The significance of soils and soil science towards realization of the United Nations Sustainable Development Goals. SOIL 2:111–128, www.soil-journal.net/2/111/2016/. https://doi.org/10.5194/soil-2-111-2016
Kyoto (1997) The Kyoto Protocol to the Convention on Climate Change, The Climate Change Secretariat, UNEP
Lal R (2010) Soil Carbon Sequestration, SOLAW Background Thematic Report - TR04B. UN Food and Agriculture Organization, Rome

Lausche B (2011) Guidelines for protected areas legislation. IUCN, Gland, Switzerland, xxvi + 370 pp
Lehner P, Rosenberg N (2017) Legal pathways to carbon neutral agriculture. Environmental Law Institute®, Washington DC
Macintosh A (2012) The carbon farming initiative: what you need to know, PrecedentAULA 70, 113 Precedent 28
Metternich G, Cowie A (2017) Kill not the goose that lays the golden egg: striving of land degradation neutrality. International Institute for Sustainable Development, SDG Knowledge Hub, 23 November 2017
Neely C, Bunning S, Wilkes A (eds) (2009) Review for evidence on dryland pastoral systems and climate change: implications and opportunities for mitigation and adaptation. Food and Agriculture Organization of the United Nations, Rome
Power M (2011) The Carbon Farming Initiative - too little, too soon? Natl Environ Law Rev 19:1
Secretariat of the Convention for Biological Diversity (2004) Addis Ababa principles and guidelines for the sustainable use of biodiversity, Montreal
Severinsen G (2010) Towards an effective legal framework for the geo-sequestration of carbon dioxide in New Zealand. Canterbury Law Rev 16:331–370
Severinsen G (2014) Constructing a legal framework for carbon capture and storage in New Zealand: approaches to legislative design. Energy Procedia 63:6629–6661
UNCCD (2017) Global land outlook, 1st edn. Bonn, Germany
UNCCD, CBD, FAO, Global Environment Facility and UNEP (2016) Draft for Consultation, Framework and Guiding Principles for a Land Degradation Indicator to Monitor and report on progress towards Target 15.3 of the Sustainable Development Goals, the strategic objectives of the Rio Conventions and other relevant targets and commitments
UNEP (United Nations Environment Program) (2018) Final Assessment of the Fourth Program for the Development and Periodic Review of Environmental Law (Montevideo IV), Evaluation Office of UN Environment
UNEP (United Nations Environment Program) (1995a) Convention on Biological Diversity
UNEP (United Nations Environment Program) (1995b) United Nations Framework Convention on Climate Change
UNFCCC (United Nations Framework Convention for Climate Change) (2002a) Marrakech Accords
UNFCCC (United Nations Framework Convention for Climate Change) (2002b) Marrakech Ministerial Declaration
United Nations (1992a) the Rio Declaration on Environment and Development
United Nations (1992b) Agenda 21
United Nations (1995) United Nations Convention to Combat Desertification
United States Geological Survey (2008) Carbon Sequestration to Mitigate Climate Change, Factsheet 2008-3097
Verschuuren J (2017) The impact of soil carbon sequestration on adaptation in Europe's agricultural sector and the potential role of regulatory instruments, Paper accepted for presentation at the 3rd European Climate Change Adaptation Conference 'Our Climate Ready Future', Glasgow, 5th–9th June 2017
Voigt C, Ferreira F (2016) Dynamic differentiation: the principles of CBDR-RC, progression and highest possible ambition in the Paris Agreement. Transnatl Environ Law 5(2):285–303
WSSD (World Summit on Environment and Development) (2002a) Plan of Implementation, United Nations
WSSD (World Summit on Environment and Development) (2002b) A Framework for Action on Agriculture, WEHAB (Water, Energy, Health, Atmosphere and Biodiversity) Working Group
WSSD (World Summit on Environment and Development) (2002c) The Johannesburg Declaration on Sustainable Development, United Nations

The manufacturer's authorised representative in the EU is Springer Nature Customer Service Centre GmbH, Europaplatz 3, 69115 Heidelberg, Germany. If you have any concerns regarding our products, please contact ProductSafety@springernature.com

Printed and bound by CPI Group (UK) Ltd, Croydon, CR0 4YY
30/09/2026
02225506-0011